Trees in a Sub-Saharan Multi-functional Landscape

Paxie W. Chirwa • Stephen Syampungani
Theodore M. Mwamba
Editors

Trees in a Sub-Saharan Multi-functional Landscape

Research, Management, and Policy

Springer

Editors
Paxie W. Chirwa
Department of Plant and Soil Sciences
Faculty of Natural and Agricultural Sciences
University of Pretoria
Pretoria, South Africa

Stephen Syampungani
Chair of Environment and Development,
Oliver R. Tambo Africa Research Chairs Initiative (ORTARChI)
Copperbelt University
Kitwe, Zambia

Theodore M. Mwamba
Faculty of Agriculture
University of Lubumbashi
Lubumbashi, Democratic Republic of the Congo

ISBN 978-3-031-69814-9 ISBN 978-3-031-69812-5 (eBook)
https://doi.org/10.1007/978-3-031-69812-5

This Springer imprint is published by the registered company Springer Nature Switzerland AG
The registered company address is: Gewerbestrasse 11, 6330 Cham, Switzerland

Foreword

Africa has suffered decades of deforestation and land degradation due to the failure of agriculture to provide either adequate food for its growing population or to enhance the livelihoods of its people. These failures reflect the adoption of inappropriate policies by colonial governments and subsequently by international development agencies with poor understanding of the different physical, social and economic needs of Africa and those of industrialized economies. The consequences of these failures have been growing hunger, malnutrition and poverty relating to the breakdown of natural, social and human capital across the continent at the landscape level. These impacts have then been exacerbated by subsequent climate change and loss of biodiversity.

This book seeks to explain the complexity of these issues. In a book related to this subject area,[1] I presented the concept of a more holistic three-step approach to addressing the failures of tropical agriculture using agroforestry to rehabilitate degraded farmland and enhance rural livelihoods. This book now adds detailed findings illustrating progress towards multifunctionality across the continent of Africa from tropical rainforest and savanna to rocky outcrops.

Very significantly, the authors of this book are scientists from around the African continent. They emphasise the progress towards the reversal of the foreign dominance of research, management and policy by adopting an integrated, people-centred and much more holistic multifunctional approach to land use across the continent. This recognizes the importance of rebuilding the intertwined strands of natural, social and human capital to achieve sustainable development. Specifically, this covers the critical need to address the failures of agricultural production and social injustice that impact the everyday lives of local people. This then can lead to beneficial impacts on the critical natural resources of vegetation, soils, water and wildlife by providing an integrated and robust socio-ecological framework to navigate and reconcile the competing anthropogenic demands of numerous stakeholders.

[1] *Multifunctional Agriculture: Achieving Sustainable Development in Africa* (2017).

Divided into 14 chapters, this book highlights the crucial role of trees, especially indigenous trees, in the rehabilitation of degraded ecosystems by their provision of environmental, cultural, and socio-economic services to agricultural production and rural development. Emphasis is given to community-based land management systems recognizing traditional knowledge, benefit sharing and collective action. Individual chapters focus on soil restoration, germplasm conservation, tree domestication and the development of markets for a wide range of tree products across the various agroecological zones of the continent. Importantly, these chapters recognize the paramount importance of local non-timber forest products and emphasise the significant roles that these trees play in the diversification of farming systems for soil health and protection, pest and pathogen control, flora and fauna conservation. They are also important for the diversification of diets and local markets. Furthermore, it is emphasised that non-timber forest products have recently emerged as 'agroforestry tree products' cultivated as new African tree crops with the virtually untapped potential to create new business opportunities. Based on ideotype selection, cultivars selected to meet the needs of farmers and different market opportunities present untapped intraspecific genetic variation in the phytochemical ingredients of these tree products. This selection then gives rise to potential new industries for Africa's economic growth.

While vital for Africa's future, all of the above also has global significance to meet the UN Sustainable Development Goals and hence the survival of the planet through enhanced social justice, the mitigation of climate change and the restoration of both wildlife habitats and critical ecosystem functions. Such synergies and 'trade-ons' generate a sustainable and harmonious future based on a sound and balanced socio-economic and ecological fabric for tropical and sub-tropical nations.

I congratulate the authors and editors of this book for taking this bold and crucial step towards the revival of Africa's sovereignty over its natural and cultural resources and livelihoods. I very much hope that policy and decision makers worldwide will recognize their leadership.

Roger R. B. Leakey

Retired Professor of Agroecology
and Sustainable Development
James Cook University, Cairns, Australia

Vice President of International Tree Foundation
Oxford, UK

Council Member of Global
EverGreening Alliance
Melbourne, VIC, Australia

Member of Earth Team/End Pandemics Alliance
San Francisco, CA, USA

Preface

Africa is largely dependent on natural resources in its quest to achieve sustainable development goals, with its woodlands and forests known to be among the most exploited ecosystems. Anthropogenic pressure has led to the progressive conversion of these woodlands and forests into patches of regrowth stands, savanna and cropland systems. Expansion with intensification of agricultural land and the unsustainable use of fuelwood biomass are among the major drivers of such landscape transformations and the associated decline of biodiversity and ecosystem services. This phenomenon is growing in magnitude and significance with emerging environmental issues including climate change, biological invasions, and land degradation arising from various land use practices, thereby jeopardizing Africa's development pathway. Agroforestry and its associated trees in a landscape formation holds promises to revert this trend, as it provides the potential for designing multifunctional agricultural landscapes that guarantee trade-offs of food security, biodiversity conservation, and other ecosystem services. Importantly, in the context of Africa, the diversity of ecological conditions and cultural heritage of the continent lends a unique scenario to the concept of multifunctional landscapes.

Drawing upon the vast experience of researchers and many other professionals in Africa, this book elucidates the roles of trees within the multifunctional landscapes of Africa, capturing the biophysical and societal diversity inherent to the region, while situating Africa's landscape multifunctionality within the broader global context. It endeavours to provide readers with a holistic understanding of trees' multifaceted roles in African landscapes, equipping them with insights and analytical tools to address landscape multifunctionality challenges effectively. The book begins by establishing the foundational concepts of multifunctional landscapes, exploring their implications for sustainable livelihoods and ecosystem services. It further delves into African woodlands' distribution, significance, threats, restoration, and conservation, offering a comprehensive overview of their multifunctional attributes. The book also traces the historical sustainability of tree-based and agroforestry systems, juxtaposing them with the modern agricultural practices introduced during the Green Revolution. This exploration highlights the challenges and opportunities presented by shifting agricultural paradigms, including the renewed emphasis on traditional food systems and the commercial potential of non-timber forest products.

The book further elaborates on the multiplicity of factors that affect production in multifunctional landscape systems in Africa. Tree-based and/or agroforestry systems to some extent were sustainable over the years because of the need to increase productivity per unit area in order to meet the growing human population. On the other hand, the Green Revolution meant the introduction of mono-crops, new so-called improved varieties, and high-input inorganic fertilizers. However, this form of production has seen a decline in yields due to economic reasons and the understanding of humankind for the need to conserve biodiversity. It has been shown in recent years that many rural communities resort to natural food systems, which act as safety nets to meet low production due to many other factors including climate variability and declining soil fertility. Hence, the ecosystems from forests provide services beyond the safety net through provision of non-timber forest products, which have now the potential for commercialization. Suddenly, everyone realized the need for traditional tree species and food crops in the landscape. These landscapes have been shown to be more diverse than human-modified landscapes where agriculture takes precedence. Indeed, there are also other food sources that are able to grow in hostile environments including rocky outcrops, an area that is not well understood.

This book is thus beneficial to undergraduate, postgraduate, and university research scholars and teachers, especially those working in areas of woodland silviculture, agroforestry, and agriculture. It will also be useful for extension agents, non-governmental organizations, natural resources managers, and policymakers.

Pretoria, South Africa — Paxie W. Chirwa

Kitwe, Zambia — Stephen Syampungani

Lubumbashi, Democratic Republic of the Congo — Theodore M. Mwamba

Acknowledgements

The gestation period for this book spanned several years until we decided it was time to overcome procrastination. The ongoing collaboration between the University of Pretoria and Copperbelt University led to the creation of our first book outline in 2022, when Prof. Stephen Syampungani was on a 2-week research visit to the University of Pretoria as an Extraordinary Professor. Further momentum was gained in 2023 when Prof. Paxie W. Chirwa took a six-month sabbatical to begin drafting the book proposal, which was ultimately submitted to Springer.

This book would not have been possible without the contributions of many eminent scientists from across sub-Saharan Africa. We acknowledge the involvement of numerous authors, as indicated by their affiliations in the respective book chapters. We also extend our gratitude to all the reviewers who meticulously reviewed the chapters to their final completion. In particular, we thank the following individuals for their valuable comments and constructive criticisms: Abasse Tougiani, Abednego Kiwia, Achille Assogbadjo, Afio Zannou, Almeida Sitoe, Andrew T. Hudak, Antoine Kalinganire, Demel F. Teketay, Edouard I. Ilunga, Emmanuel Torquebiau, Erin Sills, Evariste Fongnzossie, Festus Akinnifesi, Gudeta Sileshi, Ian Dawson, John Odindi, Lizzie Mujuru, Mesfin Gossa, Olabimpe Orubuloye, Ouedraogo Issaka, Philippe Bayen, Roger Leakey, Shem Kuyah, Simon Mng'omba, Stefan Siebert, Steven Franzel, Tajudeen O. Amusa, and Yemi Katere.

We are grateful for the support from Copperbelt University and the funding from the Oliver R. Tambo Africa Research Chairs Initiative (ORTARChI) for Environment and Development, hosted by Copperbelt University. ORTARChI is funded by Canada's International Development Research Centre (IDRC), South Africa's National Research Foundation (NRF), the Department of Science and Innovation (DSI), in partnership with the Oliver and Adelaide Tambo Foundation (OATF), and the National Science and Technology Council (NSTC) of Zambia. We also appreciate the support from the project "Agroforestry in Southern Africa: New Pathways of Innovative Land Use Systems", financed by the German Federal Ministry of Education and Research (BMF), which facilitated Dr Theodore Mwamba's visit to the University of Pretoria to work on the book project. Additionally, we thank the

University of Pretoria for its support through the Faculty of Natural and Agricultural Sciences (NAS) and the Forest Chair in the Department of Plant and Soil Sciences.

We hope the insights provided in this book will contribute to better management of multifunctional landscapes in sub-Saharan Africa.

Contents

About the Editors

Prof. Paxie W. Chirwa is a Professor of Forest Science in the Department of Plant and Soil Sciences. He is also a Forest Chair for the Forest Science Postgraduate Programme in the Faculty of Natural and Agricultural Sciences. His area of specialisation is socio-ecological systems in forests and resource use in agroforestry systems. He works in different forest ecosystems in Africa including the miombo dry forests and woodlands of southern Africa. He has supervised over 24 PhDs and 43 Masters students. He has published over 150 scientific publications in internationally peer-reviewed journals, authored 21 book chapters and edited 3 books. He has acted as Guest Editor for three special issues of forestry journals: *International Forestry Review*, *Agroforestry Systems* and *Southern Forest: A Journal of Forest Science*. He recently was one of the co-editors of the book *Miombo Woodlands in a Changing Environment: Securing the Resilience and Sustainability of People and Woodlands*. Springer, Cham. https://doi.org/10.1007/978-3-030-50104-4_1 Ribeiro N.S., Katerere Y., Chirwa P.W., Grundy I.M. (eds) (2020). He is an Associate Editor for *Agroforestry Systems Journal* (Springer) and *Journal of Forestry Research* (Springer); Editorial Advisory Board Member of *Southern Forests: A Journal of Forest Science* (Francis & Taylor); and Editor for *Environmental and Geosciences—Scientific African Journal* (Elsevier), *Malawi Journal of Agriculture, Natural Resources and Development Studies* and recently the *Carbon Footprints* journal.

Prof. Stephen Syampungani is both a restoration ecologist and environmental engineer focussing on restoration of mine wastelands and environmental risk assessment associated with various developmental activities. He holds a BSc in Forest Science (Copperbelt University), MSc in Environmental Engineering (University of Nottingham, UK) and a PhD in Forest Ecology (Stellenbosch University, South Africa). Stephen is a Professor of Restoration Ecology at Copperbelt University as well as an Extra-ordinary Professor in the Department of Plant and Soil Sciences at the University of Pretoria, South Africa. He holds a Professorial Chair of Environment and Development, the Oliver R Tambo African Research Chair Initiative (ORTARChI) Chair funded by the IDRC (Canada), National Research Foundation (NRF) and Department of Science and Innovation (DSI) of South Africa and the National Science and Technology Council of Zambia hosted by the Copperbelt University. Stephen has co-authored a number of publications in the area of mine wasteland restoration and environment risk modelling.

Dr. Theodore M. Mwamba is an agronomist and eco-physiologist with primary research interest in sustainable management of degraded ecosystems. He holds a bachelor's degree in General Agronomy from the University of Lubumbashi (DR Congo) and a PhD in Plant Stress Physiology and Regulation (Zhejiang University, China). He is a Senior Researcher at the University of Lubumbashi (DR Congo), where he has a decade of research experience in agroecosystem management. He is currently a Postdoctoral Fellow with Oliver Tambo Africa Research Chairs Initiative project (ORTARChl Chair, Environment and Development) at Copperbelt University (Zambia), where he is involved in research activities related to environmental management of metal-polluted lands in the Central African Copperbelt. His areas of research encompass land evaluation and restoration, plant growth regulation, plant germplasms management and ecotoxicological risk assessment, for which he has co-authored several papers published in SCI journals. Theodore is part of the editorial team of *Frontiers in Plant Science*, acting as Review Editor in the section Plant Nutrition.

Abbreviations

ACSM	American Congress on Surveying and Mapping
AFAP	African Fertilizer and Agribusiness Partnership
AFLP	Amplified Fragment Length Polymorphism
AFODEK	Agroforêts pour le Développement de Kipushi
AFR100	African Forest Landscape Restoration 100
AGP	Agricultural Growth Programme
AGRA	Alliance for a Green Revolution in Africa
AIP	Affordable Inputs Program
ALAP	African Landscapes Action Plan
ANR	Assisted Natural Regeneration
ARLI	Africa Resilient Landscapes Initiative
ASP	Action for Shea Parklands
ASPRS	American Society for Photography and Remote Sensing
AU-IAPSC	African Union-Inter-African Phytosanitary Council
BGCI	Botanic Gardens Conservation International
CATS Banks	Community Agroforestry Tree Seeds Banks
CBD	Convention on Biological Diversity
CBFM	Community-Based Forest Management
CBFP	Congo Basin Forest Partnership
CBI	Centre for the Promotion of Imports from developing countries
CBNRM	Community-Based Natural Resource Management
CEB	Compagnie Equatoriale des Bois
CIFOR-ICRAF	Center for International Forestry Research (CIFOR) and World Agroforestry (ICRAF)
DAP	Diammonium Phosphate
DFID	Department for International Development
DRC	Democratic Republic of Congo
EPA	Environment Protection Authority
ESA	European Space Agency
ESRI	Esri's Land Cover

EWC	Eudafano Women's Cooperative
FAO	Food and Agriculture Organization of the United Nations
FISP	Farm Input Subsidy Programme
FLO	Fairtrade Labelling Organizations
FMNR	Farmer Managed Natural Regeneration
FRA	Forest Resources Assessment
FTA	Forests, Trees and Agroforestry
GDP	Gross Domestic Product
GESS	Growth Enhancement Support Scheme
GGW	Great Green Wall
GIS	Geographic Information Systems
GLASOD	Global Assessment of Soil Degradation
GRU	Genetic Resources Unit
GSA	Global Shea Alliance
IATP	Institute for Agriculture and Trade Policy
IES	Invasive Exotic Species
IFPRI	International Food Policy Research Institute
IFTs	Indigenous Fruit Trees
IPBES	Intergovernmental Science-Policy Platform on Biodiversity and Ecosystem Services
IPCC	International Panel on Climate Change
IPM	Integrated Pest Management
IPPC	International Plant Protection Convention
IR3S/UTIAS, 2016	Integrated Research System for Sustainability Science (IR3S)/University of Tokyo Institutes of Advanced Study (UTIAS)
ISFM	Integrated soil fertility management
IUCN-WCPA	Internal Union for Conservation of Nature (IUCN)-World Commission on Protected Areas (WCPA)
KBAs	Key Biodiversity Areas
LPFN	Landscapes for People, Food and Nature
LULC	Land Use Land Cover
LVB	Lake Victoria Basin
MDGs	Millennium Development Goals
MEDC	Michigan Economic Development Corporation
MPTs	Multipurpose Trees
MTF	Multi-purpose Trees on Farmlands
MWK	Malawian Kwacha
NEPC	Nigeria Export Promotion Council
NICFI	Norwegian International Climate and Forest Initiative
NPPO	National Plant Protection Organisation
NTFPs	Non-Timber Forest Products
NTSC	National Tree Seed Centre
NWFP	Non-Wood Forest Products

OECD	Organisation for Economic Co-operation and Development
PATSPO	Provision of Adequate Tree Seed Portfolios in Ethiopia
PES	Payment for Ecosystem Services
PFM	Participatory Forest Management
PNI	Partnership for Natural Ingredients
PPM	Precision Pest Management
PPO	Plant Protection Organisation
REDD	Reducing Emissions from Deforestation and forest Degradation
REDD+	Reducing Emissions from Deforestation and forest Degradation in developing countries, and the role of conservation, sustainable management of forests, and enhancement of forest carbon stocks in developing countries
SLM	Sustainable Land Management
RIL	Reduced Impact Logging
RRC	Rural Resource Centres
SADC	South African Development Community
SETAC	Society of Environmental Toxicology and Chemistry
SOC	Soil Organic Carbon
SOM	Soil Organic Matter
SPS	Sanitary and Phytosanitary
SSA	sub-Saharan Africa
SSAL	sub-Saharan Agricultural Landscape
UNCBD	United Nations Convention on Biological Diversity
UNCCD	United Nations Convention to Combat Desertification
UNDER	United Nations Decade on Ecosystem Restoration
UNDESA	United Nations Department of Economic and Social Affairs
UNEP-WCMC	United Nations Environmental Programme (UNEP)-World Conservation Monitoring Centre (WCMC)
UNFCCC	United Nations Framework Convention on Climate Change
UNHCR	United Nations High Commissioner for Refugees
UNU-IAS	United Nations University Institute for the Advanced Study of Sustainability
USD	United States Dollars
USAID	United States Agency for International Development
TBEA	Tree-Based Ecosystem Approaches
TEK	Traditional Ecological Knowledge
TLU	Tropical Livestock Units
TLUD	Top-Lit Updraft
WIPO	World Intellectual Property Organization
WFI	Water for Injection
WFO	World Farmers' Organisation
WTO-SPS	World Trade Organization (SPS: Sanitary and Phytosanitary)
XAF	Central African CFA Franc

List of Figures

List of Tables

Chapter 1
Introduction

Paxie W. Chirwa, Stephen Syampungani, and Theodore M. Mwamba

The African continent is a natural home to a range of terrestrial biomes, providing valuable ecosystem services to humans across different landscapes. The landscapes vary in their ability to provide ecosystem services (Cheng et al. 2021). Terrestrial biomes in Africa appear in different forms ranging from tropical and sub-tropical forests to woodlands, shrublands, savannas, grasslands and deserts (Keith et al. 2022). The rich and diverse ecosystems in Africa generate goods and services that are essential in providing for the continent's food, water, energy, health and secure livelihood needs (IPBES report 2018).

The natural resources within these ecosystems generate important "landscape products" with multiple functions (García-Martín et al. 2022), supporting the livelihoods of millions of people, and sustainability across the continent. They contribute towards the well-being of communities through provision of food, energy, medicines and other ecosystem-based services (ES). Additionally, these ES offer opportunities for entrepreneurship and market integration, thereby contributing to national gross domestic products (FAO 2022). Agriculture stands out as one of the main economic pillars, sustaining the livelihoods of 70% of the population on the African continent (Uyanga et al. 2024). Therefore, the growth of the agricultural sector is

P. W. Chirwa (✉)
Department of Plant and Soil Sciences, Faculty of Natural and Agricultural Sciences,
University of Pretoria, Pretoria, South Africa
e-mail: paxie.chirwa@up.ac.za

S. Syampungani
Department of Plant and Soil Sciences, Faculty of Natural and Agricultural Sciences,
University of Pretoria, Pretoria, South Africa

Chair of Environment and Development, Oliver R. Tambo Africa Research
Chairs Initiative (ORTARChI), Copperbelt University, Kitwe, Zambia

T. M. Mwamba
Chair of Environment and Development, Oliver R. Tambo Africa Research
Chairs Initiative (ORTARChI), Copperbelt University, Kitwe, Zambia

Faculty of Agriculture, University of Lubumbashi, Lubumbashi,
Democratic Republic of the Congo

P. W. Chirwa et al. (eds.), *Trees in a Sub-Saharan Multi-functional Landscape*,
https://doi.org/10.1007/978-3-031-69812-5_1

important for enhancing food security and nutrition, thereby alleviating poverty. Similarly, non-timber forest products are vital for subsistence and trade, contributing 25–45% of the income at household level in various regions (Derebe and Alemu 2023; Leßmeister et al. 2018; Sills et al. 2011).

The livelihoods of people and terrestrial biomes in Africa are intricately linked. Garcia-Martin et al. (2022) recently identified four key groups of interrelated landscape product functions, including: (i) human links to nature; (ii) culture and identity; (iii) social capital; and (iv) nutritional sustenance and economic income. The authors argued that landscape sustainability is based on interactions between human well-being and landscape-specific services and promotes the collaboration of landscape-level actors and institutions. However, landscape sustainability needs to consider individual products, and the multiple functions, benefits and values associated with them along their supply chains.

On one hand, this human-nature link underscores the strategic importance of African landscapes in achieving sustainable development at continental level (IPBES 2018). On the other hand, it highlights the vulnerability of rural communities to undesirable changes in these landscapes (Suich et al. 2015). The reason is that people's livelihoods, including economic well-being, social capital, cultural heritage and identity, are all intricately tied to the landscapes. Land is considered an economic development asset as well as a socio-cultural resource.

Africa is generally believed to be the most resource-endowed continent in the world, with a huge proportion of the world's natural resources (UNEP 2024), both renewables and non-renewables. The continent holds 65% of the world's arable land. This natural capital underpins the continent's economy and has immense potential to transform its development through sustainable investments. Despite its large wealth of natural resources—largely underexplored, the African continent continues to experience high poverty levels, largely due to its inability to meet the food demands of a rapidly growing population. Africa's annual loss of its natural capital is estimated at USD 195 billion, through illicit financial flows, illegal mining, illegal logging, the illegal trade in wildlife, unregulated fishing and environmental degradation and loss (UNEP 2024).

Sub-Saharan Africa (SSA) is the fastest-growing region in terms of population growth globally, at an annual rate of 2.8% (Mabiso and Benfica 2019). It is also one of the most rapidly urbanizing continents, thus posing a significant challenge to achieving the second Sustainable Development Goal (zero hunger). This challenge is exacerbated by land degradation, with nearly half of the population living in ecologically fragile ecosystems such as drylands prone to frequent droughts. Currently, two-thirds of arable land in SSA is degraded to the extent that productivity from such land is very low (World Bank Group 2017). While global agricultural production has risen substantially in recent decades, yields in SSA have stagnated (World Bank Group 2017), making the region one of the most affected by food and nutrition insecurity, with 250 million people undernourished (Rege and Sones 2022).

The rate of population growth in Africa tends to increase pressure on landscape resources, leading to significant landscape degradation. The expansion and intensification of agricultural lands, unsustainable extraction of wood for fuel, and infrastructural development are key drivers of land cover changes, leading to significant conversion, fragmentation and degradation of African forests and woodlands, known to be among the most exploited ecosystems (Chirwa and Adeyemi 2020). This degradation may result in biodiversity loss and a reduced capacity of landscapes for provisioning of ecosystem services. Often, "regime shifts" emanating from such degradation, possibly coupled with high unhelpful resilience of the new regime in cases of hysteresis, can strongly influence recovery trajectories (Maes et al. 2024).

Worse still, the continent is extremely vulnerable to the impacts of climate change, with myriads of human-induced drivers. The current decade has witnessed increased temperatures and more frequent and intense precipitation events in various parts of the continent. Several scenarios have predicted a 20% expansion of drylands (World Bank Group 2017). Africa's Sahara Desert is reported to have expanded significantly since 1920, by 11–18% depending on the season since 1920 (Thomas and Nigam 2018), thus increasing the number of people exposed to ecologically sensitive and stressed environments. In this context, the future of food security and the provisioning of other ecosystem services in Africa is expected to be more challenging.

The extent of deforestation, forest degradation in SSA (Branthomme et al. 2023) and desertification necessitate concrete actions towards restoration of landscapes as well as protection of landscape resources. African countries have recognized this need and have undertaken bold continent-wide restoration initiatives to enhance landscape functionality, exemplified by tree planting projects such as the Great Green Wall Initiative (Sileshi et al. 2023), the African Resilient Landscape Initiative (ARLI), the African Forest Landscape Restoration (AFR100), and the African Landscapes Action Plan (ALAP) (Bond et al. 2019; Lewis et al. 2019). Indeed, there has been increased momentum on landscape restoration and agroforestry initiatives in the continent in recent decades (Garrity et al. 2010; Akinnifesi et al. 2010; Sileshi et al. 2014).

Innovative practices for sustainable harvesting of non-timber forest products and the creation of favourable socio-economic conditions for their exploitation are also being promoted as efforts to halt the current trend of landscape degradation (Shackleton et al. 2018). The domestication of trees, especially Africa's indigenous trees, have been advanced from studies of wild collection to clonal propagation techniques, germplasm and cultivation, to post-harvest and commercialization (Akinnifesi et al. 2008; Leakey and Akinnifesi 2017, Chaps. 9 and 10 in this volume). Based on the broad lessons learned on these various practices and initiatives (Leakey and Akinnifesi 2017), the development of an integrated and holistic form of multifunctional agriculture described as 'land maxing,' has recently been proposed as the way ahead (Leakey et al. 2022). This aims to rebuild natural, social and human capital in ways that positively impact on farm productivity, livelihoods and the local economy.

However, due to ecological and socio-economic specificities, the implementation of these efforts varies across different countries and regions, leading to varied outcomes. As editors, we acknowledge the importance of understanding the ecological attributes of African landscapes and the prevailing socio-economic systems to inform decision-making and implementation of restoration and conservation efforts. This book emphasizes regional specificities to foster contextualized actions and policy frameworks for managing tree-based multifunctional landscape resources in SSA.

Despite being a primary threat to forests and biodiversity, there is growing consensus that sustainable agricultural intensification to feed the rapidly growing population is inevitable in Africa (Chaps. 5 and 7 of this book), bringing about a dilemma in the management of African landscapes. Nonetheless, it is believed that smallholder farmers can play a crucial role as stewards of biodiversity by adopting land management practices that support forest and landscape restoration (Sileshi et al. 2008; Akinnifesi 2018). Agroforestry, in particular, holds promise for guaranteeing trade-offs of food security, biodiversity conservation, and environmental quality (Syampungani et al. 2010). This book aims at providing theoretical frameworks based on research findings and lessons on the role of agroforestry in mitigating deforestation and contextualizing the importance of trees in various farming systems, that could be useful for the restoration of forest cover in the African landscapes.

Above all, recognizing the need for multi-sectoral interventions to address poverty issues in Africa (Raffaello and Morris 2016), this book advocates for landscape management approaches that promote livelihoods diversification for rural communities. Throughout the book, emphasis is placed on landscape multifunctionality—the deliberate management of landscapes to provide multiple functions and services to society (Stürck and Verburg 2017). The book addresses the perspective of multifunctional land use as an adaptive response to socio-economic and environmental challenges faced by SSA, emphasizing the influence of existing farming systems and agroforestry practices, the diversity of services provided by tree-based non-timber forest products, the potential of untapped resources in marginal ecosystems such as rock outcrops, and the exploration of indigenous multipurpose trees through management and use of quality germplasms. These aspects are also discussed in the context of emerging environmental issues in SSA, i.e. climate change and landscape biosecurity issues.

This book has 14 chapters, covering five broad thematic areas (i) the multiple services, functions and roles of trees in multifunctional landscapes; (ii) tree-based production systems, products and their contributions to rural livelihoods; (iii) drivers of land cover changes and their impacts; (iv) approaches for managing resources to integrate trees in multifunctional landscapes; and (v) external factors affecting landscape resources management in Africa.

In terms of the structure of this book, we start with an introductory chapter (Chap. 1) presenting the general context, followed by Chap. 2 setting the tone by introducing the concept of multifunctional landscapes in the African context. Chapter 3 explores the role of trees in multifunctional landscapes, underscoring the potential impacts of trees in providing ecosystem services, while Chap. 4 highlights the unique functional traits of plant species in rocky outcrops, substantiating the view of African landscapes richness and diversity. Chapter 5 provides a comprehensive overview of the role of trees in relation to their interplay with other agroecosystem elements, while documenting the added value of tree integration in African agroecosystems. In the same vein, Chap. 6 demonstrates the importance of nitrogen-fixing trees in the SSA. Chapter 7 focuses on land use and land cover changes to highlight the need for response strategies to the current trend of landscape degradation in Africa. Chapters 8 and 9 delve into the diversity and benefits of non-timber forest products from African landscapes. Recognizing the need for suitable and healthier tree species as a crucial component of multifunctional landscape management, Chap. 10 addresses quality and sustainable tree germplasm issues, whereas Chap. 11 focuses on the management of biological invasions. Chapters 12 and 13 explore socio-ecological approaches and socio-economic considerations in management of multifunctional landscapes. Finally, Chap. 14 summarizes the book's key points and proposes future research and policy directions to secure multifunctional landscapes in Africa and contribute to sustainable development.

References

Akinnifesi FK (2018) Africa's forest landscape restoration gathers momentum. Nature Faune 32(1):3–10

Akinnifesi FK, Sileshi G, Mkonda A, Ajayi OC, Mhango J, Chilanga T (2008) Germplasm supply, propagation and nursery management of Miombo fruit trees. In: Akinnifesi FK, Leakey RRB, Ajayi OC, Sileshi G, Tchoundjeu Z, Matakala P, Kwesiga FR (eds) Indigenous fruit trees in the tropics: domestication, utilization and commercialization. CAB International Publishing, Wallingford

Akinnifesi FK, Ajayi OC, Sileshi G et al (2010) Fertiliser trees for sustainable food security in the maize-based production systems of East and Southern Africa. A review. Agron Sustain Dev 30:615–629. https://doi.org/10.1051/agro/2009058

Bond WJ, Stevens N, Midgley GF, Lehmann CER (2019) The trouble with trees: afforestation plans for Africa. Trends Ecol Evol 34:963–965

Branthomme A, Merle C, Kindgard A, Lourenço A, Ng W-T, D'Annunzio R, Shapiro A (2023) How much do large-scale and small-scale farming contribute to global deforestation? Results from a remote sensing pilot approach. FAO, Rome. https://doi.org/10.4060/cc5723en

Cheng L, Lu N, Wang M, Fu B, Xu B (2021) Alternative biome states of African terrestrial vegetation and the potential drivers: a continental-scale study. Sci Total Environ 800:149489

Chirwa PW, Adeyemi O (2020) Deforestation in Africa: implications on food and nutritional security. In: Filho LW, Azul A, Brandli L, Özuyar P, Wall T (eds) Zero hunger. Encyclopedia of the UN sustainable development goals. Springer, Cham, pp 197–211. https://doi.org/10.1007/978-3-319-69626-3

Derebe B, Alemu A (2023) Non-timber forest product types and its income contribution to rural households in the horn of Africa: a systematic review. For Sci Technol 19(3):210–220

FAO (2022) The state of the world's forests (2022) Forest pathways for green recovery and building inclusive, resilient and sustainable economies. FAO, Rome. https://doi.org/10.4060/cb9360en

García-Martín M, Huntsinger L, Ibarrola-Rivas MJ, Dámbrosio U et al (2022) Landscape products for sustainable agricultural landscapes. Nat Food 3:814–821. https://doi.org/10.1038/s43016-022-00612-w

Garrity DP, Akinnifesi FK, Ajayi OC et al (2010) Evergreen agriculture: a robust approach to sustainable food security in Africa. Food Sec 2:197–214. https://doi.org/10.1007/s12571-010-0070-7

IPBES (2018) Summary for policymakers of the assessment report on land degradation and restoration of the intergovernmental science policy platform on biodiversity and ecosystem services. Scholes R, Montanarella L, Brainich A, Barger N, ten Brink B, Cantele M et al. (eds) Secretariat of the intergovernmental science-policy platform on biodiversity and ecosystem services, Bonn, Germany, 44p

Keith DA, Ferrer-Paris JR, Nicholson E, Bishop M, Polidoro BA, Ramirez-Llodra E (2022) A function-based typology for Earth's ecosystems. Nature 610:513–518. https://doi.org/10.1038/s41586-022-05318-4

Leakey RRB, Akinnifesi FK (2017) Towards a domestication strategy for indigenous fruit trees in the tropics. In: Leakey RRB (ed) Multifunctional agriculture: achieving sustainable development in Africa. Elsevier, Academic, pp 165–176

Leakey RR, Tientcheu Avana ML, Awazi NP, Assogbadjo AE, Mabhaudhi T, Hendre PS et al (2022) The future of food: domestication and commercialization of indigenous food crops in Africa over the third decade (2012–2021). Sustain For 14(4). https://doi.org/10.3390/su14042355

Leßmeister A, Katja H, Lykke AM, Thiombiano A, Wittig R, Hahn K (2018) Contribution of non-timber forest products (NTFPs) to rural household revenues in two villages in South-Eastern Burkina Faso. Agrofor Syst 92:139–155

Lewis SL, Wheeler CE, Mitchard ETA, Koch A (2019) Regenerate natural forests to store carbon. Nature 568:25–28

Mabiso A, Benfica R (2019) The narrative on rural youth and economic opportunities in Africa: facts, myths and gaps. Papers of the 2019 rural development report, IFAD research series, 61

Maes SL, Perring MP, Cohen R, Akinnifesi FK et al (2024) Explore before you restore: incorporating complex systems thinking in ecosystem restoration. J Appl Ecol 61(5):922–939. https://doi.org/10.1111/1365-2664.14614

Raffaello C, Morris M (2016) Confronting drought in Africa's drylands: opportunities for enhancing resilience, Africa development forum series. World Bank, Washington, DC. https://doi.org/10.1596/978-1-4648-0817-3

Rege JEO, Sones K (2022) The agriculture sector in sub-Saharan Africa and the promise of biotechnology. In: Rege JEO, Sones K (eds) Agricultural biotechnology in sub-Saharan Africa. Springer, Cham, pp 1–10. https://doi.org/10.1007/978-3-031-04349-9_1

Shackleton CM, Ticktin T, Cunningham AB (2018) Non-timber forest products as ecological and biocultural keystone species. Ecol Soc 23(4):22

Sileshi G, Akinnifesi FK, Ajayi OC, Chakeredza S, Mng'omba S, Nyoka BI (2008) Towards sustainable management of soil biodiversity in agricultural landscapes in Africa. Biodiversity 9:64–67

Sileshi GW, Mafongoya PL, Akinnifesi FK, Phiri E, Chirwa P, Beedy T et al (2014) Agroforestry: fertilizer trees. Encycl Agric Food Syst 1:222–234

Sileshi GW, Dagar JC, Kuyah K, Datta A (2023) The great Green Wall initiatives and opportunities for integration of dryland agroforestry to mitigate desertification, agroforestry for sustainable intensification of agriculture in Asia and Africa, sustainable agriculture and food security. Springer Nature Singapore Pte Ltd., pp 175–206. https://doi.org/10.1007/978-981-19-4602-86

Sills E, Shanley P, Paumgarten F, de Beer J, Pierce A (2011) Evolving perspectives on non-timber forest products. In: Shackleton S, Shackleton C, Shanley P (eds) Non-timber forest products in the global contextSpringer Berlin/Heidelberg pp. 23–51

Stürck J, Verburg PH (2017) Multifunctionality at what scale? A landscape multifunctionality assessment for the European Union under conditions of land use change. Landsc Ecol 32:481–500

Suich H, Howe C, Mac G (2015) Ecosystem services and poverty alleviation: a review of the empirical links. Ecosyst Serv 12:37–147

Syampungani S, Chirwa PW, Akinnifesi FK, Ajayi OC (2010) The potential of using agroforestry as a win-win solution to climate change mitigation and adaptation and meeting food security challenges in Southern Africa. Agric J 5(2):80–88. https://doi.org/10.3923/aj.2010.80.88

Thomas N, Nigam S (2018) Twentieth-century climate change over Africa: seasonal hydroclimate trends and Sahara Desert expansion. Journal of Climate 31(9). https://doi.org/10.1175/JCLI-D-17-0187.1

UNEP report (2024) Our work in Africa. https://www.unep.org/regions/africa/our-work-africa. Accessed on 23 May 2024

Uyanga VA, Bello SF, Bosco NJ, Jimoh SO, Mbadianya IJ, Kanu UC, Okoye CO, Afriyie E, Mak-Mensah E, Agyenim- KG, Ogunyemi SO, Nkoh JN, Olamide Olasupo IO, Karikari B, Ahiakpa JK (2024) Status of agriculture and food security in post-COVID-19 Africa: impacts and lessons learned. Food Human 2:100206

World Bank Group (2017) Fighting land degradation at landscape scale: sustainable land and water management in Africa's drylands and vulnerable landscapes. International Bank for Reconstruction and Development/The World Bank, Washington, DC, 34p

Chapter 2
Trees in Multifunctional Landscapes: Definition, Classification, Systems, Structure, Functionality, Examples in Africa

Paxie W. Chirwa, Mulugheta Araia, Marie L. Avana-Tientcheu, Jonathan I. Muledi, Stephen Syampungani, Festus K. Akinnifesi, Achille E. Assogbadjo, and Eugene Loh Chia

Abstract The Chapter provides the definition and concept of multifunctional landscapes in the African context. It also highlights the merits of the different landscape approaches recognizing that no single one can address the needs of all existing African landscapes. Ultimately, the people-centred approach to multifunctional landscape management has the merit for addressing the diverse needs for biodiversity conservation, including climate change, food security and poverty alleviation. In this regard, the principles of Multifunctional Landscape approaches emphasize managing landscapes to achieve the dual purpose of multifunctionality and people-centred sustainable resource conservation. The Chapter highlights different tree-based ecosystem approaches and technologies where trees provide a broad range of ecosystem services that often go beyond their specific habitat, e.g. habitat for pollinators, corridors for wildlife mobility and contribution to the local and global climate systems. It also recognizes traditional tree-based ecosystem management practices across different African landscapes including the tropical forests, savanna woodlands and the Sahel. The selected practices presented in this chapter are neither exhaustive nor meant to be replicated or taken as a blueprint for successful multifunctional landscape approaches, but rather intended to provide a basis for innovation to integrate trees in multifunctional landscapes.

Keywords Land use · Human-modified · Tree-based ecosystem approaches · Ecosystem services · Climate change response

---

P. W. Chirwa (✉) · M. Araia
Department of Plant and Soil Sciences, Faculty of Natural and Agricultural Sciences, University of Pretoria, Pretoria, South Africa
e-mail: paxie.chirwa@up.ac.za

P. W. Chirwa et al. (eds.), *Trees in a Sub-Saharan Multi-functional Landscape*,
https://doi.org/10.1007/978-3-031-69812-5_2

2.1 Introduction

Land use and land use change have been attributed as the main causes of the decline of nature and biodiversity (Jaureguiberry et al. 2022; Brauman et al. 2019). Since 1970, the rate of expansion and intensification of monocultural land use approaches has been accelerated in response to the accelerated demand for food, animal feed, bioenergy, and fibre to the fast-growing global population assisted by the advancement of technology and trade demand (Hurtt et al. 2020). In particular, land use changes, primarily driven by agriculture, forestry, and urbanization have been shrinking, fragmenting, and degrading natural ecosystems (Tilman et al. 2017), and associated species (BGCI 2021) putting local and global biodiversity in great peril (Brauman et al. 2019). This has compromised the potential of natural forests and wooded landscapes to provide diverse ecosystem functions and services (Kumar et al. 2022).

Forests play a major role in the global biogeochemical process, influencing the soil formation, hydrological cycles, nutrient, and carbon cycles, as well as the local and global climate regulations (BGCI 2021; Hecht et al. 2016; Leakey 2014). Additionally, forests contain about half the world's carbon stocks, and more than three-quarters of the world's accessible fresh water comes from forest catchments (Newton 2021). Moreover, it is estimated that approximately 3.4–5.8 billion people worldwide rely on non-timber forest products for food, construction materials, energy, utensils, medicines, spiritual and cultural values, and income generation (Shackleton and de Vos 2022).

M. L. Avana-Tientcheu
Dèpartement de Foresterie, Facultè d'Agronomie et des Sciences Agricoles, Universitè de Dschang, Dschang, Cameroun

African Forest Forum, United Nations, Nairobi, Kenya

J. I. Muledi
Faculty of Agriculture, University of Lubumbashi, Lubumbashi, Democratic Republic of the Congo

S. Syampungani
Department of Plant and Soil Sciences, Faculty of Natural and Agricultural Sciences, University of Pretoria, Pretoria, South Africa

Chair of Environment and Development, Oliver R. Tambo Africa Research Chairs Initiative (ORTARChI), Copperbelt University, Kitwe, Zambia

F. K. Akinnifesi
Food and Agriculture Organization of the United Nations (FAO), Rome, Italy

A. E. Assogbadjo
NTFPs and Orphan Crops Unit/Lab of Applied Ecology, Faculty of Agronomic Sciences, University of Abomey-Calavi, Cotonou, Benin

E. L. Chia
Central African Regional Program, United States Forest Services International Programs, Gombe Kinshasa, Democratic Republic of Congo

Trees also play an indispensable role in the dynamics of global biodiversity (Rivers et al. 2023; BGCI 2021). However, the recent assessment of the state of globally known tree species (58,492 tree species) revealed that 0.2% of these species are extinct while about 37% are threatened or possibly threatened to extinction. The situation is not different in the tropics where about 80% of global tree species richness is found. For instance, out of 9237 tree species found in the tropical part of Africa, 0.2% and 45% are extinct or possibly threatened to extinction in their wild habitat (BGCI 2021). Unless the trend of deforestation and tree species loss is reversed, ecological collapse and the loss of ecosystem services with adverse impacts is eminent. This will aggravate the impact of climate change, biodiversity loss, poverty, food and nutritional insecurity, illegal migration, risks of future pandemics in Africa and globally.

In response, expansion of protected areas has become one of the prominent conservation strategies to conserve vulnerable and threatened species (BGCI 2021) and to bend the curve of biodiversity loss (Watson et al. 2014) by reducing direct threats. Currently, 15.4% of the global terrestrial surface has a legally protected status (UNEP-WCMS et al. 2021), and about 64% of tree species are found in at least one of the protected areas (BGCI 2021). However, most protected areas are isolated and situated in a landscape matrix with multifunctional land use, which may reduce their effective size and capacity to maintain the biodiversity that they were originally designed to conserve (Nagendra et al. 2009). Although some success stories have been reported (Schleicher et al. 2019), biodiversity loss has not been declining in many protected areas as previously anticipated (Visconti et al. 2019). Moreover, many studies have reported the negative impact of protected areas on the livelihood of local people (Oldekop and Holmes 2016) through forced displacement from their cultural landscapes (De Pourcq et al. 2017). Hence, many protected areas have been subjected to land claims by local communities and economic pressure leading to downgrading, downsizing, and degazetting of global protected areas in favour of extractive mining and local pressure for subsistence (Mascia et al. 2014).

Moreover, despite forest having been seen as the natural home of trees, nearly 1.35 billion hectares of land outside forests has about 335 million tree cover in the tropics (WRI 2024). In Africa, it was estimated that 29% of the tree cover is found outside forests (Reiner et al. 2023). Trees in these non-forest landscapes provide similar services to the forests albeit to a more limited extent than a large tract of primary forests (Thomas et al. 2021). Many studies have reported that trees in human-influenced landscapes provide a refuge for tropical biodiversity (Bhagwat et al. 2008), ensure sustainable food security (Mukhlis et al. 2022) and relieve pressure on the remaining forests (Rahman et al. 2017). Scattered trees in natural and human-modified landscapes also serve as a corridor for movement and gene flow in a fragmented environment, a habitat for some species, and provide ecosystem services that are crucial for the sustainability of rural livelihood.

The above-mentioned facts highlight the significant role trees can play in biodiversity conservation, climate change, and food and nutritional security. However, deforestation and habitat fragmentation for the purpose of expanding or intensifying a specific land use pose threats to tree species distribution and biodiversity, be it within a forest or outside the forests. This also adversely affects the associated essential ecosystem functioning of landscapes. There is mounting evidence on the relationship between biodiversity, ecosystem function (Brockerhoff et al. 2017), and the multiple services that they provide, which in turn sustain livelihoods (Felipe-Lucia et al. 2020; Soe and Yeo-Chang 2019).

Indeed, trees that are either naturally retained or planted on farmlands can help to restore vegetative cover in areas where the natural forest has been almost completely cleared. However, farmers primarily plant trees that provide them tangible benefits such as fruit, shade, poles, shelter, and restoration of soil fertility (Cook and Grut 1990). Hence, while agroforestry and forest plantations enhance biodiversity benefits better than monoculture crops and livestock farming, it cannot replace the role of natural forests in biodiversity conservation and the associated ecosystem services. Indeed, the overall contribution of agroforestry and forest plantations to biodiversity conservation in comparison with natural forests depends on (a) the type of land use that it replaces or enriches when integrated, (b) level and status of biodiversity being conserved (e.g., forest specialist and/or threatened species vis-a vis generalist species), (c) the attributes of specific agroforestry systems and practices and woodlots (e.g. parkland agroforestry, traditional silvopastoral systems, improved fallow, alley cropping etc.), (d) choices of tree species (e.g. indigenous trees versus exotic species), and (e) spatial configuration of the systems (e.g., proxy to intact forests and shrubs) (Swallow et al. 2006), belowground root architecture (Asaah et al. 2012) which may affect nutrient and water dynamics. Moreover, it has been long acknowledged that positive or negative impacts of trees in farmlands on the functioning of socio-ecological systems goes beyond the boundaries of individual farms and villages.

Based on the foregoing, traditional approaches of assigning landscapes to a single land use, be it protected areas, agroecosystems, woodlots, or settlement, have been found to be limited in capacity to fulfill multiple needs of society and nature. Such realization of trade-off between a single objective land use and the need for economic development, food and nutritional security, and conservation of forest and tree biodiversity for ecosystem services necessitated the concept of managing landscapes for multifunctional benefits (Mastrangelo et al. 2014).

2.2 The Concept of Multifunctionality

Multifunctionality at landscape level refers to the diversity and abundance of different functions and services (Stürck and Verburg 2017) that can be achieved by reconciling and synergizing competing demands of multiple stakeholders, optimizing benefits, and achieving the goals of production and conservation of tree-based land uses concurrently. Thus, understanding how a single objective land use function in a large landscape can help identify where and how to design various tree-based land uses to optimize the multifunctionality of ecosystem services they provide. A landscape framework involves looking beyond the boundaries of a small-scale land use, and to determine how the land use is affected by the external conditions, and reciprocally affects the surrounding landscape. On a broad scale, connections with the other land uses become relevant, and more comprehensive and integrated land use solutions can be developed (Lovell et al. 2021). Notwithstanding, there has been a strong support for multifunctional landscape approach, as a viable agro-ecological framework, yet it is apparent that constraints to its implementation persist (Reed et al. 2017). Hence, there is a need for research on both the theories and practices of trees (e.g. forests, trees outside of forests including agroforestry) in multifunctional landscapes for sub-Saharan Africa. This chapter presents various landscape level frameworks, principles, and examples of tree-based systems and technologies from different regions of sub-Saharan Africa as a basis for innovation to integrate trees in multifunctional landscapes.

2.3 Framing the Landscape Approaches in the African Context

In recent decades, the development of landscape frameworks has attracted the attention of many authors (Sayer et al. 2013; Ros-Tonen et al. 2014; Freeman et al. 2015), with the aim of embedding a single purpose land use planning and management within broader landscape-scale management strategies. Ideally, the intention of the approach is to create a *win-win* outcome between conservation and development initiatives in a landscape where diverse and conflicting social, economic, and environmental interests of stakeholders meet (Reed et al. 2015, 2016). Such approaches are epitomized by the "Ecosystem Approach" of the Convention on Biological Diversity (Reed et al. 2017). The ecosystem approach is a strategy for the integrated management of land, water and biotic resources that promotes conservation and sustainable use in an equitable way. Thus, the application of the ecosystem approach is adopted and promoted to reach a balance of the three objectives of the Convention: conservation; sustainable use; and the fair and equitable sharing of the benefits arising out of the utilization of genetic resources (Makino 2005).

Landscape management approaches are not entirely new to Africa (Milder et al. 2014). The approach has been advocated and implemented in the past few decades as a potent tool to tackle the inextricable challenges of agricultural and environmental restoration, food and water security, adaptation and mitigation of the negative livelihood impacts of climate change and other disruptions, health (human, livestock, wildlife) and inclusive 'green' economic growth in integrated manner (Scherr et al. 2013). However, the scope, breadth, and design of such approaches have progressively evolved in some significant ways (Milder et al. 2014), following both the emerging trends of sciences of ecosystem management and landscape ecology (Noss 1983) and in recognition of the reciprocal linkages between the livelihood needs of local communities and key drivers of deforestation, biodiversity loss and climate changes.

Milder et al. (2014) identified more than 70 approaches, including, landscape *restoration*, *biological corridors*, *bioregional planning*, *eco-agriculture*, *whole-landscape approaches*, and *diversified farming systems*. Some few years later, protected areas in multifunctional landscape, climate smart landscape, and sustainable landscape were joined to the lexicon as a contributing factor for smooth implementation of the approach (Reed et al. 2016; Mastrangelo et al. 2014). The landscape approaches described by Milder et al. (2014) have some key common features. Essentially, all the approaches recognise the importance of local context —thus, 'one-size fits all' cannot be applied (Turley 2016). Also, all the initiatives have attempted to address the optimal potential of the landscapes for social, economic, and environmental functions in an integrative manner regardless of their different starting points (Reed et al. 2016, 2020). The different starting points for those initiatives might highlight the difference in priorities of local socio-ecological issues that need to be addressed.

Moreover, integrated landscape approaches (multi-sectoral and multifunctional objectives) have been criticised for failing to achieve win-win outcomes for diverse stakeholders. Recently, there is a growing acknowledgment that the landscape approach should be based on a realistic premise having both winners and losers. Such synergies and trade-offs for the best multifunctionality options must be identified, negotiated, and accounted for (Reed et al. 2015, 2020; Freeman et al. 2015, Sayer et al. 2013). It will often be the case that optimal solutions for a stakeholder with specific interest or a specific land use objective, will be sub-optimal for another. As such, accounting for these trade-offs is fundamental to addressing the inextricable social, economic, and environmental challenges (Reed et al. 2020, 2016). By bringing together the different stakeholders operating within the landscape and attempting to identify and understand what each of their requirements and expectations are, trade-offs and synergies can be identified. Management Plans should then collaboratively aim to capitalise on the synergies while the trade-offs will enable planners to identify those losing out so that appropriate compensation (e.g. payment for ecosystem service) or alternatives can be pursued. Therefore, the landscape approach attempts, through participatory, inclusive negotiation and planning to minimize trade-offs and capitalise on synergies so that it delivers the maximum level of possible fewer losers and more winners (Sayer et al. 2015) in pursuit of multifunctionality.

2.4 The Place and Role of Trees in Multifunctional Landscapes

In recent years, the potential of managing trees in multifunctional landscapes in Africa has been taking centre stage in many of the national, regional, and international debates and treaties (Mansourian and Berrahmouni 2021). First, there has been growing global and local concern of deforestation and habitat degradation due to the ineffectiveness of sectoral approaches to conservation in Africa (Gray et al. 2016). Secondly, there has been a realisation that integrating tree communities in a whole landscape have been directly or indirectly associated with many of the previous and new integrated landscape initiatives and conceptual frameworks. These include sustainable intensification (Iiyama et al. 2018), agroecological landscapes

(Leakey 2014), climate smart agriculture (Ngoma et al. 2018), multifunctional grazing and pasture landscape management (Reid 2009), and integrated watershed management (Teka et al. 2020). There is also ample evidence that trees play an indispensable role in ecosystem functions and services that are crucial for increasing the efficiency, productivity, and sustainability of the other land uses in the landscape.

The shift in paradigm described above helps to transform agriculture and other competing land uses from a threat for forests and trees and habitats to a potential tool for protecting, conserving, and restoring forests and trees through holistic and integrated design (Gitz et al. 2022). This is particularly important where protected areas for biodiversity conservation are practically impossible to manage due to socio economic and institutional challenges. Nonetheless, the florid terminologies discussed in the above landscape approaches, the diverse tree-based ecosystem approaches, and technologies, the concept of trees in multifunctional landscapes show the need for further clarity and unification of conceptual frameworks. Thus, understanding how trees interact with the multifunctionality in landscapes of Africa to address biodiversity conservation, climate change, food security and poverty alleviation is crucial.

The most logical first step in this regard is to carry out a stocktaking of the existing tree-based land use approaches with clear and context specific terminologies, guiding principles including review of their strength and weakness from the existing multifunctional landscapes practices with an African perspective. A wide range of traditional and improved tree-based ecosystem approaches and technologies that has been practised in Africa (e.g. Leakey 2017). In this Chapter, we will provide relevant case studies of tree-based land uses and their multifunctional roles either at the land unit and/ or higher scales from a landscape management perspective. Such an approach will help to highlight the different tree-based ecosystem approaches and technologies where trees provide a broad range of ecosystem services that often goes beyond their specific habitat, e.g. habitat for pollinators, corridors for wildlife mobility and contribution to the local and global climate systems. The selected practices in this chapter are neither exhaustive nor meant to be replicated or taken as a blueprint for successful multifunctional landscape approaches, but rather intended to provide a basis for innovation to integrate trees in multifunctional landscape approaches.

2.5 Basic Terminologies and Principles of Trees in Multifunctional Landscape

At the outset, the objective of this sub-section is not to provide a glossary of the multitude of terms used in literatures related with "tree-based ecosystem approach" and multifunctional landscape approach—approaches which have emerged from different scientific realms. Nor, to discuss the principles in detail. Instead, it is to collate the relevant definitions of the key words and principles used in the chapter to keep them aligned with the objective of the chapter and for ease of understanding the theories and concepts they convey.

2.5.1 Terminologies

A ***landscape*** is a socio-ecological system that consists of a mosaic of natural and/or human-modified ecosystems, often with a characteristic configuration of topography, vegetation, land use, and settlements that is influenced by the ecological, historical, economic, and cultural processes and activities of the area (LPFN 2016). A landscape can also be defined as the interaction between human actions, and biotic and abiotic ecosystem factors that shape the physical environment. Landscape approaches are approaches based on "overarching principles, concepts and implementation rather than the retreat into the compartmentalization that has been widely used in land management for a considerable time" (FAO 2024; Sunderland 2014).

Landscape Multifunctionality: Landscape multifunctionality is the capacity of a landscape to simultaneously support multiple benefits to society from its interacting ecosystems. Landscape multifunctionality is often conceived and assessed as the joint supply of multiple ecosystem services at the landscape level (Mastrangelo et al. 2014). According to Schümann et al. (2022), maintaining or restoring landscape multifunctionality is essential to ensuring that landscapes provide a broad array of services, especially with more diverse land uses bordering each other. Multifunctional agricultural landscapes are characterized by increased connectivity between habitats, which promotes spillover of ecosystem services and their providers (Grass et al. 2019). Landscape multifunctionality is characterized by diversified land use and complex landscape structure, thereby potentially covering many, often competing interests of different stakeholder groups (Hölting et al. 2019, 2020). This warrants management strategies that increase the multifunctional use of landscapes, while sustaining natural resources for future generations. Sustainably managing landscape multifunctionality requires effective collaboration between stakeholders across spatial scales and sectors. Multifunctional landscapes require strategies that account for "land-sparing" (i.e. for instance, landscapes distinctly separated from high- yielding agriculture and natural habitats) and "land-sharing" configurations that can be harnessed to maximize synergies and minimize trade-offs. These land-sharing and land-sparing attributes are not mutually exclusive but instead exist on a continuum from measures that may be only fully considered sparing or sharing at the extreme ends (Ekroos et al. 2016).

Trees in Multifunctional Landscape: encompass forest, tree, and agroforestry approaches in multifunctional landscape (Rahman et al. 2017). Such a broad definition is used to accommodate trees that are found in forest, woodlands, agricultural lands, grasslands, steppes, and desert landscapes. In this chapter, tree is broadly defined as any perennial woody plant growing to a height of at least 2 m (Van Wyk 2013). In the absence of a universally agreed definition of trees (BGCI 2021), the aforementioned definition of a tree will go beyond the artificial distinction of trees and shrub (Van Wyk 2013) and accommodate both for their role in malfunctional landscapes.

A Landscape Approach is a multi-faceted integrated strategy that aims to bring together multiple stakeholders from multiple sectors to provide solutions at multiple scales. It can be broadly defined as a framework to address the increasingly widespread and complex environmental, economic, social, and political challenges that typically transcend traditional management boundaries (Reed et al. 2015). By ensuring the equitable and sustainable use of land, a landscape approach is a potential mechanism to alleviate poverty in an equitable manner, conserve biodiversity, safeguard forests, sustainably manage natural resources, while maintaining food production and mitigating climate change (Reed et al. 2016). A landscape approach may be undertaken by one or more stakeholders who engage in actions independently, or by multiple actors as part of a collaborative, multi-stakeholder process. This multi-stakeholder process is referred to as integrated landscape management (Denier et al. 2015).

Tree Based Ecosystem Approaches (TBEAs) refers, in this Chapter and Book, to a variety of land and forestry management systems and practices that combine trees with agricultural production in pursuit of sustained or increased productivity, enhanced ecosystem services supply, and a stronger adaptive capacity for land managers. The key criterion for a tree-based system to be classified as a TBEA is a multi-objective management strategy at the landscape scale. TBEAs include agroforestry practices (woody perennials in agricultural systems) and forestry systems that are actively managed for multiple objectives, such as food, timber and non-timber forest production, and the supply of ecosystem services (World Agroforestry Centre 2013; Iiyama et al. 2018). Trees are integrated in agroforestry-based land use by three routes: retaining remnant trees, spontaneous regenerated trees and by deliberate planting (or direct seeding (Ordonez et al. 2014)).

2.5.2 Principles of Trees in Multifunctional Landscape Approaches

The ten principles of the landscape approach (Sayer et al. 2013, Box 2.1) provide a framework from which practitioners can select and adapt to their local conditions. The principles should be considered as a ‘menu’ from which to select appropriately, depending on specific landscape contexts; not as a prescription for a spatial planning and management (Reed et al. 2016; Sayer et al. 2013).

The principles emphasize the need for managing landscapes for multifunctionality and the significance of people centred approaches to sustainable resource conservation. Both emphasize that the natural systems upon which humans depend (e.g., pollination, filtration, photosynthesis) require the proper functioning of the overall ecosystems of which these services are a part. Both approaches promote “systems-level” thinking, away from the narrow, sector-specific approaches that have, in many ways, contributed to present-day conundrums. By virtue of integrating trees with the landscape approach, the conceptual framework expands our understanding of the diverse ecosystem services derived from tree-based ecosystems (Table 2.1)—than, for example, under typical forest management plans that simply recognize the value of timber and wood products (Turley 2016).

Box 2.1: The Summary of the "Ten Principles" from Sayer et al. (2013) and Turley (2016)

Principle 1: Continual learning and adaptive management: to accommodate the dynamic and uncertain nature of landscapes, adaptive management through continuous learning and adjustment (when need) must be adopted.

Principle 2: Common concern entry point: bring together stakeholders in a transparent manner to collectively identify resource management issue to that need to be address in a short to intermediate plan while fostering shared vision and long-term objectives.

Principle3: Multiple scales: Create awareness about the process at different spatial and institutional scale that may affect local intervention.

Principle 4: Multifunctionality: Landscapes and their components have multiple uses and purposes, each of which is valued in different ways by different stakeholders. Trade-off among the multiple preference and aspirations held by stakeholders do exist, and need to be reconciled in a spatial explicit and ecosystem driven manner.

Principle 5: Multiple stakeholders: an equitable and fair engagement of multiple stakeholders are vital to obtain an optimal outcome of landscape management intervention to attain multifunctional purpose at a given time.

Principle 6: Negotiated and transparent change logic: All stakeholders should understand and accept the general logic, legitimacy, and justification for the course of actions. Communication in this regard is crucial to sustain mutual understanding and trust in pursuing the shared vision to address the common concerns.

Principle 7: Clarification of rights and responsibilities: The right and responsibilities need to be clear for good governance and management. They will also be handy managing conflicts whenever they arise.

Principle 8: Participatory and user-friendly monitoring: Information, often many kinds of information, used as a basis for the landscape management and collected for monitoring and evaluation should accessible.

Principle 9: Resilience Principle: Wholesale unplanned system changes are usually detrimental and undesirable. System-level resilience can be increased through an active recognition of threats and vulnerabilities. Actions need to be promoted that address threats and that allow recovery after perturbation through improving capacity to resist and respond.

(continued)

Box 2.1 (continued)

Principle 10: Strengthened stakeholder capacity: peoples representing different interest groups, and trusted to discharge their responsibilities should have the necessary competence- Continuous learning (Principle 1) helps in the capacity building of stakeholders. These principles emphasize that the integration of agricultural and environmental priorities will require a people-centred approach applied at landscape scales.

Table 2.1 Ecosystem services of tree in multifunctional landscape

Groups of ecosystem services	Specific benefits	Groups of ecosystem services	Specific benefits
(1) Biodiversity benefits	(a) habitat for many species; (b) ecological architecture; (c) ecological and habitat connectivity; (d) ecological services	(2) Carbon benefits	(a) Carbon uptake (b) Carbon storage
(3) Agricultural benefits	(a) pollination; (b) pest predation; (c) alternative hosts; (d) soil fertility improvements; (e) erosion control.	(4) Soil benefits	(a) enhanced soil drainage; (b) soil moisture storage; (c) increased organic matter (d) improved soil structure.
(5) Water benefits	(a) buffering the impacts of rainfall; (b); (c) recharging the moisture in the soil; (d) moderating the flow of streams; (e) erosion control.	(6) Microclimate improvements	(a) moderating urban heat island effects; (b) reduction of heat stress in agroforestry and silvo-pastoral systems; (c) evaporative cooling; (d) wind breaks.
(7) Local weather defence	(a) windbreaks; (b) shoreline protection via mangroves; (c) shade	(8) Economic benefits	(a) producing food; (b) producing timber and posts; (c) producing non-timber products,; (d) producing commercial commodities,; (e) artisanal inputs; (f) potential REDD derivatives or other offset initiatives pertaining to carbon
(9) Subsistence benefits	(a) providing food to people who live in or near forests; (b) providing fuel; (c) artisanal inputs;) providing fodder for livestock; (e) providing construction materials; (f) providing medicine	(10) Human symbolic meaning	(a) demarcation; (b) place making; (c) totems; (d) sacred groves; (e) aesthetic

(continued)

Table 2.1 (continued)

Groups of ecosystem services	Specific benefits	Groups of ecosystem services	Specific benefits
(11) Survival benefits and complex livelihood "insurance,"	(a) medicine, (b) "hunger crops"; (c) bush meat; (d) periodic extraction		

Adopted and modified from Hecht et al. (2016)

Furthermore, sustaining multifunctional values requires people centred management approaches through inclusive governance and collaboration (Favretto et al. 2021; Marques et al. 2016). Meaningful collaboration can promote the inclusion of marginalised voices, ensure appropriate actions and responses aligned to local concerns and needs, broaden the knowledge base, and bring frequently disconnected actors, sectors, and government institutions together in pursuit of a common goal. In this regard, recognition of local knowledge is important (Marques et al. 2016). Local knowledge can be a measure of community capability to make a difference in landscapes and is complementary to scientific knowledge. Berkes et al. (2000) found that traditional ecological knowledge constituted in many cases a wealth of knowledge that allowed interpretation and responding to local situation and feedback to guide resource management by communities. Such knowledge shares similarity in characters to adaptive management in terms of feedback learning, and the treatment of uncertainty/unpredictability in the landscape.

There are many promising age-long traditional governance and technical systems and practices that still exist in Africa. These include, but not limited to, taboos, totems, traditional by laws of habitats and species, traditional agroforestry practices, wetland management, culturally/traditionally protected area (Sinthumule 2023). Making this a part of the analytical, management and learning processes in landscapes would be beneficial (Marques et al. 2016). At the same time, there is a need to incorporate the current advance in governance models and technical solutions from a scientific realm as part of the conservation and development strategies at a landscape level. There are some promising polycentric governance models (e.g. Heikkila et al. 2018; Ostrom 1999) and technical solutions based on ecosystem approaches (e.g. Improved agroforestry, woodlots, integrated watershed management etc.) that may enhance the effective management for protection, conservation, and enhancement of trees in multifunctional landscapes.

2.6 Trees in Multifunctional Landscapes in Africa

2.6.1 Southern African Multifunctional Landscapes

Southern Africa has about 190 million ha of forests with an average of 32% forest cover. Forest types range from tropical moist and rainforest in the north to subtropical dry and humid forest, as well as mountain forest, in the south. Most vegetation classified by FAO as tropical forest is commonly named "woodland" in the region, for example Miombo or Mopane woodland (Di Marco et al. 2019; Chirwa et al. 2014). In Southern Africa, changes in dry land forest ecosystems and degradation are mostly driven by population growth, agricultural expansion, and energy needs. The pace, intensity and extent of deforestation and degradation are closely linked to policy, market and institutional failures that undervalue forests and woodland and promote alternative land uses that are counterproductive (Chirwa et al. 2015). Land tenure arrangements and associated equity issues are a major threat to the sustainable use of land resources. The communal land tenure system is the most widespread, in which individual property rights are weak. Hence, this poor land and tree tenure in Southern Africa is likely to have incentivised their over-exploitation. Consequently, the remaining forested areas face increasing pressure, particularly in response to high population growth rates and increasing poverty. On the positive side, there are some promising traditional and improved tree-based ecosystem systems that have been identified to conserve forest and tree species diversity and enhance agriculture production in multifunctional landscapes (Chirwa and Mahamane 2017).

2.6.1.1 Traditional Forest and Tree Management Systems

Many traditional societies in southern Africa have recognised the values and the need for conservation and management of forest and tree species diversity and their potential to sustain the provision of ecosystem services. Forest and tree species are strongly intertwined into their livelihood, cultural, emotional, spiritual, and symbolic values of their lifestyle (Araia and Chirwa 2019). Clarke (1994) highlighted the widespread importance of sacred groves for burial grounds, protection of springs and other social-cultural uses in Zimbabwe, Zambia and Mozambique. As in most cultures, both Western and non-Western (Colding and Folke 2001), social taboos, totems and traditional bylaws toward governance of resources and habitats exist in many parts of the Southern Africa (Sinthumule 2023) that protect and conserve natural ecosystems and biodiversity.

Cultural taboos as informal institutions, guide and determine human behaviour toward the natural environment (Colding and Folke 2001). For instance, in Zimbabwe, Zambia, Tanzania and Mozambique, trees growing around sacred water sources may not be felled for fear of the water drying up (van Rijsoort 2000). Individual large trees are normally used for traditional ceremonies and therefore not felled as they are thought by local community dwellers to have ties to the spirits (van Rijsoort 2000). For example, the Ficus species in many tribes in Kenya such as the Gikuyu, Meru, Luo and some clans among the Luhyia, and in some Zulu clans in South Africa have spiritual connection. In Zimbabwe, the Shangwe community conduct rainmaking ceremonies under the mubvumira, (*Kirkia acuminate*), muchakata (*Parinari curatellifolia*) and musasa (*Brachystegia speciformis*) trees that they traditionally believe to house the rain gods (Mavhura and Mushure 2019; Ngara et al. 2014). Thus, the species is culturally protected, or people refrained from cutting them down for other uses. Indigenous tree harvesting techniques in southern Africa is a key approach in managing trees in landscapes and harvesting is usually dependent on end use (Chirwa et al. 2015).

2.6.1.2 Agroforestry in Southern African Multifunctional Landscape

Southern African agroecosystems have largely remained subsistence in practice on ever-decreasing plots of lands with declining soil fertility. These farming systems have become inadequate to cope with population growth (Chirwa et al. 2015) and the adverse effect of climate change (Sheppard et al. 2020) that are experienced in the region. The landscapes in the region are characterised by traditional and improved agroforestry practices.

Traditional Agroforestry Practices

The Forest based farming system is practised characteristically in humid forests in the region and farmers practise shifting cultivation by clearing a new plot in the forest every 3–5 years and then allowed a fallow period of 7 up to 20 years. This is the basis of the Chitemene system in Zambia and the Machambas that are used to cultivate crops such as cassava, maize, beans, and potatoes in Niassa Province of Mozambique (Chidumayo 1987; Landry and Chirwa 2010).

The traditional tree legume-based system (Faidherbia albida parklands): For centuries farmers have retained a low density of trees in the parkland or two-tiered systems in the tropics, especially in the semi-arid areas in order to improve the yield of understorey crops (Akinnifesi et al. 2008). In southern Africa, farmers traditionally grow crops under scattered trees including the *Faidherbia albida*/maize system in southern and central Malawi. *F. Albida* has a unique characteristic of shedding most of its leaves during the wet season and resuming leaf growth during the dry season. This makes it possible to cultivate under its canopy with minimum shading effect on the companion crop. Substantial benefits are realized from these practices as resource-use by trees and associated crop components rarely overlap. The cultivation of crops under canopies of *F. albida* is the most notable of such traditional agroforestry practices in Southern Africa (Table 2.2).

Table 2.2 Some agroforestry systems and technologies in Southern Africa

	Agroforestry system/technologies					
Country	Improved Fallows and fertiliser tree systems	Contour Hedgerows	Conservation Agriculture	Boundary planting	Farm Forestry/ Woodlots	Traditional Systems[a]
Malawi	√	√	√	√	√	√
Mozambique						√
South Africa				√	√	
Zambia	√		√	√		√
Zimbabwe	√		√	√		
Tanzania	√				√	√

[a]Traditional Systems include agroforestry systems such as of the ngitili (traditional conserved areas) system of regeneration through controlled grazing, homesteads or homegardens and Faidherbia albida on farms in Tanzania (Viswanath and Lubina 2017)

***Transhumance silvopastural systems as adaptive resource management tradition*:** Seasonality is an important factor in the agricultural systems of the Southern African agroecosystems. It revolves around the very distinct wet and dry seasons. The availability of pasture in the late dry season is of particular importance in the management of the agroecosystems (Clarke 1994). In Zambia, transhumance grazing is still practised on the Kafue flats, where livestock graze in woodlands in the wet season and on the low-lying flood plains in the dry season (Sǿrensen 1993). Scoones (1990) also reported this pattern of grazing in Zimbabwe. This is important in the restoration of degraded land in that it allows the trees to recover through vegetative-coppice regrowth just before the start of the wet season.

Improved Agroforestry Systems and Practices

***The contribution of Fertiliser Tree*: *The case of Agroforestry practices and technologies*:** In Africa, and particularly southern Africa, the main constraint to agricultural productivity is soil nutrient deficiency. For this reason, agroforestry research in the region has focused on soil fertility replenishment (SFR) technologies over the years and the adoption and scaling-up of these practices is the main thrust of the ongoing on farm research (Akinnifesi et al. 2008; Akinnifesi et al. 2010). SFR technologies include a range of agroforestry practices (some examples in table 2.2) aimed at increasing crop productivity through growing trees (usually nitrogen-fixing), popularized as fertilizer tree-based systems, directly on agricultural land. Fertiliser tree -based systems involve soil fertility replenishment through on-farm management of nitrogen-fixing trees (Akinnifesi et al. 2010). The different fertiliser tree-based technologies that have been developed and promoted in southern Africa (Akinnifesi et al. 2008, 2010) over the last two decades include:

Hedgerow Intercropping: which is the simultaneous cultivation of two or more crops on the same field; usually, involving maize as the main crop in southern Africa and other agronomic crops as risk crops. In agroforestry based intercropping systems, species such as pigeon pea (*Cajanus cajan*), *Tephrosia vogelii, Faidherbia albida, Leucaena leucocephala*, and *Gliricidia sepium* are prominent. Gliricidia is a coppicing exotic species that is being used in the intercropping technologies throughout southern Africa.

Relay Cropping: is a system whereby nitrogen-fixing or legume trees shrubs and herbs such as *Sesbania sesban, S. macrantha, Tephrosia vogelii, Crotalaria* spp., or pigeon pea (*Cajanus cajan*), are grown as annuals and planted 3 to 5 weeks after the food crop.

Improved Fallow: has emerged as a promising alternative to traditional fallows. In an improved fallow, fast-growing, nitrogen fixing species such as *Sesbania sesban, Tephrosia vogelii, Gliricidia sepium*, and *Leucaena leucocephala* are grown for 2 to 3 years in the fallow plot after which, they are felled.

The most widely attempted system in Eastern Zambia is the improved fallow system while different forms of conservation agriculture are being promoted in Malawi, Zambia, and Zimbabwe. The most common traditional systems are common in Malawi, Mozambique and Zambia (Table 2.2).

2.6.2 East African Trees in Multifunctional Landscape

Across Eastern Africa, habitat change is largely driven by anthropogenic factors resulting both from competing uses of the land; deficient regulatory and enforcement instruments; poor or non-existent land tenure and lack of appreciation by policy makers of the intricate link between forest resource and people's livelihoods. The main human-induced habitat changes in the dry regions of East Africa include low input agriculture and especially pastoralist practices. For example, using rainfall and evapotranspiration indicators, 83% of the land is classified as arid to semi-arid, characterized by low, unreliable, and poorly distributed rainfall (Kabubo-Mariara and Karanja 2007). These areas are used for pastoral farming. In Tanzania, the unreserved areas are under heavy pressure, being converted to other land uses such as agriculture, wildlife protection, grazing, land settlement and industrial activities.

2.6.2.1 Traditional Forest and Tree Management Systems: The Case of Enclosure Practice and Its Local Governance Systems

Enclosures which are a type of land management, implemented on degraded, generally open access land in many dryland areas of East Africa are a mechanism for environmental rehabilitation with a clear biophysical impact on large parts of the formerly degraded lands (Nyberg et al. 2019). It has been shown that enclosures in Ethiopia resulted in the enhanced frequency distribution of two woody species with the recruitment of high number of woody species at lower diameter classes showing the potential of the technology to restore degraded lands (Gemechu and Dalle 2023). The most widely reported example of restoration through enclosures (*Ngitili*) and woodlots in East Africa is found in western Tanzania where it has proven very successful. Various estimations have been made by community members and researchers. For example, Kaale et al. (2003) estimated about 78,122 ha (46,593 ha individual farms and 34,206 ha communal lands) of Ngitilis from 172 villages in 2001 while Monela et al. (2005) estimated about 500,000 ha from the 833 villages. Some have estimated that there were between 500,000 and 600,000 ha of conserved areas under all interventions during the first decade of 2000 (Otsyina et al. 2008).

Enclosures are also reportedly used in Borana and Somali Region Pastoral Areas of Ethiopia. In these areas enclosures are of different types, based on the objectives and/or functions and they a driven by individual or community/group interests (Napier and Desta 2011).

2.6.2.2 Agroforestry Practices in the Restoration of Landscape

Rehabilitation of the land through agroforestry is the most common on human dominated landscapes where trees with multipurpose characteristics are used including some nitrogen fixing species for soil fertility improvement, fruit trees and timber species for wood and fibre. The most common agroforestry technologies promoted in East Africa include improved fallows in Western Kenya and rotational woodlots in dryland Western Tanzania (Table 2.3). Traditional systems include the *Faidherbia albida* parkland systems. Studies in Tanzania by Kaoneka and Solberg (1994) demonstrate that the *Shamba* people use their traditional agroforestry and intercropping systems to improve soil productivity and crops yields. This traditional agroforestry system consists of a spatial mixture of an under-storey of coffee and fruit shrub species and food crops (maize/beans and a variety of pulses), a middle storey of *Grevillea robusta* (a multipurpose species commonly used for timber, fuelwood and building poles production.

The Ngitili is also a form of traditional Agroforestry system as discussed in Sect. 2.3 and is based on the use of enclosures systems that allow the trees to either naturally regenerate or assisted to regenerate while at the same time also providing for animal fodder.

Table 2.3 Some agroforestry systems and technologies in East Africa

Country	Agroforestry system/technologies					
	Improved Fallows	Rotational Woodlots	Contour Hedgerows	Boundary planting	Home-gardens	Farm-forestry
Ethiopia					√	
Kenya	√		√	√		√
Rwanda	√		√	√		√
Tanzania		√			√	
Uganda	√		√	√	√	√

2.6.3 *Sahelian Trees in Multifunctional Landscape*

Land conversion and degradation in the Sahel is the primary environmental concern. Land conversion stems from a wide range of social and economic processes that are highly variable in both space (between countries, and between ecological regions within countries) and in time (as economic and other drivers change). Forests are removed faster than the rate they are replaced, this is if any restoration is carried out in the first place. The processes leading to land conversion in the Sahel is sometimes a consequence of multiple social, cultural, and political issues (Maisharou et al. 2015). The countries of the Sahel have been facing a persistent lack of rainfall, combined with anthropogenic factors such as monoculture agriculture, bushfires, failure or lack of fertilization, and grazing, among others. These have seriously affected ecological equilibrium, resulting in the degradation of natural resources, and a decline in agricultural production. The main identified drivers for land degradation in most Sahelian countries include increased population, expansion of economic activities dependent on natural resources, and land ownership structure. Other underlying drivers of land conversion and degradation in the Sahel region include loss of local know-how about soil conservation practices, and disappearance of services such as animal health, agricultural extension, supply of inputs and plant protection provided by the government. Tree planting has been a major approach to recover the ecological capacity of degraded lands in the Sahel, integrated as part of sustainable land management across the region (Maisharou et al. 2015).

2.6.3.1 Sustainable Land Management (SLM) Practices and Techniques in the Sahelian

The SLM concept has been considered in most Sahelian countries as an imperative for sustainable development and a tool for harmonization of interventions between the major sub-sectors of agriculture, livestock, water resources and the environment. It was unanimously accepted that SLM can promote cross-sector integration, as sought through Agenda 21 and the Poverty Reduction Strategies elaborated and adopted by most of the Sahelian countries for achieving the Millennium Development Goals (MDGs). The SLM practices and techniques are implemented in the field considering the objectives and the topography of the watershed units, which generally include plateaus, slopes, pediments, and valley bottoms. Based on the different types of soil, vegetation, and uses of these topographical units by different groups of users under different forms of land tenure, various SLM practices are implemented in the Sahel region.

There is no one-size-fits-all technical solution for the application of SLM practices and techniques. Therefore, an assessment of the watershed is made to choose a combination of techniques and practices adapted to the area's specific agro-ecological conditions (rainfall, topography, soil properties and structure, type of degradation, vegetation cover, etc.), which must also be feasible within the means of the farmers. Chirwa and Mahamane (2017) found the following are most widely used SLM practices, techniques and technologies associated with the rehabilitation of landscapes in different watersheds of Sahelian countries:

- ***Biological farming practices or crop management techniques*** such as conservation agriculture with crop rotation and intercropping, integrated pest management, tree planting and agroforestry, mulching and crop residue management.
- ***Improved pastures and rangelands techniques*** including planned grazing combined with water holes establishment, seeding grasses and woody species, control of invasive non palatable plants species, and control of bush fires through fire breaks establishment.
- ***Improvement of forest management*** with practices such as afforestation, farmers' assisted natural regeneration, shelter belts establishment, protection against wild fires and reforestation, sustainable conservation of plant diversity.
- ***Improved soil fertility management*** including utilization of crop residues, organic and humus amendment (compost, manure and green manure), integrated management of soil nutrients, environmentally friendly farming techniques.
- ***Water harvesting techniques*** for the recharge of groundwater table or improved rain water management practices through techniques as zaï (tassa), half-moon, stony bunds, ripping and sub-soiling, grass strips establishment, rehabilitation of wasteland with benches and trenches, drainage techniques.
- ***Erosion control*** using mechanical methods or structures which serve as barriers using stones or vegetation installed along contour lines, ridging/ridge tying, minimum tillage 14 for the management of water erosion; and sand dunes stabilization/fixation for the control of wind erosion.

2.6.4 Guinea-Sudanian Trees in Multifunctional Landscape

West Africa is a dynamic and fast developing region, which has experienced major biophysical and socioeconomic changes over the past several decades. The landscapes of West Africa have a long history of human settlement, exploitation, and development, with origins of agriculture dating back 4000–5000 years (Herrmann et al. 2020). Approximately 80–90% of the people in the region work in agriculture, making it the region's primary economic activity. However, one significant environmental problem affecting the area is land degradation, which has a detrimental impact on agriculture (Doso 2014). In turn, unsustainable agricultural practices in the area increase land degradation. Moreover, population growth has accelerated since the late 1960s, leading to subtle and radical land modification and conversion (Herrmann et al. 2020).

The West African vegetation is characterized by a diversity of tree species that have adapted to the region's diverse climate and environmental conditions (Sinsin et al. 2010). While the composition of tree species varies across West Africa, there are many key components that are found throughout the continent (Hawthorne and Jongkind 2006). Some of the most important tree species in West African vegetation are:

The shea tree (*Vitellaria paradoxa*) which is native to the region and is an important source of income and nutrition for local communities; the iconic baobab tree (*Adansonia digitata*) often referred to as the "Tree of Life."

The African Mahogany (*Khaya senegalensis*), a hardwood tree prized for its valuable timber; It is a fast-growing medium-sized tree up to 15–30 m in height and 1 m in diameter. Mahogany is commercially important for its a strong, tough, durable wood, and quality timber for diverse uses. It is popular for furniture, outdoor use and other woodwork due to its beauty and resistance to decay, infestation and warping. Due to illegal logging and overexploitation, it has been listed as endangered. Also, African mahogany (*Entandrophragma cylindricum*), which occurs in semi-evergreen and evergreen forests. It can grow up to 60 m and has high quality timber. It is found in countries such as Nigeria, Ghana, Côte d'Ivoire. The iroko tree (*Milicia excelsa*), a giant tropical hardwood and quality timber that can reach up to 50 m in height, is known as "king of trees". It has a wide range of habitat from deciduous to semi-deciduous or evergreen, primary or secondary forest, with an apparent preference for drier forest types, at elevations up to 1200 metres. In some countries of the region, it is normally protected as a sacred tree but mainly for its high economic value as round wood timber. Likewise, Obeche (*Triplochiton Scleroxylon*), in Central and West Africa, known as "African Whitewood." Obeche trees can reach up to 50 metres in the African rainforest.

In the dry savanna area of west Africa, are the Senegal Gum Acacia (*Acacia senegal*) producing gum arabic, a valuable substance used in various industries, including food and cosmetics. The nea tree, also known as the African rosewood (*Pterocarpus erinaceus*) is found in savannas and woodlands and used for carvings, furniture, and construction. Likewise, the kapok tree, commonly known as "ceiba" (*Ceiba pentandra*) produces cotton-like fibres in its seed pods which are used for stuffing pillows, cushions, and life jackets. In addition, many Acacia species common in West African savannas and adapted to dry conditions are known to provide valuable forage for livestock (Hawthorne and Jongkind 2006; Sinsin et al. 2010). These are just a few of the many tree species that contribute to the rich biodiversity and ecological significance of West African vegetation. These trees play essential roles in the region's ecosystems, economies, and cultures, making their conservation and sustainable management vital for the well-being of both nature and people.

The main causes of deforestation, fragmentation, and degradation in West Africa, as in other African regions, include, inter alia, demographic growth, terrorism with expanded refugees' settlements (e.g. Burkina-Faso), inappropriate soil management, shifting cultivation, insecurity in land tenure, variation of climatic conditions and intrinsic characteristics of fragile soils in diverse agroecological zones. The land degradation process is driven by both climate (direct factor) (Doso 2014), and human (indirect factors) including demographic growth, grazing pressures, agricultural practices, agricultural expansion, forest clearing and fuel wood harvesting (Dimobe et al. 2015; Padonou et al. 2017). These factors result in the reduction of vegetation cover, decrease in fallow periods and a reduction in the balance between fallow areas and cultivated fields, which are vital to maintaining soil fertility and reducing losses from erosion (Sinsin et al. 2010; Doso 2014; Dimobe et al. 2015).

Agroforestry, integrated farming and practices that restore vegetation cover are proposed as sustainable land practices in west Africa. These are expected to provide soil cover to protect soils against erosion, increase agricultural productivity and diversify farmers' sources of income (Doso 2014). The main agroforestry practice in the region is agroforestry parklands.

2.6.4.1 Parkland Agroforestry Landscape

Agroforestry practices in the Guineo-Sudanian landscape are vital for sustainable land use, biodiversity conservation, and climate resilience (Vodouhê et al. 2011). The region presents an ideal setting for agroforestry. Common agroforestry practices include parkland, livestock integration in farmlands, alley cropping, soil erosion control and water harvesting technologies, etc. In the Sudanian zone of West Africa, crops grown under a discontinuous cover of scattered trees dominate many landscapes and constitute the so-called parklands with fruit trees, nurse trees, nitrogen-fixing trees, etc. (Bayala et al. 2014).

Agroforestry parklands play an important role, through trees and shrubs providing soil cover that reduces erosion and buffers the impacts of climate change. They also provide green fodder that complements crop residues for livestock feeds, and fruits and leaves for human consumption and for income generation (Bayala et al. 2014). Being mixed agricultural systems, the interactions between these trees and crops have always been a key element determining the management options applied by farmers (Bayala et al. 2015). The commonly spared species in the parklands of the region are karité (*Vitellaria paradoxa*), néré (*Parkia biglobosa*), apple-ring acacia (*Faidherbia albida*), iroko (*Milicia excelsa*), tamarind (*Tamarindus indica*), marula (*Sclerocarya birrea*), Black Plum (*Vitex doniana*), velvet tamarind (*Dialium guineense*), etc. Protecting, conserving, and enhancing agroforestry in the Guineo-Sudanian landscape appears as an effective approach to address land degradation, promote sustainable land use, and carbon sequestration, enhance food security, and mitigate the impact of climate change (Assogbadjo et al. 2012).

2.6.5 Central Africa Trees in Multifunctional Landscape

The Central African region is vast and made up of a mix of different multifunctional landscapes spanning the Sahelian regions in the north to the tropical montane and rainforests in the South. Countries that make up the Central African region include Cameroon, Democratic Republic of Congo, Equatorial Guinea, Congo, the Central African Republic, Chad, Gabon, Burundi and Rwanda. The vegetation cover in the region is described through agroecological characteristics shaped by the agroclimatic systems. Rainfall generally decreases from the forested southern part to the Sahelian northern part. Mid-way between the forest south and the Sahelian north is found a tropical Savanna landscape.

The Central African region is home of four main agroecological zones with the most dominant being the humid tropical rainforest which also include in-land swamps and hosts the last remaining primary forests in Africa, surrounded in the sub-humid climatic zone by the semi-deciduous forests also known as moist woodland savanna and in semi-arid and arid part by savanna grassland and steppe. Each of these agroecological zones is characterized by place-based systems where people interact with forests and trees representing interdependent social-ecological relationships. As a region that is dominated by tropical humid forest agroecological systems, examples of multifunctional landscapes from this region will focus on the humid forest dominated areas.

The Convergence Plan of the Central Africa Forest Commission divides the humid forest dominated landscape space in Central Africa into 12 transboundary landscapes. The landscapes that include the Monte-Alen-Monts de Cristal; Gamba-Mavumba-Conkouati; Lope -Chailu-Louesse, Dja-Odzala-Minkebe; Sangha Tri-National; Econi-Bateke-Lefini; Lac-Tele-Lac-Tumba; Selonga-Lukenie-Sankuru, Maringa-Lapori-Wamba; Maiko-Tayna-Kahuzi-Biega; Ituri-Epulu-Aru; Virunga, cover an estimated surface area of 680,300 km^2 and cut across Cameroon, Democratic Republic of Congo, Equatorial Guinea, the republic of the Congo, the Central African Republic, Gabon, Rwanda, Uganda. These landscapes are home to thousands of plants, mammals, birds, reptiles, amphibians' species (CBFP 2005). The different tree species constitute a mosaic of land use systems that include rainforest (primary and secondary), agroforestry systems, agriculture plantations, swamp forest and mangroves etc.

2.6.5.1 Rainforest (Primary and Secondary Forests) in Multifunctional Landscapes

Landscapes in central Africa are estimated to contain <70% of intact forests in 2020 as compared to 78% (more than 210 million ha) in the year 2000 including dense evergreen rainforest, evergreen and semi-deciduous rainforest (Shapiro et al. 2023). These rainforest types are located across the landscapes in different proportions and include a very rich plant and animal biodiversity that contribute to the ecological dynamics and ecosystem services of importance to nature and people. The rainforest, that extends over six central African countries, such as Cameroon, Congo, Gabon, Equatorial Guinea, Central Africa Republic (CAR) and Democratic Republic of Congo (DRC), is the second largest continuous expanse of tropical rain forest on the planet, after the slightly larger block of the Amazon Basin (Tchatchou et al. 2015). About 30 million hectares out of the 200 million hectares of dense forests of central Africa are classified as protected areas (CBFP 2005). The primary and secondary forests constitute a greater proportion of protected areas and logging concessions for timber production and the protection of biodiversity in general

(Clark et al. 2009). Intercropping and relay cropping agroforestry systems are practised to different extents in the diverse agroecological zones found in the region.

2.6.5.2 Agroforestry Systems and Technologies

Small-scale mixed cropping systems in Central Africa always include different level of tree cover with their diversity, density and arrangement varying according to light requirement of the main crop in the production systems. This includes dispersed trees at very low density ranging from cereals and grain legumes-based systems to densely and nearly closed canopy of trees for shade tolerant roots and tuber crops such as cocoyam as well as other perennial crops, such as cocoa and coffee-based multistrata agroforestry systems. The diversity for tree species in these agroforestry systems are driven by their agroecological requirements as well as indigenous knowledge related to their uses and management practices (Sonwa et al. 2014).

The domestication research programme implemented by the World Agroforestry Centre in the last three decades has contributed to improving the diversification and intensification of useful indigenous tree species in small-scale farming systems with the aim to enhance their multifunctional potential. In the humid forest agroecosystem of Central Africa, tree species promoted include fruit trees such as *Irvingia gabonensis, I. wombolu; Dacryodes edulis, Garcinia kola, Ricinodendron heudelotii, Cola acuminata, C. nitida, Chrysophyllum albidum, Allanblackia spp.;* medicinal trees include *Prunus africana, Pausinystalia johimbe, Annickia chlorantha, Atypus affinis*; as well as melliferous trees and shrubs among which are *Vernonia spp., Alchornea cordifolia, Gambeya africana, Myrianthus arboreus, Polyscias* fulva (Leakey et al. 2022). These trees are associated to a plethora of food crops including tubers (yams, cocoyams, sweet potato and ginger), cereals such as maize and rice, vegetables, and cash crops such as coffee, cocoa, banana, oil palms (Zapfack et al. 2016).

Cocoa and coffee-based agroforestry systems are widespread mixed cropping systems in the humid forest landscapes. Over the last three decades, these systems have experienced several mutations driven by spatiotemporal dynamics of the social biophysical environment, changes in the macroeconomic context as well as agricultural policies within the region and at international level (Fogang et al. 2021). Nevertheless, several recent studies have documented these perennial based agroforestry systems as one of the most biologically diverse, climate-resilient, and socioeconomically productive farming practices addressing positive trade-off between regulation, provision, support and cultural ecosystem services (Vaast and Somarriba 2014). In these systems, the perennial crops are usually associated with both retained and cultivated indigenous tree species (timber and fruit tree species) and exotic introduced trees providing shade and other multifunctional services that contribute to sustainability of the agroforestry production systems. These diverse species help improve farmers' livelihoods, creating synergies and trade-offs between higher crop yield and the provision of ecosystem services to local households and global society (Sanial et al. 2023).

Other improved Agroforestry technologies such as improved fallows and rotational fallows, hedgerow intercropping (alley cropping/farming) using fast-growing nitrogen-fixing plants were also promoted in the humid forest and savanna agroecological zones as alternative to the slash and burn and shifting cultivation. Shifting cultivation system is considered as the main driver of forests degradation and deforestation in the region. The main nitrogen-fixing species introduced in the region for used in improved fallows and alley cropping agroforestry technologies are *Acacia auriculiformis*, *Calliandra calothyrsus*, *Cajanus cajan*, *Gliricidia sepium*, *Inga edulis*, *Tephrosia vogelii*, *Crotallaria spp*, *sesbania sesban* among others (Kanmegne et al. 2003; Keil et al. 2005; Degrande et al. 2007). Some successful experiences with indigenous species such as *Piliostigma reticulatum* and *Guiera senegalensis* have been reported in the Sahelian region that could be upscaled in the dry savanna parts of central Africa in Northern Cameroon, Chad, and Central African Republic (Bogie et al. 2019).

Other successful examples of agroforestry technologies include the smallholder Acacia-cassava based agroforestry systems adopted by farmers in the Bateke-plateau of the Kinshasa province for woodfuel supply and cassava production. A similar experiment with *Acacia auriculiformis* was reported in Haut-Katanga province as part of the AFODEK project.[1] The tree biomass production also contributes to climate change mitigation through the REDD+ mechanism. The project is expected to sequester 11–99 million tonnes of carbon dioxide equivalent over a 30-year period, and produce four million tonnes of charcoal, resulting in up to 13% of sustainably sourced woodfuel from the supply basin. This agroforestry technology is being practised in the Bateke-plateau of the Kinshasa province alongside other traditional agroforestry practices to produce several crops such as cassava, groundnuts, maize, beans, taro, and rice among, banana, vegetable, and other crops in association with oil palm, indigenous and exotic trees species providing a diversity of ecosystem services to farmers and global communities (Kachaka et al. 2020).

2.6.5.3 Swamps, Mangroves, and Peatlands for Carbon Sequestration

Swamps found in landscapes in Central Africa are made up of mangrove swamps (Buh et al. 2019; Ajonina 2022) and inland valley swamps. While the inland valley swamps are found within the humid, semi-arid and sub-humid agroecosystems and flooded by fresh water, mangrove swamps are found mainly along the Central African coast (notably the southern part of the Gulf of Guinea stretching from Cameroon's coast to Gabon's coast) and flooded by saline water. The main crop cultivated in both inland valley swamps and mangrove swamps are vegetables and rice grown in the dry season when the water level is low. The Central African region is home to five main mangrove species including the white mangroves (*Laguncularia*

[1] Agroforests for the development of Kipushi in DRC | Gret|Agroforests for the development of Kipushi in DRC | Gret

racemose and *Avicennia germinans*), the red mangroves (*Rhizophora racemose, R. harrisonii* and *R. mangle*), and an introduced species, *Nypa fruticans* (Ajonina et al. 2014; Din et al. 2016).

Research suggests that many of the swamp forests that occupy landscapes in the region are underlain by peat spanning an estimated 145,000 km^2, that store 30.6 petagrams of Carbon The region is estimated to contain the single largest peatland complex known in the tropics with a belowground carbon stocks that is equivalent to that of the aboveground tropical forests carbon stocks for the entire Congo Basin (Dargie et al. 2017). Thus, their conservation is crucial for the local and global fight against climate change.

2.7 Chapter Highlights

- Multifunctionality refers to the diversity and abundance of different landscape functions and services that can be achieved by reconciling and synergizing competing demands of different stakeholders, optimizing synergies and benefits, minimizing tradeoffs and achieving the goals of sustainable production and conservation of tree-based land uses concurrently.
- However, in an area where primary forests are highly deforested and fragmented or naturally found to be rare, trees outside of forests play important role as keystone species, a refuge for tropical biodiversity, ensuring sustainable food security, relieving pressure on the remaining forests, and providing ecosystem services that are crucial for the sustainability of rural livelihood.
- Deforestation and habitat fragmentation for a specific land use pose threats to the persistence of the tree species, be it within a forest or outside the forest.
- While agroforestry produces biodiversity benefits better than monoculture crops and livestock farming, it can only compliment rather than replace the role of primary forests in biodiversity conservation and the associated ecosystem services.
- The application of the ecosystem approach in multifunctional landscape is adopted and promoted to reach a balance of the three objectives of the Convention: conservation; sustainable use; and the fair and equitable sharing of the benefits arising out of the utilization of genetic resources;
- The role of trees in multifunctional landscapes requires a unified approach and conceptual frameworks. The aim in the end should be to provide approaches that can consolidate lessons learnt from the existing practices and accelerate the implementation of trees in multifunctional landscapes of Africa for biodiversity conservation, climate change, food security and poverty alleviation;
- The 10 principles of Multifunctional Landscape approaches emphasize the need for managing landscapes for multifunctionality of ecosystem services and the significance of people centred approaches to sustainable resource conservation;

- Trees in multifunctional landscapes from different regions of Africa are showcased ranging from traditional tree-based forest and agroforestry management systems to improved agroforestry technologies.

References

Ajonina GN (2022) Cameroon Mangroves: current status, uses, challenges, and management perspectives. In: Das SC, Pullaiah, Ashton EC (eds) Mangroves: biodiversity, livelihoods and conservation. Springer, Singapore. https://doi.org/10.1007/978-981-19-0519-3_21

Ajonina GN, Kairo J et al (2014) Assessment of mangrove carbon stocks in Cameroon, Gabon, the Republic of Congo (RoC) and the Democratic Republic of Congo (DRC) Including their potential for reducing emissions from deforestation and forest degradation (REDD+). In: Diop S, Barusseau JP, Descamps C (eds) The land/ocean interactions in the coastal zone of West and Central Africa. Estuaries of the World. Springer, Cham. https://doi.org/10.1007/978-3-319-06388-1_15

Akinnifesi FK, Chirwa PW, Ajayi OC et al (2008) Contributions of agroforestry research to livelihood of smallholder farmers in Southern Africa: 1. Taking stock of the adaptation, adoption and impact of fertilizer tree options. Agric J 3(1):58–75

Akinnifesi FK, Ajayi OC, Sileshi G et al (2010) Fertilizer trees for sustainable food security in the maize-based production systems of East and Southern Africa. A review. Agron Sustain Develop 30:615–629

Araia MG, Chirwa PW (2019) Revealing the predominance of culture over the ecological abundance of resources in shaping local people's forest and tree species use behavior: the case of the vhavenda people. South Afr Sustain 11. https://doi.org/10.3390/su11113143

Asaah EK, Wanduku TN, Tchoundjeu Z et al (2012) Do propagation methods affect the fine root architecture of African plum (Dacryodes edulis)? Trees 26:1461–1469. https://doi.org/10.1007/s00468-012-0720-9

Assogbadjo AE, Kakaï RG, Vodouhê FG et al (2012) Biodiversity and socioeconomic factors supporting farmers' choice of wild edible trees in the agroforestry systems of Benin (West Africa). Forest Policy Econ 14(1):41–49. https://doi.org/10.1016/j.forpol.2011.07.013

Bayala J, Sanou J, Teklehaimanot Z et al (2014) Parklands for buffering climate risk and sustaining agricultural production in the Sahel of West Africa. Curr Opin Environ Sustain 6:28–34. https://doi.org/10.1016/j.cosust.2013.10.004

Bayala J, Sanou J, Teklehaimanot Z et al (2015) Advances in knowledge of processes in soil–tree–crop interactions in parkland systems in the West African Sahel: A review. Agric Ecosyst Environ 205:25–35. https://doi.org/10.1016/j.agee.2015.02.018

Berkes F, Colding J, Folke C (2000) Rediscovery of traditional ecological knowledge as adaptive management. Ecol Appl 10(5):1251–1262

BGCI (2021) State of the world's trees. BGCI, Richmond

Bhagwat SA, Willis KJ, Birks HJB, Whittaker RJ (2008) Agroforestry: a refuge for tropical biodiversity? Trends Ecol Evol 23(5):261–267

Bogie NA, Bayala R, Diedhiou I et al (2019) Intercropping with two native woody shrubs improves water status and development of interplanted groundnut and pearl millet in the Sahel. Plant Soil 435:143–159. https://doi.org/10.1007/s11104-018-3882-4

Brauman KA, Garibaldi LA, Polasky S et al. (2019) Status and trends—nature's contributions to people (NCP). Global Assessment Report of the Intergovernmental Science-Policy Platform on Biodiversity and Ecosystem Services

Brockerhoff EG, Barbaro L, Castagneyrol B et al (2017) Forest biodiversity, ecosystem functioning and the provision of ecosystem services. Biodivers Conserv 26:3005–3035

Buh G, Bumtu PK, Meh A et al (2019) Space-based spatio-temporal mangrove forest dynamics and carbon stock assessment: the Ndongere mangrove, Southwest Cameroon. Conférence OSFACO: Des images satellites pour la gestion durable des territoires en Afrique, Cotonou, Benin, ⟨hal-02189377⟩ Mar 2019

CBFP (2005) The forests of The Congo basin: a preliminary assessment. The Congo Basin forest partnership. Available at: https://pfbc-cbfp.org/files/docs/Bassin%20du%20Congo/EdF/EDF2005_EN.pdf. Accessed 23 Apr 2024

Chidumayo EN (1987) Species structure in Zambian miombo woodland. J Trop Ecol 3(2):109–118

Chirwa PW, Mahamane L (2017) Overview of restoration and management practices in the degraded landscapes of the Sahelian and dryland forests and woodlands of East and southern Africa. South For J For Sci 79(2):87–94

Chirwa PW, Syampungani S, Geldenhuys CJ (2014) Managing southern African woodlands for biomass production: the potential challenges and opportunities. In: Seifert T (ed) Bioenergy from wood, Managing forest ecosystems, vol 26. Springer, Dordrecht, pp 67–87. https://doi.org/10.1007/978-94-007-7448-3_4

Chirwa PW, Larwanou M, Syampungani S et al (2015) Management and restoration practices in degraded landscapes of Southern Africa and requirements for up-scaling. Int For Rev 17(3):31–42

Clark CJ, Poulsen JR, Malonga R et al (2009) Logging concessions can extend the conservation estate for Central African tropical forests. Conserv Biol 23(5):1281–1293

Clarke J (ed) (1994) Building on indigenous natural resource management: forestry practices in Zimbabwe's communal lands. Forestry Commission, Harare

Colding J, Folke C (2001) Social taboos: "invisible" systems of local resource management and biological conservation. Ecol Appl 11(2):584–600

Cook CC, Grut M (1990) Agroforestry in Sub-Saharan Africa. Finan Develop 27:46–46

Dargie GC, Lewis SL, Lawson IT et al (2017) Age, extent and carbon storage of the central Congo Basin peatland complex. Nature 542(7639):86–90

De Pourcq K, Thomas E, Bas A et al (2017) Understanding and resolving conflict between local communities and conservation authorities in Colombia. World Dev 93:125–135

Degrande A, Asaah E et al (2007) Opportunities for and constraints to adoption of improved fallows: ICRAF's experience in the humid tropics of Cameroon. In: Advances in integrated soil fertility management in sub-saharan Africa: challenges and opportunities. Springer, Cham, pp 905–914

Denier L, Scherr S, Tchoundjeu Z et al (2015) The little sustainable landscapes book: achieving sustainable development through integrated landscape management. Global Canopy Programme, Oxford

Di Marco M, Harwood TD, Hoskins AJ et al (2019) Projecting impacts of global climate and land-use scenarios on plant biodiversity using compositional-turnover modelling. Glob Chang Biol 25(8):2763–2778

Dimobe K, Ouédraogo A et al (2015) Identification of driving factors of land degradation and deforestation in the Wildlife Reserve of Bontioli (Burkina Faso, West Africa). Global Ecol Conserv 4:559–571. https://doi.org/10.1016/j.gecco.2015.10.006

Din N, Ngo-Massou VM et al (2016) Local perception of climate change and adaptation in mangrove areas of the Cameroon Coast. J Water Res Protect 08:608–618. https://doi.org/10.4236/jwarp.2016.85050

Doso S (2014) Land degradation and agriculture in the Sahel of Africa: causes, impacts and recommendations. J Agricult Sci Appl 3(03):67–73

Ekroos J, Ödman AM, Sima S et al (2016) Sparing land for biodiversity at multiple spatial scales. Front Ecol Evol 3. https://doi.org/10.3389/fevo.2015.00145

FAO (2024) Managing landscapes for climate-smart agriculture systems. Available at: https://www.fao.org/climate-smart-agriculture-sourcebook/concept/module-a3-landscapes/a3-references/fr/. Accessed 23 Apr 2024

Favretto N, Shackleton S, Susannah MS et al (2021) Editorial for special issue: "collaboration and multi-stakeholder engagement in landscape governance and management in Africa: lessons from practice". Land 10:3. https://doi.org/10.3390/land10030285

Felipe-Lucia MR, Soliveres S, Penone C et al (2020) Land-use intensity alters networks between biodiversity, ecosystem functions, and services. Proc Natl Acad Sci 117(45):28140–28149

Fogang MK, Tientcheu MLA, Tankou C et al (2021) Spatio-temporal dynamics of land use on the expansion of coffee agroforestry systems in Cameroons production basins. Afr J Environ Sci Technol 15(11):505–518

Freeman OE, Duguma LA, Minang PA (2015) Operationalizing the integrated landscape approach in practice. Ecol Soc 20(1). https://doi.org/10.5751/ES-07175-200124

Gemechu T, Dalle G (2023) Rangeland biodiversity: status, challenges and opportunities review. J Rangeland Sci 13(3):1–9

Gitz V, Xu J, Lu Y et al (2022) Trees as hotspots: Using forests, trees, and agroforestry to foster diverse sustainable landscapes. Circular Agricult Syst 2(1):1–8

Grass I, Loos J, Baensch S et al (2019) Land-sharing/-sparing connectivity landscapes for ecosystem services and biodiversity conservation. People Nature 1(2):262–272. https://doi.org/10.1002/pan3.21

Gray E, Henninger N, Reij C et al (2016) Integrated landscape approaches for Africa's drylands. World Bank Publications

Hawthorne WD, Jongkind CCH (2006) Woody plants of Western African Forests. A guide to the forest trees, shrubs and lianes from Senegal to Ghana. Royal Botanic Gardens, Kew, UK, 1023 pp. ISBN 1842460897

Hecht SB, Pezzoli K, Saatchi S (2016) Trees have already been invented: carbon in woodlands. Collabra 2(1):24. https://doi.org/10.1525/collabra.69

Heikkila T, Villamayor-Tomas S, Garrick D (2018) Bringing polycentric systems into focus for environmental governance. Environ Policy Gov 28(4):207–211

Herrmann SM, Brandt M et al (2020) Accelerating land cover change in West Africa over four decades as population pressure increased. Commun Earth Environ 1(1):53. https://doi.org/10.1038/s43247-020-00053-y

Hölting L, Jacobs S, Frolking R et al (2019) Measuring ecosystem multifunctionality across scales. Environ Res Lett 14(12). https://doi.org/10.1088/1748-9326/ab5ccb

Hölting L, Felipe-Lucia MR, Cord AF (2020) Multifunctional landscapes, pp 128–134. https://doi.org/10.1016/B978-0-12-409548-9.12098-6

Hurtt GC, Chini L, Sahajpal R et al (2020) Harmonization of global land use change and management for the period 850–2100 (LUH2) for CMIP6. Geosci Model Dev 13(11):5425–5464

Iiyama M, Mukuralinda A, Ndayambaje JD et al (2018) Tree-based ecosystem approaches (TBEAs) as multi-functional land management strategies—Evidence from Rwanda. Sustain For 10(5). https://doi.org/10.3390/su10051360

Jaureguiberry P, Titeux N, Wiemers M et al (2022) The direct drivers of recent global anthropogenic biodiversity loss. Sci Adv 8(45):eabm9982. https://doi.org/10.1126/sciadv.abm9982

Kaale B, Mlenge W, Barrow E (2003) The potential of Ngitili for forest landscape restoration in Shinyanga region, a Tanzanian case study. In XII World Forest Congress. Food and Agricultural Organisation (FAO), Quebec City

Kabubo-Mariara J, Karanja FK (2007) The economic impact of climate change on Kenyan crop agriculture: A Ricardian approach. Glob Planet Chang 57(3–4):319–330

Kachaka ÉY, Munson AD, Gélinas N et al (2020) Adoption of an improved fallow practice using *Acacia auriculiformis* on the Batéké Plateau in The Democratic Republic of the Congo. Agrofor Syst 94:1047–1058. https://doi.org/10.1007/s10457-019-00474-5

Kanmegne J, Bayomock L, Degrande M et al (2003) Establishment of *Inga edulis* and *Calliandra calothyrsus* in improved fallow systems in southern Cameroon. Agrofor Syst 58:119–124

Kaoneka ARS, Solberg B (1994) Forestry related land use in the West Usambara Mountains, Tanzania. Agricult Ecosyst Environ 49:207–215

Keil A, Zeller M, Franzel S (2005) Improved tree fallows in smallholder maize production in Zambia: do initial testers adopt the technology? Agrofor Syst 64:225–236. https://doi.org/10.1007/s10457-004-2410-0
Kumar R, Kumar A, Saikia P (2022) Deforestation and forests degradation impacts on the environment. In: Environmental degradation: challenges and strategies for mitigation. Springer, Cham, pp 19–46
Landry J, Chirwa PW (2010) Analysis of the potential socio-economic impact of establishing plantation forestry on rural communities in Sanga District, Niassa province, Mozambique. Land Use Policy J. https://doi.org/10.1016/j.landusepol.2010.11.001
Leakey RR (2014) The role of trees in agroecology and sustainable agriculture in the tropics. Annu Rev Phytopathol 52:113–133
Leakey RRB (2017) Trees: meeting the social, economic and environmental needs of poor farmers — scoring Sustainable Development goals. In: Multifunctional agriculture: achieving sustainable development in Africa. Academic, San Diego, pp 417–420
Leakey RR, Tientcheu Avana ML, Awazi NP et al (2022) The future of food: Domestication and commercialization of indigenous food crops in Africa over the third decade (2012–2021). Sustain For 14(4). https://doi.org/10.3390/su14042355
Lovell ST, Bentrup G, Stanek E (2021) Agroforestry at the landscape level. In: Garrett HEG, Jose S, Gold MA (eds) North American Agroforestry (erd edn). American Society of Agronomy/ Wiley, pp 417–435
LPFN (2016) Integrated Landscape Management. Available at: https://peoplefoodandnature.org/about-integrated-landscape-management/. Accessed 23 Apr 2024
Maisharou A, Chirwa PW, Larwanou M et al (2015) Sustainable land management practices in the Sahel: review of practices, techniques and technologies for land restoration and strategy for up-scaling. Int For Rev 17(3):1–19. https://doi.org/10.1505/146554815816006974
Makino M (2005) The ecosystem approach of the CBD and fisheries management in Japan. Glob Environ Res English Edition 9(2). https://doi.org/10.3390/su10051360
Mansourian S, Berrahmouni N (2021) Review of forest and landscape restoration in Africa. Food and Agriculture Organisation, Rome
Marques MJ, Schwilch G, Lauterburg N et al (2016) Multifaceted impacts of sustainable land management in drylands: a review. Sustain For 8(2). https://doi.org/10.3390/su8020177
Mascia MB, Pailler S, Krithivasan R et al (2014) Protected area downgrading, downsizing, and degazettement (PADDD) in Africa, Asia, and Latin America and the Caribbean, 1900–2010. Biol Conserv 169:355–361
Mastrangelo ME, Weyland F, Villarino SH et al (2014) Concepts and methods for landscape multi-functionality and a unifying framework based on ecosystem services. Landsc Ecol 29:345–358
Mavhura E, Mushure S (2019) Forest and wildlife resource-conservation efforts based on indigenous knowledge: the case of Nharira community in Chikomba district, Zimbabwe. Forest Policy Econ 105:83–90
Milder JC, Hart AK, Minai J et al (2014) Integrated landscape initiatives for African agriculture, development, and conservation: a region-wide assessment. World Dev 54:68–80
Monela GC, Chamshama S, Mwaipopo R et al (2005) A study on the social, economic and environmental impacts of forest landscape restoration in Shinyanga Region, Tanzania. Ministry of Natural Resources and Tourism Forestry and Beekeeping Division and IUCN—The World Conservation Union Eastern Africa Regional Office
Mukhlis I, Rizaludin MS, Hidayah I (2022) Understanding socio-economic and environmental impacts of agroforestry on rural communities. Forests 13(4):556
Nagendra H, Paul S, Pareeth S, Dutt S (2009) Landscapes of protection: forest change and fragmentation in Northern West Bengal, India. Environ Manag 44:853–864
Napier A, Desta S (2011) PLI policy project review of pastoral rangeland enclosures in Ethiopia. USAID
Newton AC (2021) Ecosystem collapse and recovery. Cambridge University Press

Ngara R, Rutsate J, Mangizvo RV (2014) Shangwe indigenous knowledge systems: an ethno-metrological and ethnomusicological explication. Intl J Asian Soc Sci 4(1):81–88
Ngoma H, Angelsen A, Carter S, Roman-Cuesta RM et al. (2018) Climate-smart agriculture. Transforming REDD, p 175
Noss RF (1983) A regional landscape approach to maintain diversity. Bioscience 33(11):700–706
Nyberg G, Mureithi SM, Muricho DN et al (2019) Enclosures as a land management tool for food security in African drylands. J Land Use Sci 14(1):110–121
Oldekop JA, Holmes G (2016) A global assessment of the social and conservation outcomes of protected areas. Conserv Biol 30(1):133–141
Ordonez JC, Luedeling E, Kindt R et al (2014) Constraints and opportunities for tree diversity management along the forest transition curve to achieve multifunctional agriculture. Curr Opin Environ Sustain 6:54–60
Ostrom E (1999) Coping with tragedies of the commons. Annu Rev Polit Sci 2(1):493–535
Otsyina R, Rubanza CDK, Zahabu E (2008) Contribution of tree planting and conservation activities to carbon offsets in Shinyanga. NORAD Report
Padonou EA, Lykke AM, Bachmann Y et al (2017) Mapping changes in land use/land cover and prediction of future extension of bowé in Benin, West Africa. Land Use Policy 69:85–92. https://doi.org/10.1016/j.landusepol.2017.09.015
Rahman SA, Jacobsen JB, Healey JR et al (2017) Finding alternatives to swidden agriculture: does agroforestry improve livelihood options and reduce pressure on existing forest? Agrofor Syst 91:185–199
Reed J, Deakin L, Sunderland T (2015) What are 'integrated landscape approaches' and how effectively have they been implemented in the tropics: a systematic map protocol. Environ Evid 4(1):1–7
Reed J, Van Vianen J, Deakin EL et al (2016) Integrated landscape approaches to managing social and environmental issues in the tropics: learning from the past to guide the future. Glob Chang Biol 22(7):2540–2554
Reed J, van Vianen J, Foli S et al (2017) Trees for life: The ecosystem service contribution of trees to food production and livelihoods in the tropics. Forest Policy Econ 84:62–71
Reed J, Ickowitz A, Chervier C, Djoudi H, Moombe K, Ros-Tonen M, Yanou M, Yuliani L, Sunderland T (2020) Integrated landscape approaches in the tropics: a brief stock-take. Land Use Policy 99:104822
Reid R (2009) Trees in grazing systems. Agroforestry for natural resource management. In Nuberg I, George B, Reid R (eds), pp 219–235
Reiner F, Brandt M, Tong X et al (2023) More than one quarter of Africa's tree cover is found outside areas previously classified as forest. Nat Commun 14(1):2258
Rivers M, Newton AC, Oldfield S et al (2023) Scientists' warning to humanity on tree extinctions. Plants People Planet 5(4):466–482
Ros-Tonen MA, Mercy D, Thomas FGI (2014) From co-management to landscape governance: Whither Ghana's modified taungya system? Forests 5(12):2996–3021
Sanial F, Ruf D, Louppe M et al (2023) Local farmers shape ecosystem service provisioning in West African cocoa agroforests. Agrofor Syst 97:401–414. https://doi.org/10.1007/s10457-021-00723-6
Sayer J, Sunderland T, Ghazoul J et al (2013) Ten principles for a landscape approach to reconciling agriculture, conservation, and other competing land uses. Proc Natl Acad Sci 110(21):8349–8356
Sayer J, Margules C, Boedhihartono AK et al (2015) Landscape approaches; what are the preconditions for success? Sustain Sci 10:345–355
Scherr S, Shames S, Friedman R (2013) Defining integrated landscape management for policy makers. EcoAgriculture Policy Focus no. 10. EcoAgriculture Partners, Washington, DC
Schleicher J, Peres CA, Leader-Williams N (2019) Conservation performance of tropical protected areas: How important is management? Conserv Lett 12(5). https://doi.org/10.1111/conl.12650

Schümann H, Knierim A, Bellingrath-Kimura SD et al (2022) Working with or against multifunctional landscapes? A case study of land users' local knowledge of grassland–forest transition zones in northeastern Germany. Ecol Soc 27(1):38. https://doi.org/10.5751/ES-12970-270138

Scoones I (1990) Livestock populations and the household economy: a case study from southern Zimbabwe. PhD thesis, Imperial College, London

Shackleton CM, de Vos A (2022) How many people globally actually use non-timber forest products? For Policy Econ 135:102659. https://doi.org/10.1016/j.forpol.2021.102659

Shapiro A, d'Annunzio R, Desclée B et al (2023) Small scale agriculture continues to drive deforestation and degradation in fragmented forests in The Congo Basin (2015–2020). Land Use Policy 134:106922. https://doi.org/10.1016/j.landusepol.2023.106922

Sheppard JP, Bohn Reckziegel R, Borrass L et al (2020) Agroforestry: an appropriate and sustainable response to a changing climate in Southern Africa? Sustain For 12(17). https://doi.org/10.3390/su12176796

Sinsin B, Kampmann D, Thiombiano A, Konaté S (2010) Atlas de la Biodiversité de l'Afrique de l'Ouest. Tome I: Benin. Cotonou and Frankfurt/Main

Sinthumule NI (2023) Traditional ecological knowledge and its role in biodiversity conservation: a systematic review. Front Environ Sci 11. https://doi.org/10.3389/fenvs.2023.1164900

Soe T, Yeo-Chang Y (2019) Livelihood dependency on non-timber forest products: Implications for REDD+. Forests 10(5). https://doi.org/10.3390/f10050427

Sonwa DJ, Weise SF, Schroth G et al (2014) Plant diversity management in cocoa agroforestry systems in West and Central Africa—effects of markets and household needs. Agrofor Syst 88:1021–1034. https://doi.org/10.1007/s10457-014-9714-5

Sǿrensen C (1993) Control and sanctions over the use of forest products in the Kafue River Basin of Zambia, Rural Development Forestry Network Paper 15a. Overseas Development Institute, London

Stürck J, Verburg PH (2017) Multifunctionality at what scale? A landscape multifunctionality assessment for the European Union under conditions of land use change. Landsc Ecol 32:481–500

Sunderland T (2014) Landscape approach' defies simple definition – and that's good. Bogor, Indonesia, CIFOR. Available at: https://forestsnews.cifor.org/23834/landscape-approach-defies-simple-definition-and-thats-good?fnl=en. Accessed 12 Apr 2024

Swallow B, Boffa JM Scherr SJ (2006) The potential for agroforestry to contribute to the conservation and enhancement of landscape biodiversity. World agroforestry into the future. World Agroforestry Centre (ICRAF), Nairobi, pp 95–101

Tchatchou B, Sonwa DJ et al (2015) Deforestation and forest degradation in The Congo Basin: state of knowledge, current causes and perspectives. CIFOR, Bogor

Teka K, Haftu M, Ostwald M et al (2020) Can integrated watershed management reduce soil erosion and improve livelihoods? A study from northern Ethiopia. Int Soil Water Conserv Res 8(3):266–276

Thomas N, Baltezar P, Lagomasino D et al (2021) Trees outside forests are an underestimated resource in a country with low forest cover. Sci Rep 11(1):7919

Tilman D, Clark M, Kimmel K et al (2017) Future threats to biodiversity and pathways to their prevention. Nature 546(7656):73–81

Turley L (2016) The landscape approach: moving towards sustainable land use patterns. SSI Commentary, Report, Ottawa

UNEP-WCMC, IUCN, NGS (2021) Protected planet live report 2021. UNEP-WCMC, IUCN and NGS, Gland/Cambridge/Washington, DC

Vaast P, Somarriba S (2014) Trade-offs between crop intensification and ecosystem services: the role of agroforestry in cocoa cultivation. Agrofor Syst 88:947–956

Van Rijsoort J (2000) Non-timber forest products (NTFPs): their role in sustainable forest management in the tropics. National Reference Centre for Nature Management, International Agricultural Centre, Wageningen

Van Wyk B (2013) Field guide to trees of southern Africa. Penguin Random House South Africa

Visconti P, Butchart SH, Gauthier B et al (2019) Protected area targets post-2020. Science 364(6437):239–241

Viswanath S, Lubina PA (2017) Traditional agroforestry systems. Agrofor Anecdotal to Mod Sci:91–119

Vodouhê FG, Ousmane C et al (2011) Traditional agroforestry systems and biodiversity conservation in Benin (West Africa). Agrofor Syst 82:1–13. https://doi.org/10.1007/s10457-011-9377-4

Watson JEM, Dudley N, Segan D et al (2014) The performance and potential of protected areas. Nature 515:67–73

World Agroforestry Centre (2013) Strategy 2013–2022: transforming lives and landscapes with trees. World Agroforestry Centre

WRI (2024) Indicators of forests extent: trees outside forests. Available at: https://research.wri.org/gfr/forest-extent-indicators/trees-outside-forests. Accessed 6 Apr 2024

Zapfack L, Chimi DC, Noumi NV et al (2016) Correlation between associated trees, cocoa trees and carbon stocks potential in cocoa agroforests of Southern Cameroon. Sustain Environ 1(2):71–84

Chapter 3
Place and Roles of Trees in a Multifunctional Landscape: Trees and Environmental Services

Eméline S. P. Assèdé, Gudeta W. Sileshi, Paxie W. Chirwa, Hidirou Orou, and Stephen Syampungani

Abstract This chapter aims to highlight the role of trees in the multifunctional landscapes of sub-Saharan Africa, focusing on biodiversity, multipurpose tree management, climate change mitigation and adaptation, and tree-based approaches to the rehabilitation of degraded landscapes. Trees play a crucial role in contributing significantly to biodiversity, environmental management, and climate change mitigation in sub-Saharan Africa. Agroforestry is a sustainable system that promotes biodiversity conservation and economic benefits. In multifunctional landscapes, trees fulfil a variety of functions, ranging from supporting wildlife diversity to income generation through the production of fruit, fodder, wood, and timber. Sustainable agroforestry practices, combined with traditional knowledge, are cru-

E. S. P. Assèdé (✉)
Département Aménagement et Gestion des Ressources Naturelles, Faculté d'Agronomie, Université de Parakou, Parakou, Bénin

Department of Plant and Soil Sciences, Faculty of Natural and Agricultural Sciences, University of Pretoria, Pretoria, South Africa

G. W. Sileshi
Department of Plant Biology and Biodiversity Management, Addis Ababa University, Addis Ababa, Ethiopia

P. W. Chirwa
Department of Plant and Soil Sciences, Faculty of Natural and Agricultural Sciences, University of Pretoria, Pretoria, South Africa

H. Orou
Département Aménagement et Gestion des Ressources Naturelles, Faculté d'Agronomie, Université de Parakou, Parakou, Bénin

S. Syampungani
Department of Plant and Soil Sciences, Faculty of Natural and Agricultural Sciences, University of Pretoria, Pretoria, South Africa

Chair of Environment and Development, Oliver R. Tambo Africa Research Chairs Initiative (ORTARChI), Copperbelt University, Kitwe, Zambia

P. W. Chirwa et al. (eds.), *Trees in a Sub-Saharan Multi-functional Landscape*, https://doi.org/10.1007/978-3-031-69812-5_3

cial to the effective management of trees on farms, helping to address livelihood challenges. Trees in agroforestry systems are key players in the rehabilitation of degraded landscapes, regulating run-off water, preventing soil erosion, and maintaining water quality. In response to biodiversity loss, land degradation and climate change, a holistic approach to managing trees in multifunctional landscapes is essential to restore and strengthen the resilience of ecosystems.

Keywords Agroforestry · Biodiversity · Climate change · Landscape rehabilitation · Sustainable management · Livelihoods

3.1 Introduction

Sub-Saharan African Agricultural Landscapes (SSAL) have a long history of application of indigenous knowledge that includes tree-based cropping systems and tree management. The tree-based cropping systems support more than 500 million people in sub-Saharan Africa (SSA) with a wide range of crops and livestock integrated with trees involving a diverse set of cash income-generating activities (Assèdé et al. 2023a). Several tree management systems have been identified in SSA, with parklands singled out in the drier areas such as the Sahel and West Africa (ICRAF 2017).

Several studies have highlighted the importance of tree-based systems and their contribution to the livelihoods of local populations (Kuyah et al. 2020; Assèdé et al. 2023a). The tree-based systems in SSA offer a diversity of advantages such as the creation of marketable tree products to generate income, carbon sequestration, fuel, food, and the improvement of rural livelihoods (Kuyah et al. 2020; Muchane et al. 2020). In semi-arid areas, tree integration into agricultural system has the potential to enhance agricultural production and buffer rural livelihoods against drought (Kimaro et al. 2016, 2019; Swamila et al. 2020). The tree-crop combination can be either in spatial and or intimate mixture or in temporal rotation. Trees also provide fodder and shade for livestock as well as a range of other goods and services such as fibre, firewood and fruit (Franzel et al. 2014). Although tree introduction and management are receiving increasing attention in SSAL, the main question is how they impact biodiversity and environmental services. Additionally, there is a need to promote innovation and technology in tree management to increase uptake among farmers and policy makers for landscape rehabilitation to be successful. Studies in the last three decades observed changes in the whole agricultural system in SSA due to shortage of farmland, with associated loss of biodiversity (Gondwe et al. 2019, Assèdé et al. 2023a). The unprecedented population growth with associated increased demand for food and construction material, the increasing land and soil degradation exacerbated by climate modification have resulted in severe negative shifts in tree management systems in the SSAL (Assèdé et al. 2023a; Tripathi et al. 2019). Tree management systems in SSAL tend to be site-specific and farmer dependent considering the local biophysical circumstances and rural family motives for planting trees on farms (Liliane and Charles 2020). It is therefore important that the contribution of multifunctional landscape to improved biodiversity management and climate change mitigation and adaptation through management of multipurpose trees is well understood.

3.2 Trees and Biodiversity Improvement in Africa

3.2.1 Context and Study Design

A systematic literature search on trees in multifunctional landscapes in SAA between 1985 and 2022 was conducted using Scopus and Google Scholar search engines. To search for relevant publications, we use a combination of syntax: This review recorded 134 publications (97 and 37 from Scopus and Google Scholar, respectively). The multifunctional landscape concept in this book chapter is understood as a landscape system based on diverse and sustainable agricultural practices as well as the sustainable management of natural ecosystems with focus on the function, quality and human values.

The search covered different agroforestry zones spread across the main regions of SSA (Fig. 3.1a). Most of the studies focused on the eastern and western parts of SSA, highlighting the predominant role of trees and the growing interest in trees as an essential element of the landscape in these two regions. The central parts of SSA were relatively less represented. However, a significant research gap was noted in the southern and northern part of SSA with less than 10% of the recorded studies (Fig. 3.1b). This could be explained by a relatively low development and or research efforts on agroforestry systems in these two regions of Africa (Chap. 5). Yet, agriculture still dominates forest conversion processes, whatever the African region. But, the predominance of commercial agriculture in some countries like South Africa (Assèdé et al. 2023b) would have negatively affected the agroforestry development in southern part of Africa.

The approach used to study trees in multifunctional landscape varies with region, objective, and study context (Fig. 3.2). Around 50% of the studies on trees in multifunctional landscapes were field observations, followed by household surveys (31%) and literature reviews (10%). Experimental studies (6%) and data modelling (3%) were less represented. The observed disparity in methods used to study trees in multifunctional landscapes across SSA revealed the diversity in scientific focus and areas of interest. The paucity of experimental field studies on trees also highlights the lack of primary data on fundamental biophysical variables under real-life farmer conditions. Despite the long-term characteristic of the experimental studies on trees in farmlands, such data is urgently needed to increase our understanding of their roles in multifunctional landscapes.

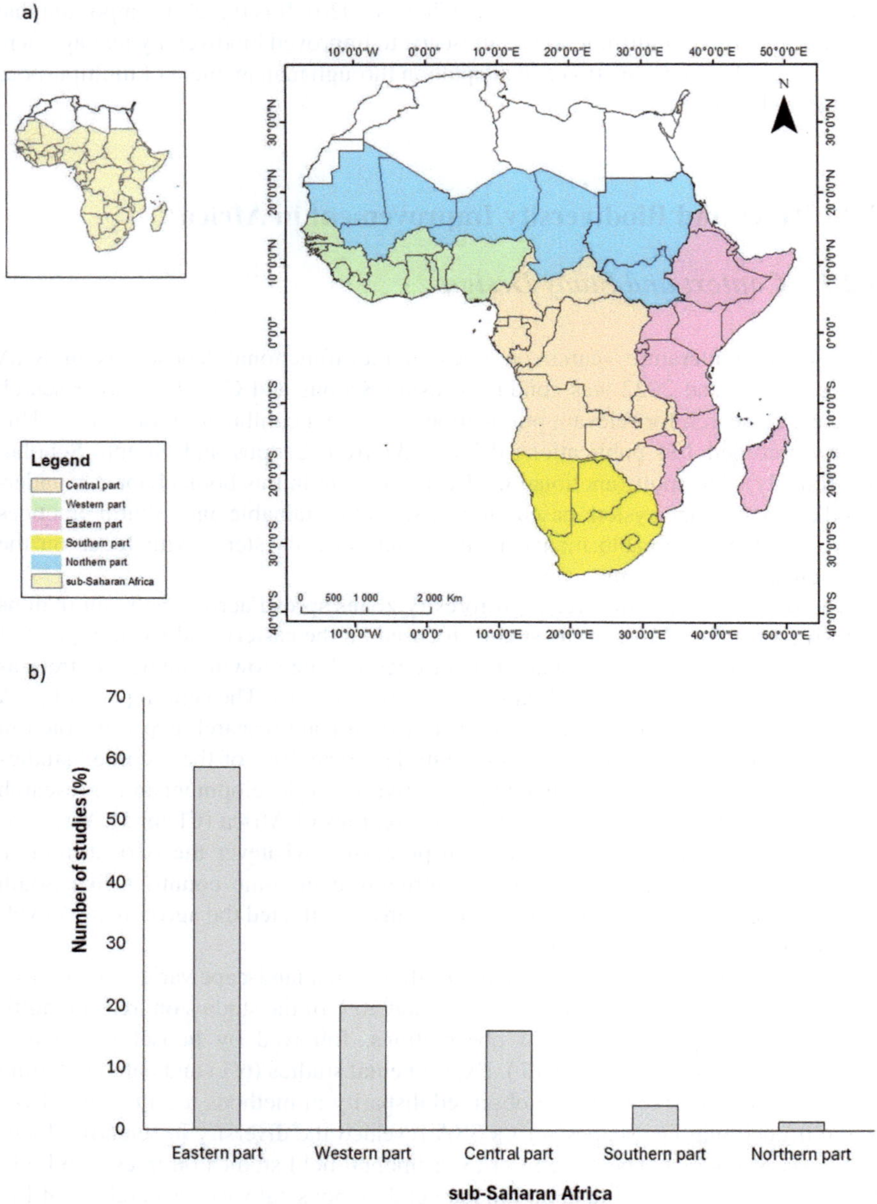

Fig. 3.1 Map of the study area showing (**a**) main regions of sub-Saharan Africa and; (**b**) distribution of studies on multifunctional landscape trees in the sub-Saharan Africa regions between 1985 and 2022

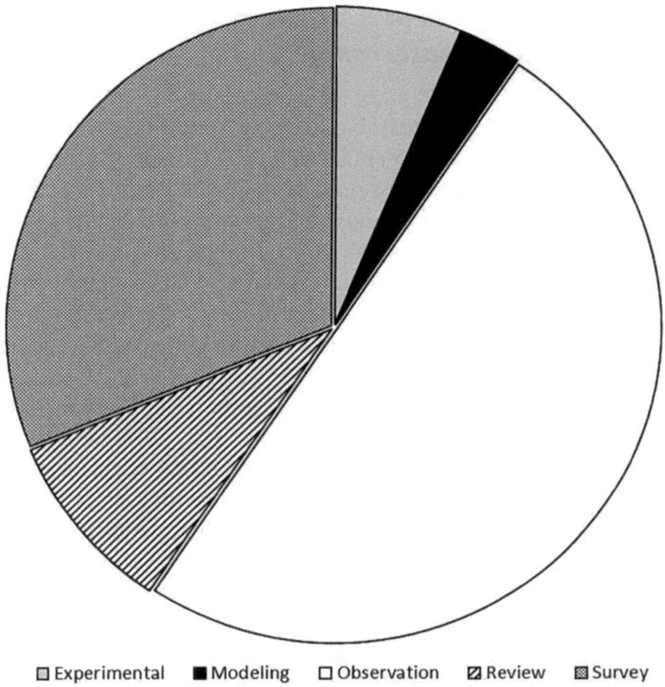

Fig. 3.2 Proportion of studies reporting trees in a sub-Saharan multifunctional landscape according to research design

3.2.2 Definition of the Concept of Agroforestry or Trees in the Agricultural Landscape

Agroforestry is broadly defined as the set of land use practices involving deliberate combination of trees (including shrubs, palms and bamboos) and agricultural crops and/or animals on the same land management unit in some form of spatial arrangement or temporal sequence such that there are significant ecological and economic interactions between the tree and agricultural components (Sinclair 1999). This definition also encompasses agroforests, which are complex agroforestry systems looking like and functioning as natural forest ecosystems but are integrated into agricultural management systems (Oke and Odebiyi 2007). Daizy et al. (2008) strengthened this definition by asserting that agroforestry is a dynamic system of natural resource management based on ecology. This is achieved through the integration of trees into agricultural lands and pastures, diversifying and supporting production to enhance social, economic, and environmental benefits.

3.2.3 *Importance of Trees in Agricultural Landscapes for Biodiversity Conservation*

Agroforestry can effectively address biodiversity conservation while simultaneously achieving social and environmental objectives (Chap. 5; Mulatu and Hunde 2020). Trees on farmlands have consistently demonstrated their capacity to improve the biodiversity within agricultural landscapes, thereby playing a key role in the conservation of biodiversity (Sistla et al. 2016; Mulatu and Hunde 2020). Studies and comprehensive analyses developed across the SSA have consistently observed a notable increase in plant diversity as a direct result of agroforestry practices. The dominant tree species in agricultural landscapes is site-specific and varies with regions across SSA (Table 3.1).

Multipurpose trees are more widely used in agroforestry systems in the Eastern part of SSA than in the Southern part (Noulekoun et al. 2017; Gebru et al. 2019; Pokwana et al. 2021; Ibrahim et al. 2022). The recorded research efforts in East Africa could be explained by the early scientific interest through important research projects (including ADDFOOD project and the Malawi Agroforestry Extension Project to address the crucial soil degradation issues in the early 1990s (Coulibaly et al. 2017). The establishment of important international organisms (including CIFORD-ICRAF) in East Africa would have greatly contributed to the agroforestry progress in this region of Africa through both technical and financial support.

The presence of trees within multifunctional landscapes also plays a crucial role in enhancing fauna diversity facilitating the interconnectedness of landscapes, their heterogeneity, and the complexity of vegetation structure (Mulatu and Hunde 2020). Additionally, trees offer nesting sites and refuge from predators, access to breeding territory, and conservation opportunities for a variety of animal species (Swallow et al. 2006). Here, we will use shade-coffee agroforestry as an example of the biodiversity benefits of agroforestry. We chose this agroforestry system because it has been evaluated for a variety of taxa including arbuscular mycorrhizae, epiphytes, pollinators, birds, and mammals (Hylander and Nemomissa 2008; Berecha et al. 2015; Buechley et al. 2015; De Beenhouwer et al. 2016; Mertens et al. 2020). According to Hylander and Nemomissa (2008), the coffee agroforests functioned as a substrate for a large number of epiphytic rainforest bryophytes and vascular plants and were of the same or higher importance in terms of harbouring forest biota than shade trees in the same system. In addition, Berecha et al. (2015) found a significant increase in the relative abundance of honeybees with increasing forest management in coffee agroforests. Buechley et al. (2015) estimated the highest bird species richness (51.00; 95% CI [44.49, 57.51]) in a shade-coffee agroforestry system compared to forest lands (19.25; 95% CI [17.82, 20.67]). Palearctic bird migrants were predominantly found in shade coffee, while more than 80% of migratory species were captured only in shade coffee. Using camera traps to compare diversity, community composition, and activity patterns of large (>2 kg) mammals between patches of natural forest and coffee forest, Mertens et al. (2020) found that species richness and diversity of large mammals were similar in coffee and natural forests.

Table 3.1 Tree species diversity in the sub-Saharan agricultural landscape

Region of sub-Saharan Africa[a]	Dominant tree species in farmlands	References
Southern part	*Acacia raddiana, Citrus sinensis, Faidherbia albida, Malus pumila, Persea americana, Prunus persica*	Pokwana et al. (2021)
Northern part	*Adansonia digitata, Balanites aegyptiaca, Combretum glutinosum, Detarium microcarpum, Faidherbia albida, Guiera senegalensis, Parkia biglobosa, Prosopis africana, Sclerocarya birrea, Terminalia spp., Vitellaria paradoxa*	Rouxel et al. (2005
Central part	*Acacia raddiana, Chrysophyllum albidum, Citrus sinensis, Dacryodes edulis, Faidherbia albida, Garcinia mangostana, Mangifera indica, Irvingia gabonensis, Persea americana, Ricinodendron heudelotii, Vitellaria paradoxa*	Etshekape et al. (2018) and Choungo Nguekeng et al. (2021)
Western part	*Acacia mangium, Acacia polyacantha, Acacia raddiana, Anarcadium occidentale, Citrus sinensis, Cola gigantea, Dialium guineense, Elaeis guineensis, Faidherbia albida, Gliricidia sepium, Lannea acida, Leucaenaleucocephala, Mangifera indica, Monotes kerstingii, Parkia biglobosa, Prosopis africana, Senna siamea, Sterculia setigera, Vitellaria paradoxa*	Tetteh et al. (2015), Damptey et al. (2021), and Assèdé et al. (2023a)
Eastern part	*Acacia melanoxylon, Acacia raddiana, Ailanthus trithysa, Albizia gummifera, Albizia harvey, Bauhinina petersiana, Brachystegia stipulata, Brachystegia utilis, Calliandra calothyrsus, Combretum apiculatum, Combretum collinum, Cordia africana, Croton macrostachyus, Dalbegiella nyassae, Diospyros kirkii, Diplorhynchus condylocarpon, Erythrina brucei, Eucalyptus camaldulensis, Eucalyptus globulus, Faidherbia albida, Faidherbia albida, Ficus sur, Ficus vasta, Vachellia negrii, Vachellia nilotica, Vachellia seyal, Gliricidia sepium, Khaya anthotheca, Lannea discolor, Leucaena leucocephala, Markhamia obtusifolia, Melia volkensii, Millettia ferruginea, Olea capensis, Olea welwitschii, Philenoptera bussei, Philenoptera violacea, Piliostigma thonningii, Psorospermum febrifugum, Pterocarpus angolensis, Pterocarpus rotundifolius, Rhamnus prinoides, Sclerocarya birrea, Senegalia senegal, Sesbania sesban, Sterculia quinqueloba, Stereospermum. Kunthianum, Tamarindus indica, Faidherbia albida, Terminalia sericea, Vachellia galpinii, Vachellia goetzei, Vachellia polyacantha, Vitex mombassae, Ziziphus abyssinica, Ziziphus spina-christi*	Noulekoun et al. (2017), Gebru et al. (2019), Nyirenda et al. (2019), Lelamo (2021), and Ibrahim et al. (2022)

[a]Refers to Fig. 3.1

In addition, in coffee-banana agroforestry, the aboveground carbon stock can reach 2.317 tons ha^{-1} (Namaalwa et al. 2019). The coffee forest land would offer a similar habitat with enough resources for food to maintain the bird diversity as high as comparable to the forest ecosystem.

In Malawi, *Macrotermes natalensis* (Haviland) and *Psammotermes allocerus* (Silvestri), were the two important woodland arthropods observed in traditional agricultural land and harvested woodland respectively (Nyirenda et al. 2019). Similarly, agroforestry systems in Zambia contribute to increasing the abundance and diversity of soil fauna (Sileshi and Mafongoya 2006; Sileshi et al. 2008). N'Woueni and Gaoue (2022) also recorded 24 arthropod species in agroforestry systems in West Africa induced by tree diversity. These results highlight the crucial role of trees in maintaining ecosystem composition.

3.2.4 Contribution of Trees in Agricultural Landscapes

Agricultural production in SSA is one of the most vulnerable production systems because it is mainly dominated by rain-fed agricultural systems which are usually associated with very low crop yield. To ensure food security, enriching the existing array of multipurpose trees within farmlands and home gardens with edible and storable crops like cereals has been identified as an efficient strategy to prevent food insecurity during periods of scarcity (Jemal et al. 2021). Trees play significant roles in various productive aspects such as providing fruit, fodder, wood, timber, and generating cash-income (Franzel et al. 2014; Kuyah et al. 2020; Assèdé et al. 2023a). Beyond these primary functions, they also offer vital agroecological services including soil fertility, erosion control, and climate change mitigation (Lelamo and Kemal 2021; Lelamo 2021). According to a diagnostic survey in Tigray region of Ethiopia, 86.4% of households acknowledge the multifaceted significance of trees, such as contributions to soil fertility improvement, regulation of runoff, amelioration of microclimates, environmental safeguarding, and provisioning of animal fodder during dry seasons (Gebru et al. 2019).

Notably, woody leguminous species offer an added advantage due to their intrinsic capability for nitrogen fixation, further enhancing their ecological importance. This further enhances soil fertility and reduces the need for synthetic fertilizers, thereby contributing to sustainable agricultural practices (Sebukyu and Mosango 2012; Sileshi et al. 2014). The maize yield in tree-based agroforestry systems is up to double (29 to 113%) in the first 3 years (Chirwa et al. 2003). In West Africa, *Gliricidia sepium* and *Cajanus cajan* fallows over 6 years can increase the maize yield by 200% compared to natural fallow (Swamila et al. 2020). A maximum crop yield of 915.9 kg/ha is obtained when tree density ranges from 20 to 40/ha (Binam et al. 2017).

The soil health benefits of agroforestry have been widely documented in SSA (Sileshi et al. 2014; Muchane et al. 2020). Agroforestry systems have been observed to increase soil organic carbon and nitrogen storage (Sileshi et al. 2008), thus improving the nourishment sources for soil microorganisms (Barrios et al. 2023). Under nutrient poor conditions, total N and SOC increased by 6–14 kg/ha/year and 2.6–194 kg/ha/year respectively (Masikati et al. 2014). Tree-based agroforestry systems can contribute between 100 kg N/ha/year and 200 kg N/ha/year (Amadalo and

Jama 2003). The trees managed in agroforestry systems improve soil texture and porosity, thus enhancing habitat quality as shelters for soil organisms, thereby fostering soil biodiversity (Sileshi et al. 2014; Barrios et al. 2023). Overall, the multifunctionality sustained by agroforestry systems consistently underlies soil erosion prevention and overall soil health, which can generate a propitious habitat for soil fauna (Muchane et al. 2020; Barrios et al. 2023). Agroforestry contributes to other ecosystem services such as regulating services (pest and diseases control, modification of microclimate, reducing wind effect) and supporting services (nutrient cycling) helping globally to build resilience.

The multifaceted role of trees extends beyond direct material gains. The tree-based crop systems have proven instrumental in addressing food security challenges faced by farmers in SSAL, improving access to essential micronutrients resources like calcium and vitamins, crucial for child growth and food security (Jemal et al. 2021; Choungo Nguekeng et al. 2021). This has led to enhanced employment opportunities and increased income, bolstering the overall resilience of farmers (Callo-Concha et al. 2019; Gebru et al. 2019). Anyanwu and Etela (2013) emphasized how livestock, an important component of farming systems in western and central parts of Africa, significantly benefits from tree management practices output, as they provide valuable fodder for sheep and goats breeders.

3.3 Multipurpose Tree Management in the Context of Multifunctional Landscape

Managing multipurpose trees in multifunctional landscapes in SSA involves tree integration into agricultural landscapes to enhance ecosystem services provision (Tscharntke et al. 2011; Pokwana et al. 2021) (Fig. 3.3). Two main management methods are generally used: planting and pruning tree branches. The perception of farmers towards tree planting on their farmlands is influenced by factors such as household energy sources and easy access to tree seedlings (Ibrahim et al. 2022).

Farmers have a long-standing tradition of maintaining on-farm trees due to benefits, including direct contributions to timber, fencing, fuelwood, and charcoal production values as highlighted by Sida et al. (2018). More than 80 edible tree species across several tree management practices including multi-storey coffee/cocoa systems, multi-purpose trees on farmlands (MPT), improved fallows, intercropping and parkland were recorded in SSAL for enhancing local household nutrition (Callo-Concha et al. 2019, Jemal et al. 2021; Assèdé et al. 2023a) (Fig. 3.4).

- **Multi-storey agroforestry systems** involve a mixture of crops and trees of different heights (e.g., coffee, cocoa, coconut) and vegetables underneath in a polyculture system. Typical examples of multi-storey agroforestry systems are the Chagga home gardens in Tanzania, the Enset-coffee homegardens in Ethiopia, and Cocoa agroforestry in West Africa (see Chap. 5 of this book). Tree-to-tree distance would be wide enough to minimize the impact on the yields of the

Fig. 3.3 *Vitellaria paradoxa* (shea tree), a multipurpose tree in agroforestry system with soy crop in West Africa (Photo Emeline Assèdé 2023)

coffee/cocoa/coconut and let sufficient light through to understory crops. Coffee, cocoa or coconut tree generally constitutes the middle-storey species in multi-storey rainforests in Western and Central parts of Africa (Djokoto et al. 2017; Sonwa et al. 2019).

- **Multi-purpose trees on farmlands** involve planted or purposely conserved trees scattered in a stand mainly for food production (see Chap. 5 of this book).
- **Improved fallows** are sequential tree fallow systems which involve deliberately planted tree/shrub (usually legumes) species as integral components in a crop-fallow rotation. It is usually called by the tree species used: "Gliricidia fallows". Using fertilizer trees, the returns to labour over 6 years in improved fallow is 17% outperform compared to natural fallow (Degrande 2001) (see Chap. 5 of this book).
- **Parkland agroforests** is a farming system developed in West Africa with *Faidherbia albida*, and indigenous fruit trees dominated by *Vitellaria paradoxa* and *Parkia biglobosa* (Boffa 1999). It is a selection and retention of naturally regenerated trees in crop fields to enhance both diet, health, and additional incomes. Trees in parklands also contribute to carbon sequestration and soil organic carbon (Bayala et al. 2018; Lelamo 2021).

Fig. 3.4 Multi-storey coffee/banana system in East Africa (Photo Emeline Assèdé 2024)

Farmer managed natural regeneration is the systematic regeneration of living tree stumps in the fields to be re-vegetated. Desired tree stumps are selected. Tree density could range 19 to 360 trees ha^{-1} (Chomba et al. 2020). For each stump, a decision is made as to how many stems will be chosen for growth. The tallest and straightest stems are selected and side branches removed to roughly half the height of the stem. The remaining stems are then culled. Regular pruning of unwanted new stems and side branches attains best results (Chomba et al. 2020). The adoption of tree pruning at various intensities rather than removing multi-purpose shade trees when crops are grown was suggested for better tree conservation in SSAL (Assèdé et al. 2021). This technique can also significantly improve biomass productivity (by 1.16%) and enhance firewood provision (Assèdé et al. 2021).

3.4 Trees as Response to Climate Change Effects

A holistic landscape management approach, encompassing diverse land-use systems like forests and mixed tree systems, has the potential to effectively diminish the impacts of climate change and amplify adaptive capacities (Dewi et al. 2013). Agroforestry practices emerge as positive contributors to the proliferation of species diversity, a main determinant of ecosystem resilience in climate change context.

The mitigation and resilience of an agro-ecosystem to climate change effects would be intimately linked to the levels of above-ground biomass (AGB) and carbon stock and the wider diversity of the agro-ecosystem (Fuchs et al. 2022). If forest ecosystems are crucial to offset greenhouse gas emissions through carbon sequestration, planting both multipurpose tree species and those more likely suited to a future climate could be seen as a potential adaptation strategy to increase resilience to climate change (Chen et al. 2023). Trees management in multifunctional landscapes has strong potential to reduce climate change impacts and increase resilience through binding both carbon stock (in biomass and soil), and microclimate modification (Dewi et al. 2013; Mbow et al. 2014; Mulatu and Hunde 2020; Fuchs et al. 2022). Studies showed that the carbon stock is closely linked to the diversity of the agroecosystem (Fuchs et al. 2022). In different parts of Africa, agroforestry systems have been shown to store more carbon than crop lands (Luedeling et al. 2011) and create resilient ecosystems that are better equipped to withstand the effects of a changing climate. In certain arid regions of Africa, where carbon stocks have declined due to land degradation, planting trees such as *Senegalia senegal* can significantly increase soil organic carbon (Bakhoum et al. 2018). In East Africa, for example, agroforestry systems have been observed to sequester large quantities of carbon through biomass accumulation (Kaonga and Coleman 2008; Luedeling et al. 2011). Studies demonstrated that improved fallows, especially those with exotic multipurpose species, sequester more carbon (2.0–6.0 Mg C ha^{-1} $year^{-1}$) than coppiced miombo woodlands, a dominant vegetation type in Southern Africa (<1.0 Mg C ha^{-1} $year^{-1}$) (Chirwa et al. 2023). These rates may depend on the tree species, site characteristics, planting densities, and the agroforestry technology used (Chirwa et al. 2023). For example, improved fallows with Sesbania tree species in Eastern Zambia sequester carbon at 26–120 Mg C ha^{-1} (Kaonga and Coleman 2008). In Zimbabwe, Nyamadzawo et al. (2005) reported carbon accumulation of 26.3 and 25.4 Mg ha^{-1} in leaves and twigs, respectively in a two-year improved fallow with *Acaciella angustissima* and *Sesbania sesban*, while different five-year fallow with rotational woodlots in tobacco growing areas in Tanzania sequestered 11.6 to 25.5 Mg C ha^{-1} (Kimaro 2009).

3.5 Tree-Based Approaches for Rehabilitation of Degraded Landscapes in SSA

According to the GLASOD (Global Assessment of Soil Degradation), there are 13 types of soil degradation including pollution, water erosion, wind erosion, overblowing, loss of nutrients, acidification, salinization, compaction and physical degradation (FAO and ITP 2015). Some evidence from SSA suggests that some agroforestry practices can reduce surface runoff and soil loss by water erosion (Muchane et al. 2020; Barrios et al. 2023). For example, agroforestry practices can reduce soil erosion rates by 50% (Muchane et al. 2020). The root systems of trees bind soil particles and reduce soil erosion during heavy rainfall. This preserves soil quality and prevents sedimentation in nearby water bodiesμ(Lelamo and Kemal 2021), which is crucial for maintaining water quality and aquatic ecosystems. In farms with improved agroforestry systems, the land use is well planned and land is effectively used. The multipurpose trees such as N-fixing trees are planted as windbreaks, shelterbelts, or around the fields on degraded lands. Trees are also intercropped with annual crops to reduce soil erosion and improve soil fertility (Assèdé et al. 2023a).

The predominant approaches to restoration in dry forests and woodlands of SSA use natural regeneration such as complete coppice, coppice with standards and selective cutting, pollarding, pruning and lopping (Chirwa et al. 2015). In the Tigray region in Ethiopia, enclosures are prevalent in livestock farming areas where they are used to encourage rehabilitation of grazing land (Chirwa et al. 2015). Also, in the Sahel, enclosures with different sustainable land management practices are used to protect young growing trees (Maisharou et al. 2015). In East Africa, enclosures are applied on degraded land in dryland areas as a mechanism to promote natural regeneration. This is mainly for environmental rehabilitation and has resulted in very visible biophysical impacts on large parts of the formerly degraded lands (Chirwa and Mahamane 2017). There is also increased adoption of forest and tree management practices and technologies that have the potential to promote rehabilitation and/or restoration of degraded forests, woodlands and parklands. For example, assisted or aided regeneration is especially prevalent in the Sahel where indigenous tree species are planted (enrichment planting) or seedlings from natural sources are protected and managed by landowners (Larwanou and Saadou 2011). Research on FMNR (Farmer Managed Natural Regeneration) indicates that old systems such as native non-browse shrubs in Ethiopia are important in facilitating regeneration of late successional tree species (Abiyu et al. 2017). The FMNR is also common in countries like Niger, with very significant gains reported in the southern part of the country (Haglund et al. 2011).

Opportunities exist for promoting forest restoration through tree planting for carbon markets in rural communities in the form of community-based REDD+ initiatives for sustainable forest management because Agroforestry is recognized as an important avenue for rehabilitation of degraded areas and enhancing ecosystem services. Natural regeneration through active involvement of local communities supported by enabling policies such as clear forest, tree and land tenure is the most successful and promising option for restoration of the large areas of degraded land. The policies promoting participatory natural resources management are in place in most African countries. However, the implementation of these regeneration activities has mostly taken a pilot project based approach, with few strategies for up scaling by the governments. In addition, enabling policies for community-based approaches, including clear cut land tenure, and equitable benefit sharing, are limited in several African countries. This constraints investments in developing, managing and using these resources, thus impairing their full contribution to rural livelihoods and their sustainability in land restoration.

3.6 Chapter Highlights

The review explores the role of trees in a multifunctional landscape in Sub-Saharan Africa, focusing on research and management approaches. Various agroforestry studies were recorded with diverse research designs, including field observations, household surveys, literature reviews, manipulative experiments, and modelling.

Agroforestry, defined as a land use system integrating trees into agricultural lands, is highlighted for its ability to address biodiversity conservation while achieving social, economic, and environmental objectives. Trees provide diverse benefits, such as fruit, fodder, wood, timber, and income to households for food security. Trees in agricultural landscapes, particularly in agroforestry systems, contribute significantly to biodiversity improvement. A positive correlation between agroforestry practices and increased plant diversity was observed, with site-specific variations across SSA.

The impact of trees extends beyond plant diversity to fauna diversity, with trees providing nesting sites, refuge from predators, breeding territory access, and conservation opportunities for various animal species. Agroforestry systems are shown to improve soil texture, porosity, and overall biodiversity. The systems also enhance soil organic carbon and nitrogen storage, preventing soil erosion and promoting overall soil health.

The chapter highlights the multifaceted significance of trees for households, acknowledging at more than 86% their contributions to soil fertility, runoff regulation, microclimate amelioration, environmental safeguarding, and animal fodder provision during dry seasons.

The management of multipurpose trees in multifunctional landscapes involves sustainable agroforestry practices, including tree integration and pruning. Farmers traditionally maintain on-farm trees due to their various benefits, such as provision of timber, fuelwood, non-timber forest products, soil fertility improvement, and crop shade. These trees enhance local household nutrition, providing timber, fuelwood, and non-timber forest products, ultimately improving livelihoods and income.

Trees are identified as crucial elements in climate change mitigation strategies. Planting multipurpose tree species, adapted to today and future climates, is considered an adaptation strategy to increase resilience and food security. Agroforestry systems are shown to sequester carbon, with diverse systems storing more carbon and creating resilient ecosystems. In arid regions, planting specific tree species is recommended to increase soil organic carbon.

Finally, tree-based approaches for the rehabilitation of degraded landscapes are discussed, emphasizing the role of agroforestry trees in regulating runoff, preventing soil erosion, and preserving water quality. Improved agroforestry systems, incorporating nitrogen-fixing species, are highlighted for their ability to reduce soil erosion and effectively utilize land.

References

Abiyu A, Teketay D, Glatzel G et al (2017) Restoration of degraded ecosystems in the Afromontane highlands of Ethiopia: comparison of plantations and natural regeneration. South For J For Sci 79:103–108. https://doi.org/10.2989/20702620.2016.1254917

Amadalo B, Jama B (2003) Improved fallows for western Kenya: an extension guideline. World Agroforestry Centre, Nairobi

Anyanwu NJ, Etela I (2013) Chemical composition and dry matter degradation characteristics of multi-purpose trees and shrubs in the humid lowlands of southeastern Nigeria. Agrofor Syst 87:747–754. https://doi.org/10.1007/s10457-012-9593-6

Assèdé ESP, Azihou FA, Biaou SSH et al (2021) Managing woodland development stages in Sudanian dry woodlands to meet local demand in fuelwood. Energy Sustain Dev 61:129–138. https://doi.org/10.1016/j.esd.2021.01.006

Assèdé ESP, Biaou SSH, Chirwa PW et al (2023a) Low-cost agroforestry technologies for climate change mitigation and adaptation in sub-Saharan Africa: a review. Bois Forets Des Tropiques 356:29–42. https://doi.org/10.19182/bft2023.356.a36908

Assèdé ESP, Orou H, Biaou SSH et al (2023b) Understanding drivers of land use and land cover change in Africa: a review. Curr Landsc Ecol Rep 8:62–72. https://doi.org/10.1007/s40823-023-00087-w

Bakhoum N, Fall D, Fall F et al (2018) *Senegalia senegal* (synonym: *Acacia senegal*), its importance to sub-Saharan Africa, and its relationship with a wide range of symbiotic soil microorganisms. S Afr J Bot 119:362–368. https://doi.org/10.1016/j.sajb.2018.10.007

Barrios E, Coe R, Place F et al (2023) Nurturing soil life through agroforestry. In: Biological approaches to regenerative soil systems, 2nd edn. CRC Press, Boca Raton, pp 265–278

Bayala J, Kalinganire A, Sileshi GW, Tondoh JE (2018) Soil organic carbon and nitrogen in agroforestry Systems in sub-Saharan Africa: a review. In: Bationo A, Ngaradoum D, Youl S et al (eds) Improving the profitability, sustainability and efficiency of nutrients through site specific fertilizer recommendations in West Africa agro-ecosystems. Springer, Cham, pp 51–61

Berecha G, Aerts R, Muys B, Honnay O (2015) Fragmentation and Management of Ethiopian Moist Evergreen Forest Drive Compositional Shifts of insect communities visiting wild Arabica coffee flowers. Environ Manag 55:373–382. https://doi.org/10.1007/s00267-014-0393-9

Binam JN, Place F, Djalal AA, Kalinganire A (2017) Effects of local institutions on the adoption of agroforestry innovations: evidence of farmer managed natural regeneration and its implications for rural livelihoods in the Sahel. Agric Econ 5:2. https://doi.org/10.1186/s40100-017-0072-2

Boffa J-M (1999) Agroforestry parklands in sub-Saharan Africa. Food and Agriculture Organization, Rome

Buechley ER, Şekercioğlu ÇH, Atickem A et al (2015) Importance of Ethiopian shade coffee farms for forest bird conservation. Biol Conserv 188:50–60. https://doi.org/10.1016/j.biocon.2015.01.011

Callo-Concha D, Mohamed O, Seyoum Aragaw H (2019) Local alternatives to local problems: the contribution of agroforestry system by-products to food and nutrition security of communities in southwestern Ethiopia. Food Stud Interdisc J 9:29–42. https://doi.org/10.18848/2160-1933/CGP/v09i01/29-42

Chen K, Wang F, Lu XL (2023) Dynamic assessment of forest assets based on ecosystem system. HSET 50:67–76. https://doi.org/10.54097/hset.v50i.8467

Chirwa PW, Mahamane L (2017) Overview of restoration and management practices in the degraded landscapes of the Sahelian and dryland forests and woodlands of East and southern Africa. South For J For Sci 79:87–94. https://doi.org/10.2989/20702620.2016.1255419

Chirwa TS, Mafongoya PL, Chintu R (2003) Mixed planted-fallows using coppicing and non-coppicing tree species for degraded Acrisols in eastern Zambia. Agrofor Syst 59:243–251. https://doi.org/10.1023/B:AGFO.0000005225.12629.61

Chirwa PW, Larwanou M, Syampungani S, Babalola FD (2015) Management and restoration practices in degraded landscapes of Southern Africa and requirements for up-scaling. Int For Rev 17:31–42

Chirwa PW, Musokwa M, Mwale SE et al (2023) Agroforestry systems for mitigating climate change and reducing carbon footprints of land-use systems in southern Africa. Carbon Footp 2(1):1

Chomba S, Sinclair, F, Savadogo et al. (2020). Opportunities and constraints for using farmer managed natural regeneration for land restoration in sub-Saharan Africa. Front For Glob Change 3:571679. https://doi.org/10.3389/ffgc.2020.571679

Choungo Nguekeng PB, Hendre P, Tchoundjeu Z et al (2021) The current state of knowledge of Shea butter tree (*Vitellaria paradoxa* C.F.Gaertner.) for nutritional value and tree improvement in west and Central Africa. Forests 12:1740. https://doi.org/10.3390/f12121740

Coulibaly JY, Chiputwa N, Nakelse T, Kundhlande G (2017) Adoption of agroforestry and the impact on household food security among farmers in Malawi. Agric Syst 155:52–69

Daizy RB, Ravinder KK, Shibu J, Harminder PS (2008) Ecological basis of agroforestry. CRC Press Taylor/Francis Group, New York, p 383

Damptey FG, De La Riva EG, Birkhofer K (2021) Trade-offs and synergies between food and fodder production and other ecosystem Services in an Actively Restored Forest, natural Forest and an agroforestry system in Ghana. Front For Glob Change 4:630959. https://doi.org/10.3389/ffgc.2021.630959

De Beenhouwer M, Geeraert L, Mertens J et al (2016) Biodiversity and carbon storage co-benefits of coffee agroforestry across a gradient of increasing management intensity in the SW Ethiopian highlands. Agric Ecosyst Environ 222:193–199. https://doi.org/10.1016/j.agee.2016.02.017

Degrande A (2001) No title found. Agrofor Syst 53:11–19. https://doi.org/10.1023/A:1012220807248

Dewi S, Van Noordwijk M, Ekadinata A, Pfund J-L (2013) Protected areas within multifunctional landscapes: squeezing out intermediate land use intensities in the tropics? Land Use Policy 30:38–56. https://doi.org/10.1016/j.landusepol.2012.02.006

Djokoto JG, Afari-Sefa V, Addo-Quaye A (2017) Vegetable diversification in cocoa-based farming systems Ghana. Agric Food Secur 6:6. https://doi.org/10.1186/s40066-016-0082-4

Etshekape PG, Atangana AR, Khasa DP (2018) Tree planting in urban and peri-urban of Kinshasa: survey of factors facilitating agroforestry adoption. Urban For Urban Green 30:12–23. https://doi.org/10.1016/j.ufug.2017.12.015

FAO, ITP (2015) État des ressources mondiales en sols (SWSR). Organisation des Nations Unies pour l'alimentation et l'agriculture et Groupe technique intergouvernemental sur les sols, Rome, Italie

Franzel S, Carsan S, Lukuyu B et al (2014) Fodder trees for improving livestock productivity and smallholder livelihoods in Africa. Curr Opin Environ Sustain 6:98–103. https://doi.org/10.1016/j.cosust.2013.11.008

Fuchs LE, Orero L, Ngoima S et al (2022) Asset-based adaptation project promotes tree and shrub diversity and above-ground carbon stocks in smallholder agroforestry Systems in Western Kenya. Front For Glob Change 4:773170. https://doi.org/10.3389/ffgc.2021.773170

Gebru BM, Wang SW, Kim SJ, Lee W-K (2019) Socio-ecological niche and factors affecting agroforestry practice adoption in different Agroecologies of southern Tigray. Ethiopia Sustain 11:3729. https://doi.org/10.3390/su11133729

Gondwe MF, Cho MA, Chirwa PW, Geldenhuys CJ (2019) Land use land cover change and the comparative impact of co-management and government-management on the forest cover in Malawi (1999–2018). J Land Use Sci 14:281–305. https://doi.org/10.1080/1747423X.2019.1706654

Haglund E, Ndjeunga J, Snook L, Pasternak D (2011) Dry land tree management for improved household livelihoods: farmer managed natural regeneration in Niger. J Environ Manag 92:1696–1705

Hylander K, Nemomissa S (2008) Home garden coffee as a repository of epiphyte biodiversity in Ethiopia. Front Ecol Environ 6:524–528. https://doi.org/10.1890/080001

Ibrahim T, Feleke E, Genete M, Bekele T (2022) Determinants and perceptions of farmers towards tree planting on farmland in northeastern Ethiopia. Trees Forests People 10:100350. https://doi.org/10.1016/j.tfp.2022.100350

ICRAF (2017) Corporate strategy 2017–2026. In: Bing. https://www.bing.com/search?pglt=43&q=ICRAF.+(2017). Accessed 27 Apr 2024

Jemal OM, Callo-Concha D, Van Noordwijk M (2021) Coffee agroforestry and the food and nutrition security of small farmers of South-Western Ethiopia. Front Sustain Food Syst 5:608868. https://doi.org/10.3389/fsufs.2021.608868

Kaonga ML, Coleman K (2008) Modelling soil organic carbon turnover in improved fallows in eastern Zambia using the RothC-26.3 model. For Ecol Manag 256:1160–1166. https://doi.org/10.1016/j.foreco.2008.06.017

Kimaro A (2009) Sequential agroforestry Systems for Improving Fuelwood Ssupply and Crop Yield in semi-arid Tanzania. University of Toronto, Toronto

Kimaro AA, Mpanda M, Rioux J et al (2016) Is conservation agriculture 'climate-smart' for maize farmers in the highlands of Tanzania? Nutr Cycl Agroecosyst 105:217–228. https://doi.org/10.1007/s10705-015-9711-8

Kimaro AA, Sererya OG, Matata P et al (2019) Understanding the multidimensionality of climate-smartness: examples from agroforestry in Tanzania. In: Rosenstock TS, Nowak A, Girvetz E (eds) The climate-smart agriculture papers. Springer, Cham, pp 153–162

Kuyah S, Sileshi GW, Luedeling E et al (2020) Potential of agroforestry to enhance livelihood security in Africa. In: Dagar JC, Gupta SR, Teketay D (eds) Agroforestry for degraded landscapes. Springer, Singapore, pp 135–167

Larwanou M, Saadou M (2011) The role of human interventions in tree dynamics and environmental rehabilitation in the Sahel zone of Niger. J Arid Environ 75:194–200

Lelamo LL (2021) A review on the indigenous multipurpose agroforestry tree species in Ethiopia: management, their productive and service roles and constraints. Heliyon 7:e07874. https://doi.org/10.1016/j.heliyon.2021.e07874

Lelamo LL, Kemal AW (2021) Distribution and regeneration status of Cordia africana (lam.) tree in agroforestry practices along Agroecology and farmers' wealth status in Sidama zone Southern Ethiopia. ujar 9:70–78. https://doi.org/10.13189/ujar2021.090302

Liliane TN, Charles MS (2020) Factors affecting yield of crops. Agron Climate Change Food Secur 9

Luedeling E, Sileshi G, Beedy T, Dietz J (2011) Carbon sequestration potential of agroforestry Systems in Africa. In: Kumar BM, Nair PKR (eds) Carbon sequestration potential of agroforestry systems. Springer, Dordrecht, pp 61–83

Maisharou A, Chirwa PW, Larwanou M et al (2015) Sustainable land management practices in the Sahel: review of practices, techniques and technologies for land restoration and strategy for up-scaling. Int For Rev 17:1–19

Masikati P, Manschadi A, Van Rooyen A, Hargreaves J (2014) Maize–mucuna rotation: an alternative technology to improve water productivity in smallholder farming systems. Agric Syst 123:62–70. https://doi.org/10.1016/j.agsy.2013.09.003

Mbow C, Smith P, Skole D et al (2014) Achieving mitigation and adaptation to climate change through sustainable agroforestry practices in Africa. Curr Opin Environ Sustain 6:8–14

Mertens JEJ, Emsens W-J, Jocqué M et al (2020) From natural forest to coffee agroforest: implications for communities of large mammals in the Ethiopian highlands. Oryx 54:715–722. https://doi.org/10.1017/S0030605318000844

Muchane MN, Sileshi GW, Gripenberg S et al (2020) Agroforestry boosts soil health in the humid and sub-humid tropics: a meta-analysis. Agric Ecosyst Environ 295:106899. https://doi.org/10.1016/j.agee.2020.106899

Mulatu K, Hunde D (2020) Agroforestry: a supplementary method for biodiversity conservation and climate change mitigation and adaptation. IJEE 5:29. https://doi.org/10.11648/j.ijee.20200503.11

N'Woueni DK, Gaoue OG (2022) Plant diversity increased arthropod diversity and crop yield in traditional agroforestry systems but has no effect on herbivory. Sustain For 14:2942. https://doi.org/10.3390/su14052942

Namaalwa J, Balaba Susan T, Ritah K, Nabanoga G (2019) Aboveground species diversity and carbon stocks in smallholder coffee agroforestry in the highlands of Uganda. In: Bamutaze Y, Kyamanywa S, Singh BR et al (eds) Agriculture and ecosystem resilience in sub Saharan Africa. Springer International Publishing, Cham, pp 403–415

Noulekoun F, Birhane E, Chude S, Zenebe A (2017) Characterization of Faidherbia albida (Del.) A. Chev. population in agroforestry parklands in the highlands of northern Ethiopia: impact of conservation, environmental factors and human disturbances. Agrofor Syst 91:123–135. https://doi.org/10.1007/s10457-016-9910-6

Nyamadzawo G, Nyamugafata P, Chikowo R et al (2005) Soil and carbon losses under rainfall simulation from two contrasting soils under maize-improved fallows rotation in eastern Zambia. In: Press CRC (ed) Soil erosion and carbon dynamics, 1st edn, Boca Raton, pp 197–206

Nyirenda H, Assédé EPS, Chirwa PW et al (2019) The effect of land use change and management on the vegetation characteristics and termite distribution in Malawian Miombo woodland agro-ecosystem. Agrofor Syst 93:2331–2343. https://doi.org/10.1007/s10457-019-00358-8

Oke DO, Odebiyi KA (2007) Traditional cocoa-based agroforestry and forest species conservation in Ondo state, Nigeria. Agric Ecosyst Environ 122:305–311. https://doi.org/10.1016/j.agee.2007.01.022

Pokwana S, Tshidzumba RP, Chirwa PW (2021) Evaluating the potential of introducing multipurpose tree species in the rural landscapes of Weza, Ugu District municipality, KwaZulu-Natal, South Africa. Trees Forests People 3:100055. https://doi.org/10.1016/j.tfp.2020.100055

Rouxel C, Barbier J, Niang A, Kaya B, Sibelet N (2005) Biodiversité spécifique ligneuse et terroirs: quelles relations? Le cas de trois villages de la région de Ségou (Mali). Bois Forets Des Tropiques 283:33–49

Sebukyu VB, Mosango M (2012) Adoption of agroforestry systems by farmers in Masaka District of Uganda. Ethnobot Res App 10:058. https://doi.org/10.17348/era.10.0.058-068

Sida TS, Baudron F, Hadgu K et al (2018) Crop vs. tree: can agronomic management reduce trade-offs in tree-crop interactions? Agric Ecosyst Environ 260:36–46. https://doi.org/10.1016/j.agee.2018.03.011

Sileshi G, Mafongoya PL (2006) Quantity and quality of organic inputs from coppicing leguminous trees influence abundance of soil macrofauna in maize crops in eastern Zambia. Biol Fertil Soils 43:333–340. https://doi.org/10.1007/s00374-006-0111-8

Sileshi G, Akinnifesi FK, Ajayi OC, Place F (2008) Meta-analysis of maize yield response to woody and herbaceous legumes in sub-Saharan Africa. Plant Soil 307:1–19. https://doi.org/10.1007/s11104-008-9547-y

Sileshi GW, Mafongoya P, Akinnifesi FK, et al (2014) Agroforestry: fertilizer trees. In: Van Alfen N (ed) Encyclopedia of agriculture and food systems, vol 1. Elsevier, San Diego, pp 222–234

Sinclair FL (1999) Une classification générale des pratiques agroforestières. Agrofor Syst 46:161–180. https://doi.org/10.1023/A:1006278928088

Sistla SA, Roddy AB, Williams NE et al (2016) Agroforestry practices promote biodiversity and natural resource diversity in Atlantic Nicaragua. PLoS One 11:e0162529. https://doi.org/10.1371/journal.pone.0162529

Sonwa DJ, Weise SF, Schroth G et al (2019) Structure of cocoa farming systems in West and Central Africa: a review. Agrofor Syst 93:2009–2025. https://doi.org/10.1007/s10457-018-0306-7

Swallow B, Boffa J-M, Scherr SJ (2006) The potential for agroforestry to contribute to the conservation and enhancement of landscape biodiversity. World agroforestry into the future, pp 95–101. World Agroforestry Centre (ICRAF), Nairobi

Swamila M, Philip D, Akyoo AM et al (2020) Gliricidia agroforestry technology adoption potential in selected dryland areas of Dodoma region. Tanzania Agriculture 10:306. https://doi.org/10.3390/agriculture10070306

Tetteh EN, Logah V, Twum-Ampofo K, Partey ST (2015) Effect of duration of reclamation on soil quality indicators of a surface–mined acid forest oxisol in South–Western Ghana. West Afr J Appl Ecol 23:63–72

Tripathi N, Hills CD, Singh RS, Atkinson CJ (2019) Biomass waste utilisation in low-carbon products: harnessing a major potential resource. npj Clim Atmos Sci 2:35. https://doi.org/10.1038/s41612-019-0093-5

Tscharntke T, Clough Y, Bhagwat SA et al (2011) Multifunctional shade-tree management in tropical agroforestry landscapes – a review: multifunctional shade-tree management. J Appl Ecol 48:619–629. https://doi.org/10.1111/j.1365-2664.2010.01939.x

Chapter 4
Distribution and Potential of Rocky Outcrop Plant Species in African Landscapes

Lackson Chama, Sylvie Andriambololonera, and Stephen Syampungani

Abstract Rocky outcrops are prominent landscape elements characterized by areas where exposed rock or boulder cover is more than 70%. They often exhibit harsh environmental conditions dominated by the presence of desiccation-tolerant plant communities. These plants exhibit several functional traits that have been shown to attract trade potential, especially in the context of their ornamental, medicinal, cultural and ecotourism values. Given their superior adaptation to growing in extremely unfriendly environmental conditions, rocky outcrop plants can potentially act as a source of valuable genetic traits for improving the performance of agricultural crops. Therefore, studying the characteristics of these plants can generate useful information in the context of increasing our understanding of their ecology, geographical distribution, conservation, economic value and their potential to enhance the adaptation of agricultural crops, especially in the face of changing climates. This is especially critical in Africa where studies on rocky outcrop ecosystems are still few, and yet the food security of the human population remains highly vulnerable to climate risks. In this chapter, we discuss the distribution of rocky outcrop ecosystems in Africa, the functional traits of rocky outcrop plant communities, their trade potential, and the possible applications of rocky outcrop plants in agriculture. Further, we discuss some of the threats faced by rocky outcrop plant species on the continent.

L. Chama (✉)
Department of Zoology and Aquatic Sciences, School of Natural Resources, Copperbelt University, Kitwe, Zambia
e-mail: lackson.chama@cbu.ac.zm

S. Andriambololonera
Missouri Botanical Garden, Madagascar Program, Antananarivo, Madagascar

S. Syampungani
Department of Plant and Soil Sciences, Faculty of Natural and Agricultural Sciences, University of Pretoria, Pretoria, South Africa

Chair of Environment and Development, Oliver R. Tambo Africa Research Chairs Initiative (ORTARChI), Copperbelt University, Kitwe, Zambia

P. W. Chirwa et al. (eds.), *Trees in a Sub-Saharan Multi-functional Landscape*,
https://doi.org/10.1007/978-3-031-69812-5_4

Keywords Distribution · Functional traits · Inselbergs · Plants · Rocky outcrops · Trade potential · Threats

4.1 Introduction

Rocky outcrops are ecosystems with geological features that have been shown to encompass a wide variety of physical environments, among which include inselbergs (Porembski et al. 2000; Fitzsimons and Michael 2017). Inselbergs are mostly dome-shaped and isolated, steep-sloped ridge, hill or small mountain rising abruptly from a plain. They may contain rock outcrops or areas with rocky outcrops that often occur in all climatic and vegetation zones of the tropics (Porembski et al. 1998; Porembski 2007; Porembski et al. 2000). They often form on geologically and geomorphologically old landscape features that are widespread on the crystalline continental shields, where these weathering-resistant outcrops tend to rise more or less abruptly above the surrounding landscape (Porembski and Barthlott 2000a; Porembski 2007, Fig. 4.1). Generally, they comprise Precambrian granites and gneisses, which are prominent landscape elements characterized by rocky areas of exposed hills or mountains called inselbergs, and where monocotyledonous mats are among the most conspicuous communities of the ecosystem (Porembski et al. 1996). The environmental conditions characterizing these ecosystems are generally harsh, with little or no soils, restricted water availability, and often with high temperatures (Porembski et al. 2000), and limited nutrient availability (Dörrstock et al.

Fig. 4.1 *Myrothamnus flabellifolius* on a rocky outcrop landscape of Muchinga escarpment, Zambia. (Photo by Stephen Syampungani)

1996; Porembski 2007). Consequently, species diversity is generally low in some rocky outcrop ecosystems and the vegetation is clearly distinct (i.e. in terms of diversity, composition and structure), and often demarcated against that of the surrounding matrix (Porembski et al. 2000; Porembski 2007).

Typically, rock outcrop vegetation is characterized by the presence of desiccation-tolerant plant communities (Porembski et al. 1996; Porembski et al. 2000). Such plants can survive cycles of dehydration and rehydration without losing viability (Hartung et al. 1998) by either keeping their photosynthetic apparatus active under water deficit or adjust to degenerate when dehydrating and rebuild when water is available again (Bewley 1979; Tuba et al. 1998; Porembski 2011). Examples of plant communities associated with these extreme edaphic and microclimatic conditions include cryptogamic crusts, ephemeral flush vegetation and monocotyledonous-mats (Porembski et al. 2000). In countries such as Ivory Coast, these vegetation types are found among several naturally occurring, extremely species-poor communities, in which competitive displacement and the low degree of disturbance prevent higher diversity (Porembski et al. 1996).

In some regions, however, rocky outcrop ecosystems have been shown to support not only high levels of species diversity (Porembski 2007; Brendonck et al. 2015), but also endemism, and have been critical in providing stable micro-climates for a diversity of species (Fitzsimons and Michael 2017). Moreover, Fitzsimons and Michael (2017), highlight that these rocky ecosystems can provide critical breeding sites for many predators, nesting sites for colonial species such as seabirds, bats and swifts, and ecological refuges for ancient lineages.

Given their superior adaptive ability to grow in extremely harsh environmental conditions, studying the genetic and evolutionary characteristics of plant communities in rocky outcrop ecosystems can be the key to identifying traits that could be utilized to improve the drought tolerance of agricultural crops. This is especially critical in Africa where studies on rocky outcrop ecosystems are still few, and yet the food security of the human population (on the continent) is highly vulnerable to the risks of climate change. In this chapter, we discuss the distribution of rocky outcrop ecosystems in Africa, the functional traits of rocky outcrop plant communities, their trade potential, and the possible applications of rocky outcrop plants in agriculture. Further, we discuss the threats facing rocky outcrop plant species on the continent.

4.2 Distribution of Rocky Outcrop Ecosystems Across Africa

Although the history of their biological exploration is relatively short, rocky outcrops have been shown to occur across all major climatic and vegetational zones, and are particularly abundant in both the Neotropical (Brazil) and Paleotropical (Africa and India) regions (Porembski 2007). In Africa, rocky outcrops have been shown to occur across regions, from the southeast and Madagascar to the western parts of the continent. The southeastern part of the continent supports rocky outcrops that are particularly rich in species diversity (Porembski 2005). For example, in Malawi and Zimbabwe, the floras of rocky outcrops are largely Zambezian with links to Madagascar (Porembski 2005, 2007). Floristically, these rocky outcrops are richer with many succulents and poikilohydric species. In fact, rocky outcrops occurring in Malawi's isolated massif of Mt. Mulanje are apparently rich in species diversity and supposedly harbour more endemics than those in West Africa (Porembski 2007). Some of the prominent endemic genera found here include *Aloe, Euphorbia, Kalanchoe* and *Utricularia.* Interesting endemics include a tree-like *Velloziaceae*, e.g. *Xerophyta splendens*, and the epiphytic *Orchidaceae,* e.g. *Polystachya johnstonii*, on stems of the preceding species. Although some closely-related southeastern rocky outcrop species are widespread and partly also found among West African rocky outcrops (e.g. the mat-forming Cyperaceae; *Coleochloa setifera* believed to be an ecological counterpart of *Afrotrilepis pilosa* in West Africa), there is a high degree of floristic divergence with vicariant taxa in Malawi (Porembski et al. 2021). Unlike on Mt. Mulanje, however, the most species rich plant communities on rocky outcrops in Zimbabwe are associated with ephemeral flush and soil-filled depressions (Seine et al. 1998), suggesting a high degree of difference in floristic communities of rocky outcrops across individual regions on the continent.

In Madagascar, rocky outcrops are a typical landscape feature of the country's high plateau. Recent research (Rabarimanarivo et al. 2019) has shown that the country's rocky outcrops are host to a diverse floristic community characterized by species-rich vascular plant genera, among which include species of *Cynorkis*, *Cyperus* and *Euphorbia.* These ecosystems are also centres of diversity for succulent plant communities, especially species of the genera *Aloe*, *Euphorbia* and *Kalanchoe* (Fischer and Theisen 2000, Rabarimanarivo et al. 2019). Further, the Madagascan rock outcrops are home to carnivorous plants (species of *Drosera* and *Utricularia*) and resurrection plants (notably species of *Xerophyta*). The Madagascan rocky outcrop habitat with the most species are ephemeral flush vegetation and monocotyledonous mats, with the latter habitat supporting a larger number of succulents and resurrection plants. In total, 724 taxa from 338 genera and 107 families have been recorded (Rabarimanarivo et al. 2019).

While both the southeast African and Madagascan rocky outcrops form centres of diversity for succulents and desiccation tolerant plants, those in the West African savanna possess a flora with comparatively few endemics. Generally, rocky outcrops in West Africa occur as hills that either form dense clusters or are isolated for hundreds of kilometres. The majority of these ecosystems occur in Guinea (around Gueckedou and Macenta), Côte d'Ivoire (Seguela and Mankono), Benin (around Dassa-Zoume and Save), Nigeria (Jos Plateau), and Angola (Amboim Plateau) (Fischer et al. 2008). While rocky outcrops located in the Madagascan rainforests have higher numbers of endemics (Parmentier and Nguema 2001), those situated in the Sudanian savannas and dry forests show a higher species richness (i.e. than their counterparts in the rainforest zone; Porembski et al. 1995). The higher number of species on these rocky outcrops could be explained by the presence of additional azonal localities (such as alluvial soils and ferricretes) and enhanced establishment opportunities for generalists, i.e. compared to a highly restrictive dispersal environment in tropical rainforest (Porembski et al. 1996).

Generally, however, rocky outcrops in West Africa are dominated by the desiccation-tolerant *Afrotrilepis pilosa* (*Cyperaceae*; Porembski et al. 1996, Porembski 2000, Müller 2008), or ephemeral flush vegetation in the savanna zones, while small fringing monocotyledonous mats are suspected to be predominant in tropical rainforests (Porembski 2000). In some regions of West Africa, such as Mt. Niangbo in Ivory Coast, nearly 200 species of vascular plants have been recorded, albeit its rocky outcrops are dominated by ephemerals (Porembski et al. 1996). Other than ephemeral communities, rocky outcrops in this region are also covered by plants such as lichens, bear *Afrotrilepis* pilosa mats, patches of forest or related communities, and many carnivorous species (Porembski and Barthlott 1993). Besides the regions and countries mentioned above, rocky outcrops have been shown to occur in several other parts of the continent, including South Africa (Foden and Potter 2005, Mucina and Rutherford 2006, Fischer et al. 2008), Botswana, Burundi, Eswatini, Ethiopia, Lesotho, Namibia and Rwanda (Porembski et al. 1997). Rocky outcrop ecosystems are also present in the Central African Copperbelt, a mineral-rich region stretching from Democratic Republic of Congo to Zambia. This region is host to some of the world's largest concentrations of mineral deposits (Yoshida et al. 2004), and is associated with rocky outcrops enriched with copper and cobalt deposits. These rocky outcrops support unique plant communities arising from deterministic species selection by environmental filters such as metal toxicity and random processes, including dispersal limitation in insular habitats (Ilunga et al. 2015). Here, about 32 strict endemic plant species (mostly metallophytes) and roughly 23 broad endemics occur on metal-rich soils (Faucon et al. 2010), most of which are associated with rocky outcrops. Several other studies have revealed a mosaic of plant communities differing in species assemblages within and among Cu-Co outcrops (Saad et al. 2012; Ilunga et al. 2013, 2015; Séleck et al. 2013), effectively suggesting that these unique ecosystems might be widespread across the entire continent.

4.3 Functional Traits of Rocky Outcrop Plants

Several functional traits have been identified across rocky outcrop plants on the African continent. Among these include carnivorous, geophytic, graminoid, resurrection, succulent and therophytic traits (Porembski 2005). Carnivorous plants are those which absorb nutrients from dead animals with the help of some morphological, physiological and behavioural features that enables them to attract, trap and digest their prey (Kulkarni et al. 2023). Examples of carnivorous plants include those belonging to the genus *Utricularia* and *Drosera*. Utriculariaceae are native to southern Africa, where they can be found across several countries, namely Angola, Eswatini, Lesotho, Madagascar, Namibia, and South Africa. It also forms part of the fynbos in the Western Cape of South Africa (Manning and Paterson-Jones 2007).

This genus often grows in damp, sandy or peaty soils among mosses by streams or wet depressions at altitudes from near sea level to 1200 m (Taylor 1989).

Drosera (of the family Droseraceae) is one of the largest genera of carnivorous plants. With over 190 species (McPherson 2010), members of this Genus survive by luring, capturing and digesting insects with the aid of stalked mucilaginous glands covering their leaf surfaces. In this case, they use insects to supplement the poor mineral nutrition of the soil in which the plants often grow. Although the origins of the genus are thought to have been in Africa or Australia (Rivadavia et al. 2003), various species of *Drosera* are widely spread and are now native in every continent except Antarctica (McPherson 2008). While Australia is believed to be the centre of diversity, with roughly 50% of all known species of *Drosera*, at least 20 species of this genus occur across the African continent (Rivadavia et al. 2003), of which six have been recorded for Madagascar (Fleischmann et al. 2020).

Geophytes are plants with underground or tuberous subterranean storage organs or structures (modified leaf bases, roots, or stems) that allows them to survive unfavourable weather conditions (e.g. fires) or to flower at times when essential resources like water, warmth or sun are not highly abundant in their environment (Kulkarni et al. 2023). Their tuberous storage organs (classified as bulbs, corms, rhizomes, stem tubers, root tubers, and enlarged hypocotyls) are often filled with stored food reserves (e.g. carbohydrates, proteins), nutrients (e.g. mineral salts) and water, which provides for rapid regrowth when favourable conditions return. Because of these functional attributes, geophytes are plants with the capability to survive arid environmental conditions by dying back (surviving as underground storage organs) and re-sprout in favourable conditions (Transeau 1994). In Africa, geophytes have been shown to exist across rocky outcrops in the western and southern regions of the continent. Here, they often occur on rocky outcrops either as a ground layer of wetter forest type (e.g. in the Sudano-Zambezian ecoregion), or forest undergrowth (e.g. across most of West Africa; Porembski 1996). The common geophytes on the continent include Araceae (e.g. *Anchomanes difformis*), Anthericaceae (e.g. *Chlorophytum togoense*), Orchidaceae (e.g. *Habenaria gabonensis*) and Cyperaceae (e.g. *Mariscus dubius*). *Mariscus dubius* has particularly been shown to be the most dominant rocky outcrop geophytes on the continent (Porembski 1996).

Graminoids are grass-like plants that show desiccation tolerance and/or mat formation, often exhibiting C4 photosynthesis and wind pollination characteristics (Kulkarni et al. 2023). Among these plants include those belonging to the families Poaceae (e.g. *Cymbopogon plurinodis, Aristida canescens, Eragrostis curvula, Setaria sphacelata, Heteropogon contortus, Elionurus muticus,* Fuls et al. 1992) and Cyperaceae (e.g. *Carex, Cyperus, Tetraria* species, Schroeder et al. 1993, Archer 2005). The Poaceae have a characteristically wide degree of ecological adaptation, spreading from eastern to southern Africa (Avoseh et al. 2015). Comprising at least 144 species, the genus *Cymbopogon* are particularly widely distributed and are adapted to growing in moderate and extremely harsh climatic conditions such as rocky outcrops (Padalia et al. 2011; Avoseh et al. 2015).

The family Cyperaceae comprise over 100 genera and more than 5000 species worldwide among which include rhizomatous sedges and other grass-like species

(Archer 2005). Its largest genus is *Carex,* comprising roughly 2000 species worldwide, followed by *Cyperus* with at least 550 species, both of which are widely distributed across the world (Archer 1998). East and southern Africa are the centres of diversity for the genus Cyperus with up to 200 species occurring in these regions as a whole (Simpson and Muasya 2004; Archer 2005; Archer and Craven 2004). Several other species of the family Cyperaceae occur across the continent, including those of *Tetraria,* which is an endemic genus to the winter rainfall region of the Western Cape, South Africa, where up to 50 species have been recorded (Hyde et al. 2024; Bean and Johns 2005). Plants of the Cyperaceae exhibit several attributes such as early maturation, rapid growth, short juvenile period, survival and the ability to produce seed under adverse environmental conditions, high photosynthetic rate (C4 photosynthesis) and increased water-use efficiency (Klingman et al. 1982; Stuckey and Barkley 1993; Muenscher 1955; Rejmanek 1996). These and many other attributes have been critical in defining their adaptation to grow in adverse conditions such as sandstones and rocky outcrops.

Succulents are annual ephemeral or perennial plants with organs (roots, stems, and leaves) modified to store water, consequently enabling them to deal with drought-like conditions that follow the growing season and during dry spells across seasons (Griffiths and Males 2017; Kulkarni et al. 2023). Examples of succulent plants common across African rocky outcrops include species of the genera *Aloe*, *Crassula*, *Euphorbia* and *Kalanchoe.* These and many other genera extend their range from Madagascar and southern Africa, to eastern and western African regions (Barthlott and Porembski 1996). Rocky outcrops in Madagascar, southern and eastern Africa are centres of diversity for succulents and desiccation tolerant plants (Barthlott and Porembski 1996, Burke 2013, Cock 2015). For example, of the over 600 species of the Aloe genus recognised worldwide, approximately 290 are in Southern Africa, followed by Eastern Africa with roughly 200 and Madagascar with over 50 species (Reynolds 2004; Cock 2015), albeit not all these species grow on rocky outcrops.

Euphorbia is one of the largest and most diverse genera of flowering plants in the world, with roughly 2000 species (Carter and Eggli 1997). At least 850 of these are succulent, of which over 700 are found only in Africa and Madagascar where they are widespread across different habitats from forests to deserts and rocky outcrops (Archer 2005). The genus *Kalanchoe,* which is well known for its ornamental and traditional medicinal values, comprise approximately 150 known species globally (Akulova-Barlow 2009; Walker 2019; Vargas et al. 2022). The majority of these species are in Madagascar and continental Africa (Descoings 2003). Madagascar is particularly the centre of diversity for *Kalanchoe*, with at least 60 species (Allorge-Boiteaij 1996; Clifton 2003; Klein et al. 2021). Southern and Eastern Africa accounts for another 56 of the known species worldwide (Allorge-Boiteaij 1996).

Resurrection plants are any poikilohydric or desiccation tolerant vascular plants that can survive extreme dehydration, for several months or years, but recover without suffering any physiological consequences when water is present (Porembski 2011; Oliver et al. 2020). They are a particularly remarkable group of plants as they can remain alive in their dried state for several years, albeit they rehydrate and become fully functional within 48 h upon receiving water (Scott 2000). Although the number of known resurrection plants could be well-over 1000 species globally, only about 300 of these are known to grow on rocky outcrops (Porembski and Barthlott 2000b; Porembski 2011). These include algae, mosses, ferns, and lycopods, as well as angiosperm plants (both monocots and dicots), albeit not gymnosperms (Lyall and Gechev 2020; Oliver et al. 2020). Although resurrection plants are geographically found in almost all continents, Africa (especially Southern Africa), is particularly rich in these species (Lyall and Gechev 2020, Oliver et al. 2020). Examples of resurrection plants common on rocky outcrops in mainland Africa and Madagascar include those from the monocotyledonous genus *Xerophyta*, a member of the Velloziaceae family (WFO 2024). There are approximately 70 species of this genus on the continent, 45 of which are in mainland Africa and 25 in Madagascar (Behnke et al. 2013). Most of the *Xerophyta* species exhibit poikilochlorophyllous characteristics, suggesting that they lose their chlorophyll during desiccation as protective mechanisms that are fully activated in their leaves (Mello-Silva et al. 2011, Behnke et al. 2013, Farrant et al. 2015, Wanga et al. 2024). Some species of this genus, such as *Xerophyta splendens* have a restricted range, known to grow only on steep rocky slopes of Mt. Mulanje (in Malawi), where they are locally abundant and widespread, and on Mt. Namuli (in Mozambique) where they are apparently rare and localised (Strugnell 2006; Timberlake et al. 2009; Harris et al. 2011). Other species include *Xerophyta retinervis* which is restricted to South Africa and some parts of Eswatini. South Africa alone is host to at least 50 species of the genus *Xerophyta*.

Therophytes are annual plants that complete their life cycle in a short period when conditions are favourable and survive harsh conditions as seeds (Kulkarni et al. 2023). In this case, therophytes have evolved to die off after reproduction and then survive unfavourable seasons (e.g. winter, drought) in the seedbank. The seeds germinate when conditions in their environment become favourable (Porembski 1996). These functional attributes have enabled therophytes to successfully grow in predominantly harsh habitats such as rock outcrops and ferricretes or edaphically dry sites in wet forests. These plants have also been shown to exhibit weedy tendencies, especially along roadsides and on fallow lands (Porembski 1996). Therophytes have been shown to be the most dominant and important floristic life-form in the context of both species richness and abundance on rocky outcrops across Africa (Porembski 1996). Notable therophytes found here include plants belonging to the families Commelinaceae (e.g. *Cyanotis lanata*), Lamiaceae (e.g. *Aeollanthus pubescens*), and Scrophulariaceae (e.g. *Lindernia exilis*) (Porembski 1996; Porembski 2005; Porembski and Barthlott 2000b).

All plants described above exhibit key attributes shaping their adaptation to grow in predominantly harsh environmental conditions underpinning rocky outcrop ecosystems. In a world where biodiversity is increasingly facing threats from climate change, these attributes are going to be crucial in driving the survival of these plant communities. Future studies should focus on establishing not only the species diversity, but also the genotypes characterizing these plants (e.g. Barbará et al. 2008; Gonçalves-Oliveira et al. 2017) to better appreciate their adaptation to grow in such ecosystems. Other functional types that occur across Africa (but not discussed above) include hemiparasites and legumes plants.

4.4 Trade Potential of Rocky Outcrop Plants of Africa

Despite their unique growth form attributes, rocky outcrop plants attract enormous trade potential across Africa and globally. Their value is mainly centred on the ornamental, medicinal, cultural and ecotourism attributes that these plants exhibit (Surendranath et al. 2012; Rabarimanarivo et al. 2019). For example, some carnivorous plant species (especially from the genus *Drosera*) have increasingly become valuable ornamentals mainly due to the beauty of their glistening traps (Rice 2006).

Across the world, there are 250 species of *Drosera*, 30 of which occur on the African continent (Rjosk et al. 2022). The relatively stringent environmental requirements of these plants have somewhat made it difficult to cultivate them for commercial purposes. Research has shown that a few (e.g. *D. aliciae, D. capensis* and *D. spatulata*) of the hardiest varieties have, nonetheless, made their way into the mainstream nursery trade where they are cultivated (Simpson 2010) and sold as ornamentals on markets across several countries (Rice 2006). Similarly, several species of geophytes have also been shown to contribute significantly to the global ornamental industry resulting from commercialized bulb and flower production, including outdoor and forced fresh-cut flowers, potted plants and for landscaping, including private gardening (Surendranath et al. 2012). Although each of these markets produce a high percentage of their own cut flowers, they also import considerable quantities from across the world, including African countries (e.g. Madagascar and South Africa), albeit some of these are illegally harvested and sold (Rabarimanarivo et al. 2019).

Besides their ornamental potential, rocky outcrop plants are also in demand due to their medical properties. For example, several *Drosera* species (e.g. *D. anglica, D. intermedia, D. linearis, D. madagascariensis* and *D. rotundifolia)* have been used as medical herbs from as early as the twelfth century (Altmeyer and Bacharach-Buhles 2024). These plants have long been recommended by herbalists for the treatment of dry coughs, and inflammatory diseases such as bronchitis, whooping cough, asthma and bronchial cramps (Schilcher and Elzer 1993; Ghate et al. 2016), and several other health related conditions (Crowder et al. 1990; Hoffmann 2003; Wolf et al. 2006; Vogel 2023). Although carnivorous plant trade has its primary origins in East Africa and Madagascar, a few of them are now cultivated throughout the world, as they are increasingly threatened by extinction in the wild (https://wa.kaiserpermanente.org/kbase/topic.jhtml?docId=hn-2171001).

Drosera plants also have been shown to produce mucilage which is an important material used in nano technology or biomaterials research (Zhang et al. 2010; Gaddam et al. 2021). For example, research on the adhesive mucilages of these species (e.g. *D. capensis* and *D. spatulata*) observed several networks of nanofibers and nanoparticles of various sizes within the residues and identified some key components of biological salts comprised of calcium, magnesium, and chlorine (Zhang et al. 2010). Apparently, these nanoparticles can increase the viscosity and stickiness of the mucilage, in turn increasing the effectiveness of the traps used by these carnivorous plants. Even more importantly, the adhesive and elastic qualities of the mucilage can provide a suitable substitute for the attachment of living cells, and this is especially important in tissue engineering. For example, during tissue implant surgery (e.g. replacement hip or an organ transplant), the use of *Drosera* mucilage can drastically improve the rate of recovery and decrease the potential for rejection because living tissues can effectively attach and grow on it. And because this mucilage can stretch to nearly a million times its original size and is readily available for use, it can be an extremely cost-efficient source of biomaterial (Gaddam et al. 2021).

Drosera plants have several other uses that highlight their economic attributes and therefore trade potential. These plants are used as a source of food (Barthlott et al. 2004), dyes for the textile industry (Yusuf et al. 2017; Plantarara 2001), and as a liqueur made from fresh leaves of *D. capensis*, *D. spatulata*, and *D. rotundifolia* (Plantarara 2001). Similarly, resurrection plants such as *Xerophyta retinervis* have many traditional applications. For example, their roots are smoked to relieve asthma, while smoke from the whole plant is used to stop nosebleeds (Johnson 2008). *X. retinervis* exhibit tough, fibrous stems and these are apparently used extensively in traditional home craft for making brushes and mats, and to make rope for hut and screen building (Johnson 2008). Further, outcrop plants are a popular tourist attraction. For example, rocky outcrops have been shown to attract at least 15,000 ecotourists per year in Madagascar, and the money raised from ecotourism activities contribute to the conservation of the local flora and fauna, as well as to the protection of the local people's cultural heritage (e.g. tombs; Rabarimanarivo et al. 2019).

4.5 Potential Applications of Rocky Outcrop Plants in Agriculture

Several studies on rocky outcrop plants have focused on the genome and the mechanisms that underlie desiccation tolerance in these species (Farrant et al. 2015; Costa et al. 2017; Radermacher et al. 2019; Wanga et al. 2019, 2021). This can be key in tackling food security (Wanga et al. 2024), especially the decreasing sustainability of rain-fed agriculture amidst the global warming threats. Undeniably, agriculture remains the cornerstone of food production, and therefore critical for achieving worldwide food security (Smutka et al. 2009; Otsuka 2013; Smutka et al. 2015; Wegren and Elvestad 2018). With the global human population projected to rise to over 9 billion by 2050 (UNDESA 2017), feeding this population will undoubtedly demand an increase in the overall food production from the current production scale (FAO 2009). However, attaining the goal of increased food production would mean finding ways to confront, and combat several threats facing the agricultural sector. The threat posed by climate change is especially likely to affect food security at the global, regional, and local level. The projected increases in temperature, alterations in the frequency and severity of droughts and floods, could cause crop failures (Kim and Mendelsohn 2023), and consequently reduce agricultural productivity, disrupt food availability, reduce access to food, and affect food quality (Ziska et al. 2016; Brown et al. 2015). Therefore, the future of agricultural productivity will largely depend on our ability to identify and introduce crops that exhibit superior and suitable traits to grow in changing climates. One way to do this will be to invest in crop improvement through the alteration of the genetics of plant populations by introducing traits that will facilitate the continued productivity of agricultural crops in the face of the threats posed by climate change.

If utilized for this purpose, rocky outcrop plants can potentially act as a source of valuable genetic traits that can improve the performance of agricultural crops in a world of changing climates. As already discussed above, these plants are adapted to growing in environmental conditions that are generally harsh, with little or no soils, restricted water availability and often with high temperatures (Porembski et al. 2000; Kulkarni et al. 2023). In fact, some of these plants (e.g. therophytes) exhibit attributes that enable them to complete their life cycles in a short period of time, i.e. when conditions are favourable. Such functional attributes have been critical in supporting the successful establishment and productivity of these plants in hostile conditions such as rocky outcrops and ferricretes or other edaphically dry sites (Porembski 1996). Therefore, agricultural crops could potentially benefit from receiving such genetic traits and this could be crucial in the context of establishing resilient food production systems.

Given the many valuable benefits they provide (e.g. as ornamental or medicinal value), research should also exploit the potential for large-scale domestic cultivation of rocky outcrop plants (through agriculture) for commercial purposes, i.e. besides using them as a source of genetic traits. This will not only increase the benefits generated from these plants, but also uplift the global efforts to protect rocky outcrop plants and their ecosystems (e.g. Kaiser and Jennings 2001; Gaston and Fuller 2008), especially in the face of heightened anthropogenic threats.

4.6 Threats to Rocky Outcrop Plant Species

Irrespective of their unique ecological importance, rocky outcrops in Africa are facing numerous threats, notably human impacts such as the large scale-extension of farmland (Barthlott et al. 1993) and anthropogenic fires. Other threats result from the collection of plants for ornamental and medicinal purposes (e.g. *Myrothamnus flabellifolius* in Zambia and the endemic *M. moschatus* in Madagascar). Many rocky outcrops, particularly in commodity production landscapes, face a variety of threats including large-scale extension of farmland (Porembski and Barthlott 1993), soil compaction and erosion caused by livestock; nutrient enrichment and weed invasion; introduced predators; and physical damage caused by recreational and quarrying activities (Fitzsimons and Michael 2017). Rocky outcrops are quarried (Fig. 4.2)

Fig. 4.2 Production landscape in Ankazobe, Madagascar, a threat to rocky outcrop plant species (Photo by Stefan Porembski)

and used for landscaping, constructing buildings and roads. Some rocky types are used to make pavers, benchtops and sculptures. In peri-urban areas, rocky outcrops are destroyed or removed to make way for urban development.

Additionally, mineral exploitation in Africa especially in the Central African Copperbelt has also been observed to impact on the rocky outcrop species, as some of these ecosystems are associated with mineral deposits (Ilunga et al. 2013). In the Democratic Republic of Congo, most Katangan copper outcrops have been allotted to mining companies and this could result in irreversible damage of the rocky outcrop ecosystems (Ilunga et al. 2013).

4.7 Chapter Highlights

This chapter has highlighted that most rocky outcrop plants are desiccation-tolerant, suggesting they can survive cycles of dehydration and rehydration without losing viability (Hartung et al. 1998). These plant communities are widespread on rocky outcrops across the African continent, albeit their species richness appear to be particularly higher in Madagascar and southeastern Africa. The chapter has also revealed that the rocky outcrop plant species are associated with some mineral deposits, especially in the Central African Copperbelt, notwithstanding the damaging consequences posed by mineral exploration and mining activities on the integrity of these ecosystems. Plants that are associated with rocky outcrop ecosystems display several functional characteristics, including carnivorous, geophytic, graminoid, resurrection, succulent and therophytic traits (Porembski 2005). Some of these traits are responsible for shaping the trade potential characterizing these plants (e.g. ornamental, medicinal, nanotechnology, etc.). The adaptation of rocky outcrop plants to grow in ecologically harsh conditions suggests that they pose genetic traits that could potentially be transferred to agricultural crops to increase the latter's adaptation to climate change. If successful, this has the potential to contribute towards increasing agricultural productivity and ensure the attainment of global food security, especially in the face of projected human population growth and increased demand for agricultural produce. Besides, research should exploit the potential for large-scale domestic cultivation of rocky outcrop plants for commercial purposes. This will not only increase the benefits generated from these plants, but also uplift global efforts to protect rocky outcrop plants, especially in the face of heightened anthropogenic threats.

References

Akulova-Barlow Z (2009) Kalanchoe. Natl Cact Succ J 81(6):268–276. https://doi.org/10.2985/015.081.0601

Allorge-Boiteaij L (1996) Madagascar Center of Speciation and Origin of the genus Kalanchoe (Crassulaceae). Biogéographie de Madagascar:137–145

Altmeyer P, Bacharach-Buhles M. (2024) Drosera herba. Altmeyers medical encyclopedia. Available on: https://www.altmeyers.org/en/naturopathy/drosera-herba-143953

Archer C (1998) Grasses and grass-like plants. PlantLife 18:14–16

Archer RH (2005) National herbarium Pretoria. South Africa National Biodiversity Institute. Accessed on 6th April 2024 from: https://pza.sanbi.org/euphorbia#

Archer C, Craven P (2004) Cyperaceae in Namibia. SABONET News 9:18–28

Avoseh O, Oyedeji O, Rungqu P, Nkeh-Chungag B, Oyedeji A (2015) Cymbopogon species, ethnopharmacology, phytochemistry and the pharmacological importance. Molecules 20(5):7438–7453. https://doi.org/10.3390/molecules20057438

Barbará T, Lexer C, Martinelli G, Mayo S, Fay MF, Heuertz M (2008) Within-population spatial genetic structure in four naturally fragmented species of a neotropical inselberg radiation, Alcantarea imperialis, A. Geniculata, A. Glaziouana and A. Regina (Bromeliaceae). Heredity 101:285–296. https://doi.org/10.1038/hdy.2008.65

Barthlott W, Porembski S (1996) Biodiversity of arid islands in tropical Africa: the succulents of inselbergs. In: van der Maesen LJG, van der Burgt XM, van Medenbach de Rooy JM (eds) The biodiversity of African plants. Springer, Dordrecht. https://doi.org/10.1007/978-94-009-0285-5_9

Barthlott W, Gröger A, Porembski S (1993) Some remarks on the vegetation of tropical inselberg: diversity and ecological differentiation. Biogeographica 69:105–124

Barthlott W, Porembski S, Seine R, Theisen I (2004) Karnivoren Biologie und Kultur fleischfressender Pflanzen. Verlag Eugen Ulmer, Stuttgart (Hohenheim). ISBN 9783800141449

Bean A, Johns A (2005) Stellenbosch to Hermanus South African wild flower guide 5. BotSoc, Cape Town

Behnke HD, Hummel E, Hillmer S, Sauer-Gürth H, Gonzalez J, Wink M (2013) A revision of African Velloziaceae based on leaf anatomy characters and rbcL nucleotide sequences. Bot J Linn Soc 172(1):22–94. https://doi.org/10.1111/boj.12018

Bewley JD (1979) Physiological aspects of desiccation tolerance. Annu Rev Plant Physiol 30:195–238. https://doi.org/10.1146/annurev.pp.30.060179.001211

Brendonck L, Jocqué M, Tuytens K, Timms BV, Vanschoenwinkel B (2015) Hydrological stability drives both local and regional diversity patterns in rock pool metacom munities. Oikos 124:741–749. https://doi.org/10.1111/oik.01710

Brown ME, Antle JM, Backlund P, Carr E.R, Easterling WE, Walsh MK, Ammann C, Attavanich W, Barrett CB, Bellemare MF, Dancheck V, Funk C, Grace K, Ingram JSI, Jiang H, Maletta H, Mata T, Murray A, Ngugi M, Ojima D, O'Neill B, Tebaldi C (2015) Climate change, global food security, and the U.S. food system (146 pages). Available online at http://www.usda.gov/oce/climate_change/FoodSecurity2015Assessment/FullAssessment.pdf

Burke A (2013) Succulent plants on arid inselbergs. Flora 208:321–329. https://doi.org/10.1016/j.flora.2013.05.001

Carter S, Eggli U (1997) The cites checklist of succulent *Euphorbia taxa* (Euphorbiaceae). German Federal Agency for Nature Conservation

Clifton R (2003) Madagascar's plants: checklist of the endemic (and indigenous) flowering plants, 2nd ed

Cock IE (2015) The genus aloe: Phytochemistry and therapeutic uses including treatments for gastrointestinal conditions and chronic inflammation. Prog Drug Res 70:179–235. https://doi.org/10.1007/978-3-0348-0927-6_6

Costa MCD, Artur MAS, Maia J, Jonkheer E, Derks MFL, Nijveen H, Williams B, Mundree SG, Jiménez-Gómez JM, Hesselink T, Schijlen EGWM, Ligterink W, Oliver MJ, Farrant JM, Hilhorst JWM (2017) A footprint of desiccation tolerance in the genome of Xerophyta viscosa. Nat Plants 3:1–10. https://doi.org/10.1038/nplants.2017.38

Crowder AA, Pearson MC, Grubb PJ, Langlois PH (1990) Drosera L J Ecol 78(1):233–267. https://doi.org/10.2307/2261048

Descoings B (2003) Kalanchoe. In: Eggli U (ed) Crassulaceae. Illustrated handbook of succulent plants. Springer-Verlag, Berlin/Heidelberg/New York, pp 143–181

Dörrstock S, Porembski S, Barthlott W (1996) Ephemeral flush vegetation on inselbergs in the Ivory Coast (West Africa). Candollea 51:407–419

FAO (2009) The State of Food and Agriculture. Food and Agriculture Organization of the United Nations. Viale delle Terme di Caracalla, 00153, Rome, Italy

Farrant JM, Cooper K, Hilgart A, Abdalla KO, Bentley J, Thomson JA, Dace HJW, Peton N, Mundree SG, Rafudeen MS (2015) A molecular physiological review of vegetative desiccation tolerance in the resurrection plant Xerophyta viscosa (Baker). Planta 242:407–426. https://doi.org/10.1007/s00425-015-2320-6

Faucon MP, Meersseman A, Shutcha MN, Mahy G, Luhembwe MN, Malaisse F, Meerts P (2010) Copper endemism in the Congolese flora: a database of copper affinity and conservational value of cuprophytes. Plant Ecol Evol 143:5–18. https://doi.org/10.5091/plecevo.2010.411

Fischer E, Theisen I (2000) Vegetation of Malagasy inselbergs. In: Porembski S, Barthlott W (eds) Inselbergs, Ecol Stud, vol 146. Springer, Berlin, Heidelberg, pp 259–276. https://doi.org/10.1007/978-3-642-59773-2_13

Fischer J, Brosi B, Daily GC, Ehrlich PR, Goldman R, Goldstein J, Lindenmayer DB, Manning AD, Mooney HA, Pejchar L, Ranganathan J, Tallis H (2008) Should agricultural policies encourage land sparing or wildlife-friendly farming? Front Ecol Environ 6(7):380–385. https://doi.org/10.1890/070019

Fitzsimons JA, Michael DR (2017) Rocky outcrops: a hard road in the conservation of critical habitats. Biol Conserv 211(B):36–44. https://doi.org/10.1016/j.biocon.2016.11.019

Fleischmann AS, Rakotoarivelo NH, Roccia A, Gonella PM, Andriamiarisoa LR, Razanatsima A, Rakotoarivony F (2020) A new and endemic species of Drosera (Droseraceae) from Madagascar. Plant Ecol Evol Journal 153(2):283–291. https://doi.org/10.5091/plecevo.2020.1705

Foden W; Potter L. (2005) Salsola glabrescens Burtt Davy. National assessment: Red list of South African plants version 2017.1. http://redlist.sanbi.org/species.php?species=3590-56

Fuls ER, Bredenkamp GJ, Van Rooyen N, Theron GK (1992) The physical environment and major plant communities of the Vredefort-Kroonstad-Lindley-Heilbron area, northern Orange Free State. S Afr J Bot 58:317–325. https://doi.org/10.1016/S0254-6299(16)30817-1

Gaddam SA, Kotakadi VS, Subramanyam GK, Penchalaneni J, Challagundla VN, Dvr SG, Pasupuleti VR (2021) Multifaceted phytogenic silver nanoparticles by an insectivorous plant Drosera spatulata Labill var. bakoensis and its potential therapeutic applications. Sci Rep 11(1):21969. https://doi.org/10.1038/s41598-021-01281-8

Gaston KJ, Fuller R (2008) Commonness, population depletion and conservation biology. Trends Ecol Evol 23(1):14–19. https://doi.org/10.1016/j.tree.2007.11.001

Ghate NB, Das A, Chaudhuri D, Panja S, Mandal N (2016) Sundew plant, a potential source of anti-inflammatory agents, selectively induces G2/M arrest and apoptosis in MCF-7 cells through upregulation of p53 and Bax/Bcl-2 ratio. Cell Death Discov 2:15062. https://doi.org/10.1038/cddiscovery.2015.62

Gonçalves-Oliveira RC, Wöhrmann T, Benko-Iseppon AM, Krapp F, Alves M, Wanderley MDGL, Weising K (2017) Population genetic structure of the rock outcrop species Encholirium spectabile (Bromeliaceae): the role of pollination vs. seed dispersal and evolutionary implications. Am J Bot 104(6):868–878. https://doi.org/10.3732/ajb.1600410

Griffiths H, Males J (2017) Succulent plants. Curr Biol 27(17):890–896. https://doi.org/10.1016/j.cub.2017.03.021

Harris T, Darbyshire I, Polhill R (2011) New species and range extensions from Mt Namuli, Mt Mabu and Mt Chiperone in northern Mozambique. Kew Bull 66(2):241–251. http://www.jstor.org/stable/23044677

Hartung W, Schiller P, Dietz KJ (1998) Physiology of Poikilohydric plants. In: Behnke HD, Esser K, Kadereit JW, Lüttge U, Runge M (eds) Progress in botany, vol 59. Springer, Berlin/Heidelberg. https://doi.org/10.1007/978-3-642-80446-5_11

Hoffmann D (2003) Medical herbalism: the science and practice of herbal medicine. Simon and Schuster. ISBN 978-1-59477-890-2

Hyde MA, Wursten BT, Ballings P, Palgrave CM (2024) Flora of Zimbabwe: Genus page: Tetraria. https://www.zimbabweflora.co.zw/speciesdata/genus.php?genus_id=289. Retrieved 4 Apr 2024

Ilunga EI, Séleck M, Colinet G, Faucon MP, Meerts P, Mahy G (2013) Small-scale diversity of plant communities and distribution of species niches on a copper rock outcrop in Upper Katanga, D.R. Congo. Plant Ecol Evol 146(2):173–182. https://doi.org/10.5091/plecevo.2013.816

Ilunga EI, Mahy G, Piqueray J, Séleck M, Shutcha MN, Meerts P, Faucon MP (2015) Plant functional traits as a promising tool for the ecological restoration of degraded tropical metal-rich habitats and revegetation of metal-rich bare soils: a case study in copper vegetation of Katanga, DRC. Ecol Eng 82:214–221. https://doi.org/10.1016/j.ecoleng.2015.04.084

Johnson I (2008) Xerophyta retinervis Baker. Retrieved from South African National Biodiversity Institute (SANBI). http://pza.sanbi.org/xerophyta-retinervis

Kaiser MJ, Jennings S (2001) Ecosystem perspectives on conserving targeted and non-targeted species. In: Reynolds JD, Mace GM, Redford KH, Robinson JG (eds) Conservation of exploited species. Cambridge University Press, Cambridge, pp 343–369

Kim SM, Mendelsohn R (2023) Climate change to increase crop failure in U.S. Environ Res Lett 18:014014. https://doi.org/10.1088/1748-9326/acac41

Klein DP, Shtein R, Nusbaumer L, Callmander MW (2021) Kalanchoe darainensis (Crassulaceae), a new species from northeastern Madagascar. Candollea 76:117–123. https://doi.org/10.15553/c2021v761a12

Klingman GC, Ashton FM, Noordhoff LJ (1982) Weed science: principles and practices, 2nd edn. Wiley, New York

Kulkarni A, Shetti R, Shigwan BK, Vijayan S, Datar MN (2023) The relationship between vegetation, plant functional diversity and environment on rock outcrops in the Western Ghats. India Plant Ecol Divers 16(1–2):87–98. https://doi.org/10.1080/17550874.2023.2255999

Lyall R, Gechev T (2020) Multi-omics insights into the evolution of angiosperm resurrection plants. Annu Plant Rev Online. https://doi.org/10.1002/9781119312994.apr0730

Manning JC, Paterson-Jones C (2007) Field guide to fynbos. Struik, Cape Town

McPherson SR (2008) Glistening carnivores. Redfern Natural History Productions Ltd, Poole

McPherson SR (2010) Carnivorous plants and their habitats. Redfern Natural History Productions Ltd, Poole
Mello-Silva R, Santos DY, Salatino ML, Motta LB, Cattai MB, Sasaki D, Lovo J, Pita PB, Rocini C, Rodrigues CD, Zarrei M, Chase MW (2011) Five vicariant genera from Gondwana: the Velloziaceae as shown by molecules and morphology [corrected]. Ann Bot 108(1):87–102. https://doi.org/10.1093/aob/mcr188
Mucina L, Rutherford MC (eds) (2006) The vegetation of South Africa, Lesotho and Swaziland. South African National Biodiversity Institute, Pretoria
Muenscher WC (1955) Weeds, 2nd edn. Cornell University Press, Ithaca
Müller J (2008) Herbaceous and non-inundated vegetation of Sahelian inselbergs in Burkina Faso. Candollea 63(1):57–79
Oliver MJ, Farrant JM, Hilhorst HWM, Mundree S, Williams B, Bewley JD (2020) Desiccation tolerance: avoiding cellular damage during drying and rehydration. Annu Rev Plant Biol 71:435–460. https://doi.org/10.1146/annurev-arplant-071219-105542
Otsuka K (2013) Food insecurity, income inequality, and the changing comparative advantage in world agriculture. Agric Econ 44(1):7–18. https://doi.org/10.1111/agec.12046
Padalia RC, Verma RS, Chanotiya CS, Yadav A (2011) Chemical fingerprinting of the fragrant volatiles of nineteen indian cultivars of Cymbopogon Spreng (Poaceae). Rec Nat Prod 5:290–299
Parmentier I, Nguema N (2001) La vegetación de los inselbergs de Rio Muni. In: Aedo C, Morales R, Telleria MT, Velayos M (eds) Botanicay Botanicos en Guinea Ecuatorial. Real Jardin Botanico. Agencia Espanola de Cooperaci6n lnternacional, Madrid, pp 201–23
Plantarara (2001) Artzneimittle, Tee, und Likör aus fleischfressenden Pflanzen Archived 2006-06-18 at the Wayback Machine
Porembski S (1996) Notes on the vegetation of inselbergs in Malawi. Flora 191(1):1–8. https://doi.org/10.1016/S0367-2530(17)30685-0
Porembski S (2000) The invasibility of tropical granite outcrops ("inselbergs") by exotic weeds. J Roy Soc Western Australia 83:131–134
Porembski S (2005) Floristic diversity of African and south American inselbergs: a comparative analysis. Acta Bot Gallica 152(4):573–580. https://doi.org/10.1080/12538078.2005.10515515
Porembski S (2007) Tropical inselbergs: habitat types, adaptive strategies and diversity patterns. Braz J Bot 30:579–586. https://doi.org/10.1590/S0100-84042007000400004
Porembski S (2011) Evolution, diversity, and habitats of poikilohydrous vascular plants. In: Lüttge U, Beck E, Bartels D (eds) Plant desiccation tolerance, Ecological studies. Springer, Berlin, pp 139–156
Porembski S, Barthlott W (1993) Okogeographische Differenzierung und Diversitat der vegetation von Inselbergen in der Elfenbeinkiuste. Animal-plant interactions in: Barthlott W, Naumann CM, Schmidt- Loske K, Schuchmann KL(eds) Tropical environments. Results of the annual meeting of the German Society for Tropical Ecology, Bonn, pp 149–158
Porembski S, Barthlott W (2000a) Granitic and gneissic outcrops (inselbergs) as centers of diversity for desiccationtolerant vascular plants. Plant Ecol 151:19–28. https://doi.org/10.1023/A:1026565817218
Porembski S, Barthlott W (eds) (2000b) Inselbergs. Biotic diversity of isolated rock outcrops in tropical and temperate regions. Ecol Stud:146. https://doi.org/10.1046/j.1366-9516.2001.00123.x-i2
Porembski S, Brown G, Barthlott W (1995) An inverted latitudinal gradient of plant diversity in shallow depressions on lvorian inselbergs. Vegetatio 117:151–163. https://doi.org/10.1007/BF00045506
Porembski S, Brown G, Barthlott W (1996) A species-poor tropical sedge community: Afrotrilepis pilosa mats on inselbergs in West Africa. Nord J Bot 16:239–245. https://doi.org/10.1111/j.1756-1051.1996.tb00222.x
Porembski S, Seine R, Barthlott W (1997) Inselberg vegetation and biodiversity of granite outcrops. J R Soc West Aust 80:193–199
Porembski S, Martinelli G, Ohlemüller R, Barthlott W (1998) Diversity and ecology of saxicolous vegetation mats on inselbergs in the Brazilian Atlantic rainforest. Biodiv Res 4:107–119. https://doi.org/10.1046/j.1365-2699.1998.00013.x

Porembski S, Becker U, Seine R (2000) Islands on islands: habitats on inselbergs. In: Porembski S, Barthlott W (ed) Inselbergs. Biotic diversity of isolated rock outcrops in tropical and temperate regions. Ecol Stud 146:49–67

Porembski S, Rexroth J, Weising K, Bondi L, Mello-Silva R, Centeno DC, Datar MN, Watve A, Thiombano A, Tindano E, Rabarimanarivo MN, de Paula LFA (2021) An overview on desiccation-tolerant mat-forming monocotyledons on tropical inselbergs. Flora 285:0367–2530. https://doi.org/10.1016/j.flora.2021.151953

Rabarimanarivo MN, Ramandimbisoa B, Rakotoarivelo N, Phillipson PB, Andriambololonera S, Callmander MW, Porembski S (2019) The extraordinary botanical diversity of Malagasy inselbergs. Candollea 74:65–83. https://doi.org/10.15553/c2019v741a8

Radermacher AL, du Toit SF, Farrant JM (2019) Desiccation-driven senescence in the resurrection plant *Xerophyta schlechteri* (Baker) N.L. Menezes: comparison of anatomical, ultrastructural, and metabolic responses between senescent and non-senescent tissues. Front Plant Sci 10:1396. https://doi.org/10.3389/fpls.2019.01396

Rejmanek M (1996) A theory of seed plant invasiveness: the first sketch. Biol Conserv 78:171–181

Reynolds T (2004) Aloe chemistry. In: Reynolds T (ed) Aloes the genus aloe. CRC Press, Boca Raton, pp 39–74

Rice B (2006) Growing carnivorous plants. Timber Press, Portland

Rivadavia F, Kondo K, Kato M, Hasebe M (2003) Phylogeny of the sundews, Drosera (Droseraceae), based on chloroplast rbcL and nuclear 18S ribosomal DNA sequences. Am J Bot 90(1):123–130. https://doi.org/10.3732/ajb.90.1.123

Rjosk A, Neinhuis C, Monizi M, Lautenschläger T (2022) Synopsis of the genus Drosera (Droseraceae) in Angola and The Democratic Republic of the Congo. Blumea 67(1):1–14. https://doi.org/10.3767/blumea.2022.67.01.03

Saad L, Parmentier I, Colinet G, Malaisse F, Faucon MP, Meerts P, Mahy G (2012) Investigating the vegetation-soil relationships on the copper-cobalt rock outcrops of Katanga (D.R. Congo), an essential step in a biodiversity conservation plan. Restor Ecol 20:405–415. https://doi.org/10.1111/j.1526-100X.2011.00786.x

Schilcher H, Elzer M (1993) Drosera (sundew): a proven antitussive. Zeitschrift für Phytotherapie 14(50):4

Schroeder J, Thomas SH, Murray L (1993) Yellow and purple nutsedge and Chile peppers host southern root-knot nematode. Weed Sci 41:150–156. https://doi.org/10.1017/S0043174500057738

Scott P (2000) Resurrection plants and the secrets of eternal leaf. Ann Bot 85(2):159–166. https://doi.org/10.1006/anbo.1999.1006

Seine R, Becker U, Porembski S, Follmann G, Barthlott W (1998) Vegetation of inselbergs in Zimbabwe. Edinb J Bot 55:267–293. https://doi.org/10.1017/S0960428600002195

Séleck M, Bizoux JP, Colinet G, Faucon MP, Guillaume A, Meerts P, Piqueray J, Mahy G (2013) Chemical soil factors influencing plant assemblages along copper-cobalt gradients: implications for conservation and restoration. Plant Soil 373:455–469. https://doi.org/10.1007/s11104-013-1819-5

Simpson MG (2010) Diversity and classification of flowering plants: eudicots (Chapter 8). In: Plant systematics, 2nd edn, pp 275–448. doi: https://doi.org/10.1016/B978-0-12-374380-0.50008-7

Simpson DA, Muasya AM (2004) Three new species of Cyperus (Cyperaceae) from Eastern and Southern Africa. Kew Bull 59(4):593–598. https://doi.org/10.2307/4110917

Smutka L, Steininger M, Miffek O (2009) World agricultural production and consumption. Agris Online Pap Econ Informat 1(2)

Smutka L, Steininger M, Maitah M, Škubna O (2015) The Czech agrarian foreign trade-ten years after the EU accession. Agrarian Perspectives XXIV. Global Agribusiness and the Rural Economy. In: Proceedings of the 24th international scientific conference, Czech University of Life Sciences Prague, Faculty of Economics and Management, , Prague, Czech Republic, pp 385–392

Strugnell AM (2006) A checklist of the spermatophytes of Mount Mulanje, Malawi. Scripta Botanica Belgica 34:5–194

Stuckey RL, Barkley TM (1993) Weeds. In: Flora of North America editorial committee (editor), Flora of North America north of Mexico 1. Oxford University Press, New York, pp 193–198

Surendranath R, SharathKumar M, Ganga M (2012) Scenario and scope of ornamental geophytes. Intl Conf Agric Horticult Sci 14-15. https://doi.org/10.4172/2168-9881.S1.004

Taylor P (1989) The genus Utricularia – a taxonomic monograph. Kew bulletin additional series XIV, London

Timberlake J, Dowsett-Lemaire F, Bayliss J, Alves T, Baena S, Bento C, Cook K, Francisco J, Harris T, Smith P, de Sousa C (2009) Mt Namuli, Mozambique: biodiversity and conservation. Royal Botanic Gardens, Kew/London

Transeau EN (1994) General Botany. Discovery Publishing House. ISBN9788171412525. https://books.google.co.zm/books?id=0-TiglQPN_UC

Tuba Z, Protor CF, Csintalan Z (1998) Ecophysiological responses of homoiochlorophyllous and poikilochlorophyllous desiccation tolerant plants: a comparison and an ecological perspective. J Plant Growth Regul 24:211–217. https://doi.org/10.1023/A:1005951908229

UNDESA (2017) World population prospects: the 2017 revision. UNDESA, New York

Vargas A, Herrera I, Nualart N, Guézou A, Gómez-Bellver C, Freire E, Jaramillo Díaz P, López-Pujol J (2022) The genus Kalanchoe (Crassulaceae) in Ecuador: from gardens to the wild. Plants (Basel) 11(13):1746. https://doi.org/10.3390/plants11131746

Vogel A(2023) Drosera rotundifolia L. A round-leafed sundew. Plant Encyclopaedia. Available online at https://www.avogel.com/plant-encyclopaedia/drosera_rotundifolia.php

Walker CC (2019) Two shrubby Madagascan Kalanchoes. New Zealand Cactus Succulent J:72(1). Published online at https://oro.open.ac.UK/59187/1/Cactus%20Journal%20Vol72%20No1%20C%20Walker.pdf. Accessed 10 April 2024

Wanga VO, Dong X, Oulo MA, Munyao JN, Mkala EM, Kirika P, Gituru RW, Hu GW (2019) The complete chloroplast genome sequence of Xerophyta spekei (Velloziaceae). Mitochondrial DNA B Resour 5(1):100–101. https://doi.org/10.1080/23802359.2019.1698365. PMID: 33366441

Wanga VO, Dong X, Oulo MA, Mkala EM, Yang JX, Onjalalaina GE, Gichua MK, Kirika PM, Gituru RW, Hu GW, Wang GF (2021) Complete chloroplast genomes of Acanthochlamys bracteata (China) and Xerophyta (Africa) (Velloziaceae): comparative genomics and phylogenomic placement front. Plant Sci 12:1–16. https://doi.org/10.3389/fpls.2021.691833

Wanga VO, Ngarega GK, Oulo MA, Mkala EM, Ngumbau VM, Onjalalaina GE, Odago WO, Nanjala C, Ochieng CO, Gichua MK, Gituru RW, Hu GW (2024) Projected impacts of climate change on the habitat of Xerophyta species in Africa. Plant Divers 46(1):91–100. https://doi.org/10.1016/j.pld.2023.05.001

Wegren SK, Elvestad C (2018) Russia's food self-sufficiency and food security: an assessment. Post-Communist Econ 30(5):565–587. https://doi.org/10.1080/14631377.2018.1470854

WFO (2024) Xerophyta splendens (Rendle) N.L.Menezes. Published on the online at http://www.worldfloraonline.org/taxon/wfo-0000751961. Accessed 8 Apr 2024

Wolf E, Gage E, Cooper DJ (2006) *Drosera rotundifolia* L. (roundleaf sundew): a technical conservation assessment (PDF). USDA Forest Service, Rocky Mountain Region

Yoshida M, Kokonyangi J, Batumike JI (2004) The Central Africa Copperbelt in southeastern Democratic Republic of Congo: IGCP-450 field workshop on Proterozoic sediment-hosted base metal deposits of West Gondwana. Gondwana Res 7:640–643. https://doi.org/10.1016/S1342-937X(05)70820-0

Yusuf M, Shabbir M, Mohammad F (2017) Natural colorants: historical, processing and sustainable prospects. Nat Prod Bioprospect 7(1):123–145. https://doi.org/10.1007/s13659-017-0119-9

Zhang M, Lenaghan SC, Xia L, Dong L, He W, Henson WR, Fan X (2010) Nanofibers and nanoparticles from the insect capturing adhesive of the Sundew (Drosera) for cell attachment. J Nanobiotechnol 8(20):20. https://doi.org/10.1186/1477-3155-8-20

Ziska L, Crimmins A, Auclair A, DeGrasse S, Garofalo JF, Khan AS, Loladze I, Pérez de León AA, Showler A, Thurston J, Walls I (2016) Ch. 7: food safety, nutrition, and distribution. In: The impacts of climate change on human health in the United States: a scientific assessment. U.S. Global Change Research Program, Washington, DC, pp 189–216. https://doi.org/10.7930/J0ZP4417

Chapter 5
Trees and Agricultural Production Systems

Gudeta W. Sileshi, Festus K. Akinnifesi, Emeline Assede, Simon A. Mng'omba, Paxie W. Chirwa, Shem Kuyah, Bikila Warkineh, and Emelda M. Hachoofwe

Abstract While agriculture is one of the greatest causes of deforestation and the associated biodiversity loss in Africa, smallholder agriculture also holds the key to conservation of the remaining biodiversity, increasing tree cover and reducing greenhouse emissions through tree planting activities. However, tree planting by smallholders and traditional agroforestry practices have not been well documented across the range of African farming systems. Using the farming systems lense, here we provide insights into how farmers are adapting tree planting to fit household needs and emerging market opportunities. We also highlight a number of important but poorly documented traditional agroforestry practices developed in each farming system. We conclude that tree planting and agroforestry projects in Africa have not been appropriately contextualized within a farming systems framework and farmers

G. W. Sileshi (✉) · B. Warkineh
Department of Plant Biology and Biodiversity Management, Addis Ababa University, Addis Ababa, Ethiopia

F. K. Akinnifesi
Food and Agriculture Organization of the United Nations (FAO), Rome, Italy

E. Assede
Laboratoire d'Ecologie, Botanique et Biologie Végétale, Bénin Faculté d'Agronomie, Université de Parakou, Parakou, Bénin

Department of Plant and Soil Sciences, Faculty of Natural and Agricultural Sciences, University of Pretoria, Pretoria, South Africa

S. A. Mng'omba
Lilongwe University of Agriculture and Natural Resources (LUANAR), Lilongwe, Malawi

P. W. Chirwa
Department of Plant and Soil Sciences, Faculty of Natural and Agricultural Sciences, University of Pretoria, Pretoria, South Africa

S. Kuyah
Department of Botany, Jomo Kenyatta University of Agriculture and Technology (JKUAT), Nairobi, Kenya

E. M. Hachoofwe
School of Natural Resources, The Copperbelt University, Kitwe, Zambia

P. W. Chirwa et al. (eds.), *Trees in a Sub-Saharan Multi-functional Landscape*,
https://doi.org/10.1007/978-3-031-69812-5_5

livelihood choices. This has led to the misconception that agroforestry technologies developed in one context could be successfully deployed in another. We argue that the failure of many tree planting projects to realize their potential arises from the lack of appreciation of the diversity of farming systems and farmers' livelihood choices. We strongly recommend support for agroforestry practices through incentives such as payment for ecosystem services with specific focus on farmers livelihood choices.

Keywords Agroforestry · Biodiversity · Farming systems · Landscape restoration · Livelihoods

5.1 Introduction

Driven by the need to satisfy the demands of a rapidly growing global population, agricultural land is expanding at the expense of natural forests, grasslands and other lands (Johnson et al. 2014; Zabel et al. 2019). Several analyses have demonstrated the role of agricultural expansion as the main driver of deforestation, biodiversity loss, and habitat fragmentation underpinning the global biodiversity crisis (Branthomme et al. 2023; Curtis et al. 2018; Gibbs et al. 2010; Johnson et al. 2014; Masolele et al. 2024; Zabel et al. 2019). For example, between 1980 and 2000 more than 55% of new agricultural land came from conversion of intact forests, and another 28% came from disturbed forests across the tropics (Gibbs et al. 2010). According to Curtis et al. (2018), permanent land use change for production of agricultural and forestry commodities, wildfire, and urbanization are the primary drivers of forest loss. The Global Forest Resources Assessment of 2020 has shown that agriculture has driven almost 90% of deforestation over the period 2000–2018 globally (Branthomme et al. 2023). Forest conversion to cropland and grassland by small-scale farming accounted for 68% of agriculture-driven deforestation during 2000–2018 globally, of which 40% was conversion for cropland and 28% for livestock farming (Branthomme et al. 2023). Cropland expansion by small-scale and large-scale farming was responsible for 71% and 29% of deforestation, respectively (Table 5.1, Fig. 5.1a). Likewise, 64% of deforestation due to livestock farming was attributed to small-scale farming and 36% to large-scale farming (Fig. 5.1b).

The extent and spatial distributions of deforestation caused by expansion of cropland and livestock farming in the period 2000–2018 across Africa are shown in Table 5.1 and Figs. 5.1. According to the FRA 2020 analysis, the total forest area converted to small-scale agricultural land is over 44 million ha in the period 2000–2018 (Table 5.1). Deforestation due to small-scale conversion to cropland affected 36.7 million ha, while conversion to livestock farming affected 7.5 million ha. Conversion to large-scale crop and livestock farming affected about 1.4 million ha in the same period (Branthomme et al. 2023). Deforestation due to large-scale farming was observed only in a few areas such as in the southern coast of Western Africa (Branthomme et al. 2023). Large-scale farming affecting forests mainly

Table 5.1 Agriculture-driven deforestation across Africa in the period 2000–2018 based on data from FRA 2020

Category	Forest area converted to cropland (million ha)	Forest area converted to livestock farming (million ha)	Total forest area converted to agriculture (million ha)
Small-scale	36.7 (97)[a]	7.5 (95)[a]	44.2
Large-scale including oil palm	1.0 (3)[a]	0.4 (5)[a]	1.4
Africa total	37.7 (82.7)[b]	7.9 (17.3)[b]	45.6
World total (% share of Africa)	85.9 (44)	66.6 (12)	152.5 (30)

[a]Figures in parentheses represent % of the total area of 37.7 and 7.9 million ha forest land converted to cropland and livestock farming in Africa

[b]Figures in parentheses represent % of the total area of 45.6 million ha agriculture-driven deforestation in 2000–2018

involved tree crops such as cocoa (*Theobroma cacao*), oil palm (*Elaeis guineensis*), rubber (*Havea brasilensis*), coffee (*Coffea* spp.) and to some extent food crops such as maize, rice and cassava. A strong coupling between cocoa production and deforestation in West Africa has been a growing concern over the past few decades. For example, in Côte d'Ivoire, the world's leading cocoa producer, more than 90% of the forest has disappeared (Kouassi et al. 2023). Recent analysis by Kalischek et al. (2023) shows that cocoa cultivation is an underlying driver of over 37% of forest loss in protected areas in Côte d'Ivoire and over 13% in Ghana. Similarly, analysis of data for the period 2000–2020 show dramatic decrease in forest cover and increase in forest fragmentation (53.8%) in the Congo Basin (Ma et al. 2023). Even where the forest is not completely lost, substantial forest fragmentation exists, and this is a major driver of loss of biodiversity and ecosystem services (Haddad et al. 2015; Ma et al. 2023).

In the past, shifting cultivation (swidden, slash and burn) has often been cited as the main driver of forest disturbance in sub-Saharan Africa (Fig. 5.1, Curtis et al. 2018), but emerging evidence suggests that this characterization is associated with misconceptions about shifting cultivation (see also Nath et al. 2022). For example, the map by Curtis et al. (2018, Fig. 5.1 in this chapter), gives the impression that forests in much of sub-Saharan Africa are subjected to shifting cultivation. This misconception diminishes the role of the real drivers of deforestation in Africa. In reality, shifting cultivation is practised mainly in the humid tropical regions of Africa, and only a few such as the forest-based farming systems are amenable to shifting cultivation. In a typical sequence of shifting cultivation, the forest is clear cut and burnt and crops are planted. After 2–3 years of cultivation, the fields are abandoned and the farmer clears another piece of forest. The abandoned field is left to fallow for several years to decades to rebuild fertility before the farmer returns to it to start the process again. This is not practical any more in many parts of Africa. Even in areas where shifting cultivation is feasible, its role in deforestation has been widely debated (see references in Nath et al. 2022). For example, in some areas in

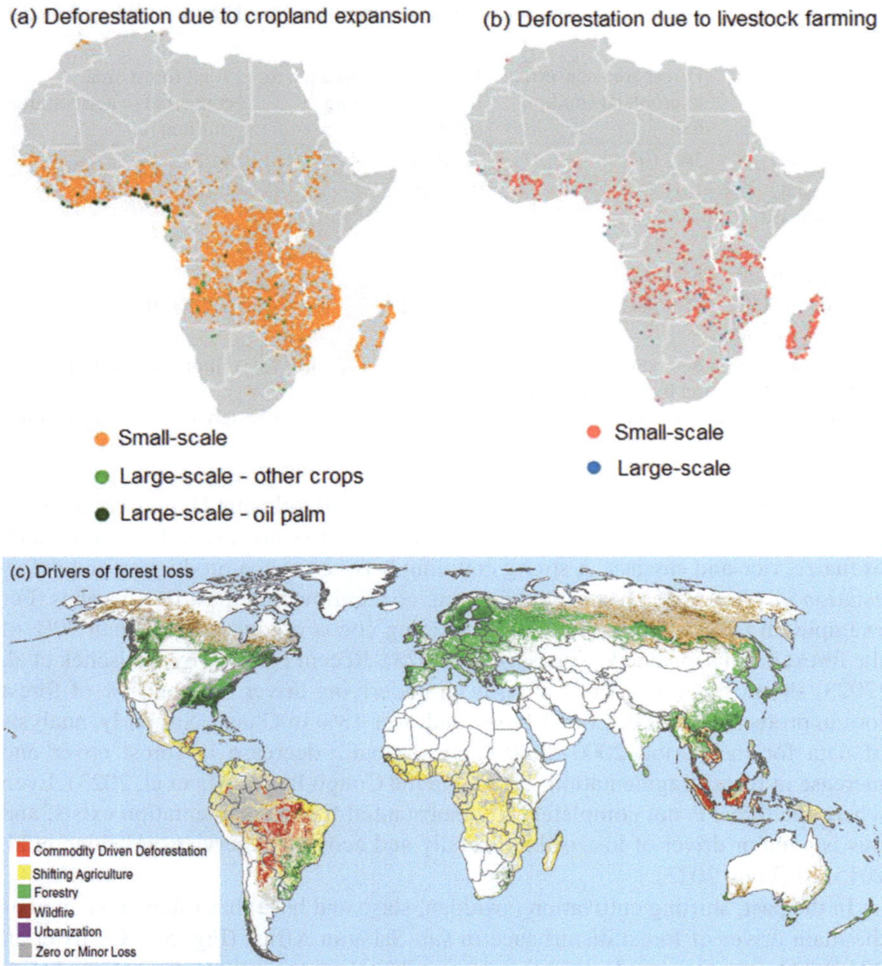

Fig. 5.1 Distribution of deforestation driven by small-scale and large-scale cropland expansion (**a**) and livestock farming (**b**) in the period 2000–2018 (adapted from Branthomme et al. 2023), and (**c**) drivers of forest loss in the years 2001–2015 according to Curti et al. (2018)

the Democratic Republic of Congo, shifting cultivation was shown to have a smaller footprint than other activities (Molinario et al. 2015).

Although the real culprit is the permanent conversion of forests for commercial plantations such as cocoa, oil palm, rubber in the humid forests, and tobacco and cashew in dry forests (Masolele et al. 2024), this problem is rarely acknowledged. For example, expansion of commodity crops such as cacao, oil palm and rubber is now the dominant drivers of forest loss in the humid regions of western and central

Africa (Masolele et al. 2024; Ordway et al. 2017). Hotspots of deforestation for cashew and tobacco plantations dominate in the dry forests of western and south-eastern Africa (Geist 1999; Masolele et al. 2024). In tobacco producing areas, deforestation takes place both directly, as a result of land clearing for tobacco cultivation, and indirectly, through wood used for curing and construction of barns (Geist 1999; Jew et al. 2017; Jimu et al. 2017). To meet the wood demand for tobacco curing, 140,000 ha of miombo woodlands are cleared annually, accounting for 4–26% of the deforestation in the miombo eco-region (Geist 1999). A more recent analysis shows that tobacco production accounts for 15% of deforestation in Zimbabwe and 26% in Malawi (Jimu et al. 2017). In post-forest landscapes, agriculture and related activities including grazing and fire often have negative effect on the regeneration of trees. For example, in the Sahel, negative associations have been found between livestock grazing and tree regeneration (Lohbeck et al. 2020).

While agricultural expansion is currently considered as one of the greatest threats to forests and biodiversity in Africa, smallholder agriculture also holds the key to conservation of the remaining biodiversity on agricultural land because smallholder farmers are the ultimate land managers and stewards of biodiversity (Sileshi et al. 2008a). In that sense, supporting and incentivizing smallholders to adopt sustainable agricultural practices can play a crucial role in the fight against deforestation. For example, in areas undergoing forest degradation, sustainable intensification on the existing cropland can support forest landscape restoration (De Pinto et al. 2017). However, the relationship between agricultural expansion and deforestation is not often nuanced. The negative impacts of high food prices, causing acute hunger, severe malnutrition and food insecurity, can amplify the humanitarian appeal of agricultural expansion to the society. Therefore, it is crucial to understand the context in which agriculture is contributing to deforestation and forest degradation, and to explore opportunities for sustainable intensification of agriculture to halt or reverse the current trend of forest loss.

One of the approaches to forest and landscape restoration has been through the promotion of agroforestry and other tree planting initiatives. However, these approaches and their contributions have been poorly contextualized to be in tandem with prevailing farming systems and farmers livelihood choices in Africa. The diversity of agroforestry practices and the context in which they can be promoted has also not been examined from the perspective of farming systems. Therefore, the objectives of this chapter are to: (1) synthesize the state of knowledge on the place and role of trees in African farming systems; (2) identify and highlight traditional and modern agroforestry practices in the different farming systems; and (3) highlight the role that multifunctional landscapes with trees can play in increasing productivity, climate change adaptation and mitigation. Here we undertook a critical review of the literature from Africa to meet our objectives.

5.2 Role of Trees in African Farming Systems

In the following passages we will use a farming systems lens to highlight the place and roles of trees in agriculture and livelihoods across Africa. We chose the farming systems approach because it allows characterization and analysis of systems within which smallholder households live and make resource management decisions (Dixon et al. 2020). In the past, inadequate consideration of a farming systems focus has led to the misconception that tree planting and agroforestry projects piloted in one setting could be successfully deployed in another setting. As a result many tree planting and agroforestry projects have failed to realize their potential. A farming system is defined as a population of farm households that as a group have broadly similar patterns of resource use, livelihoods, consumption, constraints and opportunities, and for which similar bundles of development strategies and interventions would be appropriate (Dixon et al. 2020). Such systems share broadly similar agro-ecological and market access conditions (Dixon et al. 2020). Dixon et al. (2020) identified 15 farming systems in sub-Saharan Africa (Fig. 5.2, Table 5.2). These farming systems are highly interconnected as materials (e.g., crop seeds, livestock, manure, etc.) and knowledge are shared and exchanged among the populations residing in the different farming systems. Indigenous knowledge of farmers, pastoralists and forest dwellers remains highly relevant in the management of cropland, rangeland, livestock and trees.

Trees and agroforestry practices are an integral part of the various farming systems in Africa (Table 5.3). As such trees and agricultural activities coexist not only in larger landscape contexts but also on single farms. African farmers and forest dwellers either integrate tree species in productive niches on their farms or manage forest resources (Miller et al. 2018). Farmers also plant trees to demarcate and assert informal claims to land. For example, a third of smallholder farmers in five African countries reported cultivating trees on their farms; with fruit trees and cash crops such as coffee, cocoa, and cashew nuts and timber trees being the most frequently planted trees (Miller et al. 2018). The contribution of trees to household income was estimated at 17% of total gross income among tree crop growers (Miller et al. 2018).

Many of the agricultural production systems take some form of agroforestry, where products and services of varying commercial values are generated from trees and crops. In this section, we will identify and describe agroforestry systems as low-value, medium-value and high-value based on the value of the products or commodities produced (Table 5.3). An agroforestry practice is deemed to be of low value when it produces products mainly for subsistence use, but of medium value when commodities are produced for sale in local or domestic markets. An agroforestry practice is said to be of high value when it involves export commodities where the value chain employs a large number of people. Some tree commodities have value chains providing a source of employment and livelihoods for a large number of Africans (Minang et al. 2021). For example, in Ethiopia coffee is still largely integrated into agroforestry systems and supports the livelihoods of over 15 million farmers and generates 25% of the country's export earnings (Moat et al. 2017).

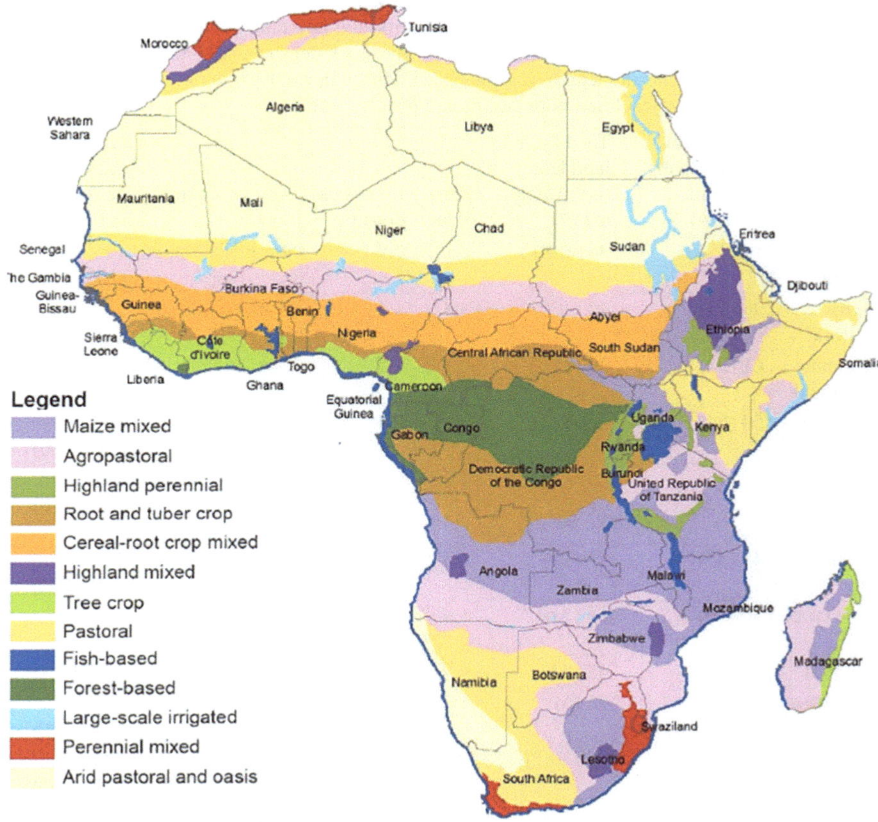

Fig. 5.2 Map of farming systems of Africa (Source: Dixon et al. 2020). The island and the urban and peri-urban farming systems were not mapped by the time of production of this map

Similarly, in Burundi the coffee value chain is a source of employment for 0.6–1.6 million people and livelihood to about a third of the population. Over 70% of the global cocoa supply originates from smallholder farmers in West Africa (Sassen et al. 2022), and over two million smallholder farmers are engaged in cocoa cultivation in Ghana and Cote D'Ivoire alone (Kalischek et al. 2023). Cocoa farmers are estimated at 600,000 in Cameroon and 300,000 in Nigeria. In Nigeria, the palm oil value chain alone employs about 1.8 million people (Minang et al. 2021). Therefore, we believe the distinction between low, medium and high value commodities in agroforestry is important to inform policy and investment in research and development.

African agriculture and the conditions in which trees are managed and commodities are produced is also heterogeneous, complex and dynamic. The diversity of soils, climate, institutions, markets and agricultural services also results in diverse

Table 5.2 Farming systems in Africa, agricultural population (in million people), extreme poverty (in % of rural population), total system land area (in million ha), cultivated land area (in million ha), farm size (cultivated area in ha per household), livestock density (TLU[a] per cultivated area), major agroecological zone and length of growing session (in days). Farming systems are arranged in descending order of agricultural populations and poverty levels

Farming system	Agricultural population[b] (extreme poverty)	Total system land area (cultivated area)[b]	Farm size	Livestock density	Agroecological zone[c]	Length of growing period (range in days)
Maize mixed	107 (56)	361 (40)	2.1	0.9	Warm subhumid	196 (150–240)
Agro-pastoral	98 (53)	443 (68)	3.8	1.1	Warm semiarid	130 (75–165)
Highland perennial	61 (51)	39 (8)	0.7	2.5	Cool subhumid	267 (240–330)
Root and tuber crop	50 (54)	236 (23)	2.5	0.4	Warm humid	269 (210–300)
Cereal-root crop mixed	43 (55)	202 (34)	4.3	0.9	Warm subhumid	187 (150–210)
Highland mixed	45 (37)	53 (13)	1.6	2.2	Cool subhumid	183 (150–270)
Tree crop	30 (40)	64 (7.6)	1.4	0.4	Warm humid	299 (270–330)
Pastoral	38 (40)	488 (33)	4.7	1.3	Warm arid	64 (30–90)
Fish-based	22 (48)	75 (5)	1.2	8.6	Warm subhumid	197 (150–330)
Forest-based	12 (52)	135 (0.7)	0.3	0.7	Warm humid	343 (330–365)
Irrigated	48 (NA)	46 (12)	NA	1.9	Warm arid	54 (0–90)
Arid pastoral-oases	8 (27)	915 (0.4)	0.3	67.0	Warm arid	12 (0–30)
Perennial mixed	12 (22)	37 (13)	5.2	0.4	Warm semiarid or subhumid	185 (150–240)
Island	1.5 (40)	1 (0.2)	0.8	NA	Warm subhumid	Variable
Urban and peri-urban	NA	NA	NA	NA	Variable	Variable

[a]TLU (Tropical Livestock Units) are livestock numbers converted to a common unit using conversion factors of 0.7 for cattle, 0.1 for sheep and goat, 0.2 for pig and 0.1 for chicken
[b]All figures are estimates by 2015 obtained from different chapters in Dixon et al. (2020)
[c]Since all farming systems (except perennial mixed) fall in the tropics, we removed the prefix "Tropical" to reduce clutter. *NA* estimate not available

land use zones and farming systems (Dixon et al. 2020). For the same reason, the benefits of trees and agroforestry practices are context-specific. If appropriately contextualized in farming systems and deployed in the right place, we believe agroforestry practices and other tree planting initiatives can provide important opportunities for strengthening the synergies between agriculture and forestry in providing

Table 5.3 Traditional and introduced agroforestry systems in African farming systems

Farming system	Access to services[a]	Dominant agroforestry practices	Value of commodities	References
Maize mixed	Medium	Agroforestry parkland; Improved fallows; Alley cropping; Intercropping with trees; Fodder banks; Rotational woodlots	Low-medium	Akinnifesi et al. (2010) Boffa (1999) Dubiez et al. (2019) Kachaka et al. (2020) Kwesiga and Coe (1994) Nyadzi et al. (2003) Sileshi et al. (2014) Tian et al. (2005) Kang et al. (1990)
Agro-pastoral	Low-medium	Agroforestry parklands	Low-medium	Boffa (1999)
Highland perennial	High	Home gardens; Coffee agroforestry; Tea agroforestry; Spice crop agroforestry; Rotational woodlots Alnus alley cropping	High	Aerts et al. (2011) Cyamweshi et al. 2023) Hemp and Hemp (2009) Negawo and Beyene (2017) Reyes et al. (2009) Winter (2009)
Root and tuber crop	Low-medium	Cassava-based agroforestry Yam-based agroforestry	Low-medium	Adjei-Nsiah et al. (2020)
Cereal-root crop mixed	Medium-high	Alley cropping	Low-medium	Kang et al. (1990)
Highland mixed	Low-medium	Cocoa agroforestry; Coffee agroforestry; Khat agroforestry	High	Jagoret et al. (2017) Feyisa and Aune (2003)
Tree crop	High	Cocoa agroforestry; Coffee agroforestry	High	Asare et al. (2014) Jagoret et al. (2017) Saj et al. (2017)
Pastoral	Low	Agroforestry parkland	Low-medium	Boffa (1999)
Fish-based	Medium-high	NA	NA	NA
Forest-based	Low	Improved fallows; Taungya Homegardens	Medium-high	Akinbisoye et al. (2014) Bandi et al. (2022) Kachaka et al. (2020)
Irrigated	Medium-high	NA	NA	NA

(continued)

Table 5.3 (continued)

Farming system	Access to services[a]	Dominant agroforestry practices	Value of commodities	References
Arid pastoral-oases	Very low	Oasis agroforestry	Medium-high	Santoro (2023)
Perennial mixed	High	Olive agroforestry	High	Temani et al. (2021)
Island	Medium	NA	NA	NA
Urban and peri-urban	Medium-high	Home gardens; Fodder banks	Medium-high	Chakeredza et al. (2007) Lee-smith et al. (2020)

[a]Information summarized from Dixon et al. (2020)
NA Information not available

various ecosystem services including food production, water regulation, biodiversity conservation, climate change adaptation and mitigation. In the following sections, we will briefly describe the farming systems and the role of trees and agroforestry in each of them.

5.2.1 Maize Mixed Farming System

The maize mixed farming system covers the subhumid areas of East, Central and Southern Africa, mainly the plateaus and highlands (800–1500 metres) in the Democratic Republic of Congo (DRC), Ethiopia, Kenya, Lesotho, Madagascar, Malawi, Mozambique, South Africa, Swaziland, Tanzania, Zambia and Zimbabwe (Fig. 5.2). This farming system has the highest agricultural population, with an annual growth rate of 2.7%. Out of a system land area of 361 million ha, the cultivated area occupies 40 million ha of which only 1.9 million ha is irrigated (Table 5.1) and 90% is under smallholders (Blackie et al. 2020). It is also characterized by medium access to services including markets (Blackie et al. 2020). The length of the growing season is on average 196 days (Table 5.1). Livestock density is relatively low. The livelihood sources include maize, legumes, cassava, tobacco, cotton, cattle, goats, poultry and off-farm work (Dixon et al. 2020).

Trees are an integral part of this farming system. Traditional agroforestry practices on smallholder farms involve planting maize and other food crops between scattered fruit trees (e.g., mango, citrus, and indigenous fruit trees), and fertilizer trees (e.g., *Faidherbia albida* and *Vachellia* species) (Akinnifesi et al. 2010; Sileshi 2016). Tree production on the larger and better endowed farms is becoming more widespread, in part to enhance household fuelwood supplies. Based on several decades of research, different agroforestry practices suited to this farming system have also been developed and promoted to address both soil fertility decline and shortage of animal feed. The key agroforestry practices include fertilizer tree systems promoted in Malawi, Zambia and Zimbabwe, fodder banks in Kenya and Zimbabwe, rotational woodlots in Tanzania and fruit trees in most countries (see Sect. 5.2).

5.2.2 *Agropastoral Farming System*

The agropastoral farming system is characterized by a mixed crop and livestock production in semi-arid areas covering 25 African countries (Fig. 5.2, Boffa et al. 2020). It is characterized by low access to services (e.g., markets) and lack of basic infrastructure (e.g., roads and electricity). Rainfall is unimodal in western and southern Africa but bimodal in eastern Africa. Drought is a regular phenomenon, and the length of growing period is 130 days with a total annual rainfall of 300–800 mm. Livestock density is relatively high, where rangelands and fallow lands provide livestock forage. Livelihoods are derived from sorghum, maize, pearl millet, pulses, sesame, cattle, sheep, goats, poultry and off-farm work. The sustainability of the agropastoral system relies on the maintenance of woody perennials in agriculture (Boffa et al. 2020). In the Sahelian, Eastern and Southern African subsystems, farmers grow crops under the canopy of scattered trees and shrubs of varying density, a system called agroforestry parklands (Boffa 1999). Livestock form a significant component of these farming systems, and hence the parklands are typically agrosilvipastoral systems (Boffa 1999). Agroforestry parklands, a form of traditional agroforestry, involve a mixture of perennial species characterized by one or a few dominant tree species and many secondary species whose composition varies according to local ecology and farmer preferences (Boffa 1999). For example, the Sahelian parklands are dominated by *Faidherbia albida*, *Parkia biglobosa* and *Vitellaria paradoxa* under which sorghum, millet, groundnut and sesame are grown (Boffa 1999). In most regions *Faidherbia* forms dense stands under which soils are fertile and crops and grass grow abundantly (Sileshi 2016). *Faidherbia* sheds its leaves at the onset of the wet season, permitting cultivation of sorghum and millet on the soil enriched by litterfall, manure and deep-capture of nutrients (Boffa 1999; Sileshi 2016). The perennials in agroforestry parklands are rarely planted but are derived from natural regenerations managed by farmers on agricultural land (see farmer-managed natural regeneration under Sect. 5.3).

5.2.3 *Highland Perennial Farming System*

This is a mixed farming system characterized by perennial crops including banana, plantains, enset (*Enset ventricosum*), coffee and tea in humid East African highlands with elevations above 1400 m asl (Lynam 2020). The highland perennial farming system covers the cool humid areas of Burundi, DRC, Ethiopia, Kenya, Rwanda, Tanzania and Uganda (Fig. 5.2). Access to services including markets is medium to high (Lynam 2020). Rainfall is bimodal, and the length of growing period is on average 267 days. Since the highlands have long rainy periods, agriculture is not usually water-limited. The relatively fertile soils and agroecological conditions make this farming system suitable for horticultural crops, tree crops and dairy farming (Lynam 2020). As a result, this farming system has experienced diversification into high value cash crops such as coffee and smallholder tea especially in central Kenya and northern Tanzania. Livelihoods are derived from diversified cropping including maize, cassava, sweet potato, beans, cereals, horticultural crops, livestock and poultry (Lynam 2020). Cattle are also an integral part of the system, with a recent shift to higher value dairy animals and a corresponding increase in forage production (Lynam 2020).

Over the centuries, communities have developed unique traditional agroforestry practices such as home gardens, where tree crops such as coffee are grown under shaded trees. In the Southern Ethiopian subsystem, the enset-coffee home gardens, a multi-storey agroforestry with high crop diversity predominates. Here, home gardens are mainly composed of Arabica coffee (*Coffea arabica*), enset (*Ensete ventricosum*), banana (*Musa paradisiaca*), avocado (*Persea americana*), mango (*Mangifera indica*) and Khat (*Catha edulis*) (Lynam 2020; Negash et al. 2022). In the Jimma zone of southwest Ethiopia, Arabica coffee is indigenous to the understorey of the moist evergreen montane rainforests. Here, patches of Afromontane forests are interspersed with managed coffee stands, which may be considered as agroforests of high conservation value and economic significance (Aerts et al. 2011). Another version exists where semi-wild coffee is grown in forest fragments where farmers manage the climax forest canopy and undergrowth to reduce the shading of coffee (Aerts et al. 2011). This is somewhat similar to rustic coffee in Mexico.

In the Albertine Rift subsystem including eastern DRC, Rwanda, Burundi, southwestern Uganda and the Kagera region of Tanzania, coffee and East African highland banana (*Musa acuminata*) are intercropped with annuals such as beans and root crops (Lynam 2020) in complex traditional agroforestry systems. In Sironko and Mbale districts of Eastern Uganda, coffee agroforestry is the main agricultural production system (Negawo and Beyene 2017). In Mubende in Uganda, fast-growing trees are planted in large scale Robusta coffee plantations. Modern agroforestry practices (e.g., alley cropping, scattered planting) with fertilizer trees such as alder (*Alnus acuminata*) have been widely promoted on farms to restore degraded land in the highlands of Rwanda (Cyamweshi et al. 2023).

In the Northern Tanzanian subsystem, the Chagga home garden is the most complex and well documented traditional agroforestry practice (Hemp and Hemp 2009; Winter 2009). This multi-storey agroforestry practice consists of highland banana grown under tall shade trees, coffee bushes under the banana plants, pulses, vegetables and root crops under the coffee bushes (Winter 2009). Livestock are stall-fed and manure collected from stalls is used to fertilize the gardens. With over 500 plant species, of which 400 are not cultivated, the home gardens host high native species biodiversity (Hemp and Hemp 2009). Another notable example in this subsystem is the spice crop agroforestry involving black pepper (*Piper nigrum*) and cardamom (*Elettaria cardamomum*) in the East Usambara Mountains of northeastern Tanzania

(Hall et al. 2011; Reyes et al. 2009). Cardamom is a valuable source of income for over 60% of the highland farmers (Reyes et al. 2009). The cardamom agroforestry involves the thinning of canopy trees to create only 50% shade and the complete clearance of the lower strata of a once natural forest (Hall et al. 2011). The feasibility of producing black pepper and cardamom in home garden agroforestry has also been explored. Pepper and cardamom intercropped with grevillea were shown to produce 3.9 and 2.3 times more than in monoculture stands (Reyes et al. 2009).

5.2.4 Root and Tuber Crop Farming System

This farming system is characterized by traditional agroforestry consisting of a mixture of root and tuber crops (mainly cassava, yam and cocoyam, potatoes), tree crops (e.g., oil palm, cocoa, rubber, cashew and mangoes) and cereals mainly rice, maize, sorghum and millet (Adjei-Nsiah et al. 2020). This farming system is found in tropical humid forest and the forest-savannah transitional zones of west and central Africa across 13 countries stretching from South Sudan to Sierra Leone (Fig. 5.2). Access to services is medium. Out of a system area of 236 million ha, only 23 million ha is cultivated and 1% under irrigation (Table 5.1). The length of growing period is on average 269 days and an annual rainfall of 1000–2500 mm is received. The agricultural population by 2015 was estimated at 50 million. The average farm size is 2.3 ha per household. Extreme poverty affects 54% of the rural population, representing 10.9% of rural poor people in SSA. Livestock density is 0.4 TLU per ha of cultivated area (Table 5.3), which is low due to prevalence of diseases (Adjei-Nsiah et al. 2020). Livelihoods are derived principally from cassava, yams and legumes. Other livelihood sources include cereals (e.g., maize, rice, sorghum and millet) and legumes (groundnut, cowpea and bambara nut), sugarcane, coffee, cocoa, plantain, banana, sesame and citrus cultivated in agroforestry systems. In the more humid, low altitude zones the major tree crops include oil palm, cocoa and rubber, commonly intercropped with cassava and cocoyam until the tree canopy closes (Adjei-Nsiah et al. 2020). Here cassava-based and yam-based agroforestry practices are common. In the sub-humid areas, cashew and mangoes are cultivated in traditional agroforestry systems. The fan palm (*Borassus aethiopum*) and African locust bean (*Parkia biglobosa*) are also important sources of cash and nutrition in the forest-savannah transition zone (Termote et al. 2022).

5.2.5 Cereal Root Crop Farming System

The cereal-root crop mixed farming system has one of the most suitable agroecological conditions in Africa, where tropical cereals (maize, sorghum, millet and upland rice), root and tubers (cassava, yams and sweet potato), legumes and vegetables are grown, usually as mixed intercrops (Kassam et al. 2020). This farming is practised under rainfed conditions, with limited use of fertilizers. The cereal root crop farming system is found in the rain forest and warm subhumid savannah zones in West and Central Africa (Fig. 5.2). In most of the countries, this farming system is regarded as the 'grain belt' and the 'bread basket' (Kassam et al. 2020). Access to services is medium. Livelihoods are derived principally from sorghum, maize, millet, cassava, yams, legumes and cattle. Cotton is more common in the north and root crops in the south (Kassam et al. 2020).

Apart from the traditional agroforestry systems, such as the home garden and scattered trees on farms, the main introduced agroforestry practice at scale was alley farming. Alley farming (hedgerow intercropping) is an agroforestry practice where trees are grown in rows and arable crops (cereals, legumes or horticultural crops) are cultivated in the alleys between the tree rows. For example, cassava can be planted in alleys between hedgerows of leguminous trees such as *Leucaena leucocephala* and *Gliricidia sepium*.

5.2.6 Highland Mixed Farming System

The highland mixed farming systems is an extensive crop-livestock system dominated by highland cereals such as wheat, barley and teff, cool season legumes such as peas, lentils and broad beans, potatoes and livestock (Amede and Lemenih 2020). It is found in tropical cool subhumid and mountainous areas with altitudes above 1700 m in Angola, Cameroon, Eritrea, Ethiopia, Lesotho, Nigeria, Morocco, Mozambique and Zimbabwe (Fig. 5.2). Access to services is low to medium. There is strong interdependency between crops and livestock production. Cattle are kept mostly for draught power, while sheep and goats are kept for generating cash from sales of live animals, wool and mohair. Livelihoods are derived predominantly from wheat and barley, teff, peas, lentils, broad beans, rape, potatoes, sheep, goats, cattle and poultry. High value tree crops such as coffee, cocoa, peach, apple and other fruits are produced in some parts of this farming system depending on the subsystems (Amede and Lemenih 2020). This farming system is heavily constrained by land degradation, high population density, small landholdings, and poor infrastructure. Livestock feed is another major constraint across the farming system. Up to 70% of the feed comes from crop residues. Although agroforestry practices involving fodder trees can play a role in alleviating feed shortages, these have been promoted only to a limited extent in this farming system. Other tree species such as *Eucalyptus* species and Khat (*Catha edulis*) are commonly grown on cropland across the subsystems in Ethiopia (Alemie (2009). Khat is a small evergreen tree cultivated for production of fresh leaves that are chewed as a stimulant (euphoric properties) in East Africa and the Arabian Peninsula. Khat is Ethiopia's second largest export commodity after coffee (Feyisa and Aune 2003; Negash et al. 2022). Khat production has been traditionally integrated within agroforestry systems involving maize or sorghum in the Ethiopian highlands (Feyisa and Aune 2003, Negash et al. 2022). According to Feyisa and Aune (2003) khat-maize agroforestry is 2.7 times more profitable per hectare than maize monocropping. The labour requirements for maize production are 50% lower in the khat-maize agroforestry than in maize monocropping (Feyisa and Aune 2003). As a result, it is a rapidly expanding perennial commodity crop in the Ethiopian highlands (Feyisa and Aune 2003, Negash et al. 2022). In Cameroon and rainforest zone of Nigeria, cocoa cultivation is based on complex agroforestry systems where cocoa is associated with forest trees, banana/plantain or fruit tree species that farmers preserve and/or plant after partial forest clearing (Jagoret et al. 2017, 2020).

5.2.7 Tree Crop Farming System

The tree crop farming system is dominated by tree crops, notably cocoa, coffee, oil palm and rubber production in tropical warm humid areas of Cameroon, Côte d'Ivoire, Ghana, Liberia, Nigeria, Sierra Leone and Togo in west and central Africa and Madagascar in southern Africa (Gockowski 2020, Fig. 5.2). Access to services including markets is high. Livelihoods are derived mainly from cocoa, coffee, oil palm, rubber, yams, maize, cassava and citrus (Gockowski 2020). This farming system is distinguished from other farming systems by a greater reliance on global markets for tree products (Gockowski 2020). Traditional agroforestry practices where food crops are cultivated between tree crops mainly for subsistence are common. Livestock production involves mainly goats, sheep, pigs and chickens; cattle are scarce due to the prevalence of trypanosomiasis.

Cocoa, oil palm, coffee and rubber are the main tree commodities in this farming system. In Cameroon, for example, cocoa agroforestry involves growing cocoa under the shade of larger trees mostly from native forest trees, sometimes thinned out to provide space for cocoa seedlings to be planted, or, to a lesser extent, trees and other perennials (banana, plantain) specially planted to provide shade (Hervé and Albanie 2021; Jagoret et al. 2017, 2020; Saj et al. 2017). The native trees that are retained and the new, exotic trees planted, provide not just shade, but are also used to provide extra income (Hervé and Albanie 2021). This mixture of shade trees and shrubs creates a multi-storey canopy similar in structure and function to a forest-like agroforest system. Cocoa agroforestry has raised a lot of hope and is now on the political agenda as a solution to rehabilitate highly degraded forests and reconcile agriculture with forestry in post-forest landscapes. Long-term sustainability could be achieved if the basal area share of the cocoa stand does not exceed 40% of the total basal area of cocoa agroforestry systems (Saj et al. 2017).

5.2.8 Pastoral Farming Systems

The pastoral farming system is characterized by transhumant (annual migration of livestock herders) or extensive pastoralism dominated by cattle production in tropical warm arid to semi-arid areas (de Leeuw et al. 2020). This farming system extends across 13 African countries (Fig. 5.2). Crop production is limited due to the short growing period (30–90 days). Access to services is also low. Livelihoods are derived mainly from cattle, camels, sheep, goats and some cereal crops. Four subsystems have been recognized based on geographic location and annual migration into more humid farming systems: (1) Sahelian, (2) East African, (3) Southern African and (4) Northern African subsystems (de Leeuw et al. 2020). In all subsystems, the extent of rangelands is declining due to land use change towards crop production (de Leeuw et al. 2020), overgrazing and bush encroachment due to the spread of local woody species and/or invasive alien species such as *Prosopis juliflora* in grasslands (Birhane et al. 2017).

The traditional agroforestry practices rely on farmer assisted natural regeneration of dryland trees used as sources of fruits and livestock fodder. Fruit trees such as *Adansonia digitata*, *Balanites aegyptiaca, Cordeauxia edulis*, *Sclerocarya birrea*, *Tamarindas indica*, *Vitellaria paradoxa*, *Ziziphus auritani*, and *Ziziphus mucronate* play a key role as sources of fruits and nuts as well as fodder for livestock (Kalinganire et al. 2008). In some regions, enclosures are enforced using fences and community by-laws or social fencing. For example, in Kenya, pastoralist communities in West Pokot and Baringo use enclosures to allow regeneration of trees and grass by setting aside fractions of the land as reverse for the dry season (Wairore et al. 2016). Unfortunately, there is a dearth of information on the traditional agroforestry practices and other tree planting intervention in the pastoral farming system because of poor documentation.

5.2.9 Forest-Based Farming System

The forest-based farming system is confined to tropical forests, forest-savannah mosaics and monsoon climate zones characterized by a year-long growing period and a tropical warm humid climate (Hauser et al. 2020, Table 5.2). This system constitutes a transition phase from forest to permanent cultivation or fallowing in the humid forest zones of the Democratic Republic of Congo, Republic of the Congo, Cameroon, Central African Republic, Equatorial Guinea, Gabon and parts of Liberia and Côte d'Ivoire (Fig. 5.2). Due to the high risk of trypanosomiasis, there are only few cattle and small ruminants. Access to services is low (Hauser et al. 2020). Livelihoods are largely derived from subsistence food crops including cassava, maize, beans, coco yams, taro, banana, plantain, and hunting and gathering forest products. Other livelihoods come from perennial tree crops such as cocoa and palm oil (Hauser et al. 2020). This farming system encompasses the Congo Basin, which contains the oldest, densest and most ecologically significant forest with the greatest role in the stability of the global climate. Although the Amazon covers a larger area of primary rainforest, trees in the Congo Basin being taller and more resilient to climate change than the Amazon rainforest, soak up some 1.2 billion tonnes of carbon dioxide annually and store one third more carbon over the same area of land than those of the Amazon (Cuni-Sanchez et al. 2021).

Two subsystems are recognized based on the pattern of forest clearing, namely, the forest mosaic subsystem and contiguous forest fringe subsystem. Deforestation due to shifting cultivation, establishment of plantations and activities associated with logging operations are the key challenge in this farming system (Kengoum et al. 2020). The high demand for food, wood energy and timber in the rapidly growing urban centres and the low agricultural productivity on existing agricultural land result in a very strong pressure on the forest resources (Hauser et al. 2020). Another key driver of change in this farming system is foreign investment in large-scale plantations (e.g., oil palm) and large-scale agriculture (e.g., rice cultivation) (Rulli et al. 2013). As such, this farming system needs urgent attention for landscape restoration in tandem with increasing productivity through practices such as agroforestry. This is especially the case in the DRC where increasing productivity in the supply basins can reduce the demand for expanding cropland into intact forest areas. Increased integration of trees into cropping systems through agroforestry can concomitantly increase access to tree products such as fruits, firewood, timber, building materials, animal fodder and leaf biomass that can be used as a fertilizer input (Kuyah et al. 2019; Sileshi et al. 2014). While traditional agroforestry practices (e.g. homegardens, taungya, etc) are poorly documented, new practices such as improved fallows and fertilizer tree systems have been proposed as alternatives to shifting cultivation.

5.2.10 Large-Scale Irrigated Farming Systems

The large-scale irrigated farming system is centred around water collected in dams from rivers and applied over a large area that can be physically irrigated (Williams et al. 2020). This system is located in low rainfall areas in Chad, Egypt, Mali, Niger, Nigeria, Somalia, Sudan and Zambia (Fig. 5.2). One of the greatest threats to this system is irrigation-induced salinization leading to land abandonment. This not only threatens the productivity of land within irrigated areas but also other farming systems downstream. There is no documentary evidence of agroforestry practices and other tree planting projects. The role of trees and agroforestry in land reclamation and opportunities for sustainable use of abandoned land needs to be explored in this farming system.

5.2.11 Arid Pastoral and Oases Farming System

The arid pastoral and oases farming system is located in the heart of the vast arid and hyper-arid areas encompassing 17 countries across Africa (Chaibou and Bonnet 2020, Fig. 5.2). The length of the growing period is less than 30 days (Table 5.3). With over 67 TLU per ha of cultivated area, this farming system has the highest livestock density. Availability of permanent water sources makes utilization of the vast expanses of highly unpredictable grazing lands possible (Chaibou and Bonnet

2020). The livelihood sources include date palms, cattle, small ruminants, irrigated crops and vegetables (Chaibou and Bonnet 2020).

This farming system is characterized by two closely interconnected components; arid pastoralism and scattered oasis farming. The arid pastoral component is characterized by extensive livestock production (mainly camels and goats) on wide expanses of drylands surrounding oases. The oases component is made up of dense vegetation dotted with small intensively cultivated plots using groundwater from wells. Traditional oases represent a unique form of agroforestry systems developed through the centuries by combining trees, crops and livestock (Santoro 2023). The traditional oasis agroforestry systems in Northern Africa are complex agrosilvipastoral systems consisting of a three-layer vertical structure, with the upper stratum occupied by date palms, the intermediate stratum composed olive trees and fruit trees (e.g., pomegranate, oranges, apricot and figs), and the lower stratum composed of cereals (e.g., sorghum, wheat, maize, rice and barley), vegetables and fodder crops such as alfalfa (Santoro 2023). This system is well integrated with animal husbandry, where fodder crops supply livestock feed and the livestock in turn provide manure and workforce for transport, water extraction and ploughing (Santoro 2023). The oasis agroforestry system in the northern part of the Sahara is dominated by the doum palm (*Hyphaene thebaica*) and the date palm (*Phoenix dactylifera*) and fruit trees such as mandarin, mango, grapefruit, pomegranate, and olives. The oasis agroforestry in Northeastern Ethiopia and Djibouti involves planting date palm by pastoralists (Betemariyam and Kefalew 2022). In Niger, Chad, Mauritania and Sudan the oasis agroforestry consists of desert date (*Balanites aegyptiaca*), jujube (*Ziziphus mauritiana*) and *Acacia nilotica*. Fruit tree production is increasing in importance (Chaibou and Bonnet 2020, Fig. 5.2). While some documentation exists on traditional agroforestry practices in northern Africa, information is virtually lacking on agroforestry in arid pastoral and oases farming system in eastern and southern Africa.

5.2.12 Perennial Mixed Farming System

The perennial mixed farming system is based on commercial trees (e.g., olives, fruit trees and forestry), grapes, sugarcane and cereals, pulses and livestock confined to the southern and northern extremes of the African continent (Mudhara et al. 2020). The southern subsystem is characterized by subtropical/tropical warm semi-arid climate and a growing period of 165 days, whereas the northern African subsystem has subtropical warm subhumid agroecology with 207 days (Mudhara et al. 2020). This farming system is distinguished from other African farming systems by the Mediterranean climate and access to relatively good domestic and export markets. This farming system has three recognizable subsystems: (1) the sugarcane and production forest subsystem in Swaziland and South Africa; (2) the deciduous fruit and wine grape subsystem in the Western Cape region of South Africa; and (3) the north African rainfed mixed subsystem in northern Algeria, Morocco and Tunisia

(Mudhara et al. 2020). The sugarcane-forestry subsystem consists of a mixture of smallholder farms, large commercial plantations where high value crops such as fruits, grapes, sugarcane are mostly grown under irrigation. Here forestry plantations (mainly pines, *Eucalyptus* species and *Acacia mearnsii*) are established on the steep slopes, and sugarcane plantation on better quality land. In the deciduous fruit and wine grape subsystem of the Western Cape, grapes, deciduous fruits and citrus are grown on irrigated commercial farms.

In the North African subsystem, most farmers engage in rainfed crop and livestock production involving winter cereals (e.g., wheat, barley), pulses (e.g., chickpeas, lentils, medics), olives, grapes and livestock mainly cattle, sheep and goats (Mudhara et al. 2020). Traditional olive-based agroforestry systems with seasonally open grazing and annual cropping is one of the outstanding examples of agroforestry in this farming system (Temani et al. 2021). Olive agroforestry is a widespread traditional practice in Algeria (Katsoulis et al. 2022), Morocco (Ben zineb et al. 2022; Temani et al. 2021) and Tunizia (Leauthaud et al. 2022). Olive trees are usually planted in lines, but they can also be planted in a scattered pattern (Katsoulis et al. 2022). Several species of fruit trees (e.g., fig, carob, quince), cereals (e.g., wheat, barley), grain legumes (e.g., faba bean, chickpea) and medicinal and aromatic plants are grown in association with olive trees (Katsoulis et al. 2022; Temani et al. 2021). The olive agroforestry systems in these arid environments are more productive and produce high-quality grains than sole crops and trees (Ben zineb et al. 2022; Katsoulis et al. 2022; Temani et al. 2021). For example, in north-eastern Morocco, olive agroforestry improved grain weight and the protein content of durum wheat and faba bean (Temani et al. 2021). In another study in northern Morocco, the olive agroforestry generated higher ecological stability than the sole crop under stressful climate conditions (Ben zineb et al. 2022). The study also showed that olive agroforestry enhanced network complexity of the soil microbiome leading to more stable beneficial microbial communities, especially in severe drought (Ben zineb et al. 2022). In Algeria, chamomile and anise grown as understory crops in olive agroforestry systems yielded high quality essential oils meeting commercial specifications (Katsoulis et al. 2022).

5.2.13 Urban and Peri-urban Farming Systems

Urban and peri-urban farming systems are located within cities, or on their fringes, and are gaining increasing significance due to the growing numbers of urban households involved (Lee-smith et al. 2020). However, this farming system has not been fully mapped and essential statistics were unavailable by the time of writing this chapter. This farming system is characterized by two subsystems, the backyard subsystems and open space subsystems (Lee-smith et al. 2020). With 50–67% of farming households, the backyard subsystem is the most common, and it consists of home gardens, rooftop and institutional gardens in inner urban areas. The home garden is essentially an urban agroforestry practice providing both subsistence and the opportunity for urban enterprise. The open space subsystem is more prevalent in peri-urban areas dominated by small-scale operators producing vegetables and small-scale livestock as well as large-scale glasshouses producing fruits, flowers and vegetables (Lee-smith et al. 2020).

The urban agroforestry practices consist of a mixture of fruit and nut trees, berry shrubs, and crops designed as multifunctional polycultures to supply food and a broad array of other ecosystem services (Lovell 2020). Fruit trees such as avocados, bananas, guava, mango, papaya, citrus fruits are often grown in the home gardens in many urban areas across Africa. With the expanding urban populations, there is a growing demand for both fruit and ornamental trees giving rise to flourishing tree nursery enterprises in most African cities (Lee-smith et al. 2020). In cities like Niamey in Niger, agroforestry systems in urban and peri-urban home gardens have a high richness and diversity of species (Lee-smith et al. 2020). Widespread adoption of urban agroforestry is likely to be constrained by availability and access to land. This is particularly the case where property values are high or future development is anticipated (Lovell 2020).

5.2.14 *Fish-Based Farming Systems*

The fish-based farming system is a mixed fishing and farming system where households derive 30–50% of their income from small-scale fishing but also engage in other livelihood activities including forestry, livestock production, hunting and gathering (Hamerlynck et al. 2020). This farming system is found along coasts, lakes and rivers across Africa, in a wide range of climatic zones (Fig. 5.2). Information is virtually lacking on agroforestry practices and tree-based commodities in these two farming systems. We believe that is because of poor documentation of the practices.

5.2.15 *Island Farming Systems*

The island farming system is characterized by mixed fisheries, tree crops, sugarcane, cereal and tuber crops, fruit, vegetables, spices, livestock and some forestry in Island states including Cabo Verde, the Comoros, Mauritius, Sao Tome and Principe, and the Seychelles (Mudhara et al. 2020). As in the Fish-based farming system, information is virtually lacking on agroforestry practices and tree-based commodities in these farming systems.

In each of the farming systems described above, isolated and sacred trees protected by farm households exist and these often contribute to biodiversity conservation. The role that such trees play in ecosystem functioning has been recently highlighted in Sileshi and Nath (2023).

5.3 Innovative Tree-Based Technologies for Diversifying Production

In this section we briefly describe tree-based technologies in Africa under four main categories, namely, fertilizer trees, fodder trees, fruit and nut trees and timber trees (Table 5.4). Innovative technologies here broadly cover introduced agroforestry practices designed or adapted by researchers for specific contexts.

5.3.1 Fertilizer Tree Systems

Fertilizer tree systems involve the utilization of nitrogen-fixing leguminous trees in cereal production systems to improve the availability of nitrogen to crops, while also providing additional benefits such as rehabilitation of degraded land and provision of wood (Akinnifesi et al. 2010; Sileshi et al. 2014, 2020a, b, 2023a, b). These systems include improved fallows, relay cropping, intercropping, alley farming (hedgerow cropping), evergreen agriculture and biomass transfer. Each of these systems have been described in great detail in earlier publications (Akinnifesi et al. 2010; Kang et al. 1990; Kuntashula et al. 2004; Sileshi et al. 2014, 2020a, 2023a, b). Therefore, we will refrain from detailed descriptions of the practices, but we will provide brief discussion of their applicability and farming systems in which they are likely to be adopted.

The improved fallows technology was specifically developed as an alternative to shifting cultivation (slash-and-burn agriculture) and the traditional grass fallows that take a long time to rehabilitate the soil (see references in Sileshi et al. 2020a). This technology was widely tested in the maize mixed farming system in southern and eastern Africa (Akinnifesi et al. 2010; Sileshi et al. 2014) and to some degree in the forest-based farming system in DR Congo (Kachaka et al. 2020). However, the adoption of improved fallows by smallholder farmers has been slow despite intensive promotion especially in southern and eastern Africa. This practice holds some potential for adoption in the maize mixed farming system where land holding sizes are large and, in the forest-based farming systems, where traditional shifting cultivation is still practised. It is unlikely that improved fallows can be adopted in other farming systems where land holding sizes are small (see Table 5.2) and shifting cultivation is not traditionally practised.

Intercropping with trees is a common traditional practice. An innovative practice involves intercropping Gliricidia with maize, which was widely tested in the maize mixed farming system in Malawi and Zambia (Akinnifesi et al. 2010, Sileshi et al. 2014). Gliricidia-maize intercropping holds a great potential for adoption in the maize mixed farming system where land holding sizes are small. Another example involves growing annual crops under scattered trees of species such as *Alnus acuminata*, *Grevillea robusta* and in the humid tropical highlands of Rwanda and Kenya (Sileshi et al. 2020a). As indicated in Sect. 5.2.3, this practice holds a great potential for adoption in the highland perennial farming system.

Table 5.4 Categories of recent and innovative agroforestry practices and the farming systems where they have been tested

Categories	Innovative practices and tree species	Farming systems	Predominant regions	References
Fertilizer trees	Improved fallows and relay cropping with *Sesbania sesban*, *Tephrosia* spp. and pigeon pea (*Cajanus cajan*), *Acacia* spp.	Maize mixed	East, Central and Southern Africa	Akinnifesi et al. (2010), Kachaka et al. (2020) Kwesiga and Coe (1994) Sileshi et al. (2014)
	Intercropping with *Gliricidia* and *Alnus*,	Maize mixed	Southern Africa	Akinnifesi et al. (2010) Sileshi et al. (2020b)
	Alley cropping with *Alnus* and *Leucaena* spp	Cereal root crop	West and Central	Kang et al. (1990)
	Biomass transfer using any of the nitrogen-fixing species		Eastern and Southern Africa	Cyamweshi et al. (2023) Kuntashula et al. (2004) Sileshi et al. (2011)
Fodder trees	Pure stands of *Calliandra, Leucaena, Acacia, Vachelia spp* planted as hedges or on farm boundaries	Urban and peri-urban	East, Central and Southern Africa	Chakeredza et al. (2007)
Fruit trees	Domestication of indigenous species	Almost all farming systems	Eastern, Western and Southern Africa	Akinnifesi et al. (2008) Leakey et al. (2022)
	Cocoa agroforestry	Tree crop and forest-based		Asare et al. (2014) Kouassi et al. (2023) Saj et al. (2017)
Timber and fuelwood trees	Taungya with *Cedrela odorata*, *Gmelina arborea*, *Khaya* spp. and *Tectona grandis*; Scattered planting of *Grevillea robusta*, *Eucalyptus* and *Pinus* spp. on cropland	Forest based Highland mixed Perennial mixed	West Africa East Africa Southern Africa	Akinbisoye et al. (2014) Kalame et al. (2011)
	Rotational woodlots (*Acacia crassicarpa, A. polyacantha, Leucaena* spp.)	Maize mixed	East Africa; Central Africa	Dubiez et al. (2019) Nyadzi et al. (2003)

Alley farming was tested and developed in the humid regions of Africa as an alternative to the traditional shifting cultivation (slash and burn agriculture) (Kang et al. 1990; Tian et al. 2005). Where livestock is part of the farming system, alley farming supplies quality fodder for livestock. The provision of fuelwood is also an important benefit.

Biomass transfer involves moving tree pruning and litter produced in one part of the farm (e.g., farm boundaries, in fodder banks, fallows, alley farming) to another part such as a vegetable garden (Sileshi et al. 2014). Biomass transfer using *Alnus, Calliandra*, *Gliricidia*, *Leucaena* and *Tithonia* pruning has been widely tested on maize and vegetable crops in East Africa and southern Africa and significant increases in crop productivity and soil fertility has been demonstrated (Cyamweshi et al. 2023; Kuntashula et al. 2004; Sileshi et al. 2011).

5.3.2 Fodder Trees

Fodder trees and shrubs are often grown in fodder banks. In the agropastoral and pastoral farming systems in the Sahel, trees such as *Balanites aegyptiaca*, *Acacia polyacantha* and *Faidherbia albida* are managed in grazing lands to provide fodder, shade and shelter for the animals. These trees also allow luxurious growth of forage under their canopies because of the positive effect on soil moisture and soil fertility (Boffa 1999). Fodder trees and shrubs can be planted scattered on cropland, along boundary lines, as hedges around the homestead and along pathways, along terrace edges on sloping land and permanent contour bunds. The fodder bank technology capitalizes on the use of leguminous trees and shrubs managed for production of fodder as protein supplements for livestock (Chakeredza et al. 2007). Tree species commonly planted in fodder banks include species in the genera *Acacia*, *Calliandra*, *Flemingia*, *Gliricidia*, *Leucaena*, Sesbania and *Vachelia* spp. Planting the trees in double rows at wider spacing (50 cm apart with 1–1.5 m in between double rows) has been recommended to improve fodder productivity (Chakeredza et al. 2007). Where land is scarce, farmers may plant trees in single rows. Once the trees established, grass may be allowed to grow between the tree rows. Studies from Tanzania, Kenya and Uganda have demonstrated the profitability of fodder banks in zero-grazing, cut and carry systems (Chakeredza et al. 2007). Fodder trees hold significant potential for adoption in the agropastoral, highland perennial, highland mixed, arid pastoral an urban and peri-urban farming systems given the large livestock population (see Table 5.2) and the huge demand for feed.

5.3.3 Fruit and Nut Trees

Exotic and indigenous fruit trees managed in agroforestry systems play an important role in the farming systems in Africa. The common exotic fruit trees are avocado, mango, citrus fruits, cashew, deciduous fruits, olives (National Academy of Sciences. 2008). Olives and deciduous fruit trees (apples, pears, peaches, nectarines, plums, prunes and apricots) are the dominant sources of export commodities in the perennial mixed farming system. Some of the indigenous fruit tree species such as cocoa and oil palm are export crops from the tree crop and forest-based farming systems. In West Africa, cocoa agroforestry is a traditional practice that is now being promoted as an alternative to monoculture cocoa plantations (Kouassi et al. 2023).

Other fruit tree species play a key role in meeting the nutritional requirement of households and income generation (Akinnifesi et al. 2008; Ickowitz et al. 2014; Sileshi et al. 2023c). Their uses as fodder, timber and making small tools and medicinal value have also been well-documented (Akinnifesi et al. 2008). Sileshi et al. (2023c) provides a list of priority species in each African region and discusses in detail the potential and challenges for their domestication and research needs. Over the years, priority species have been jointly identified by researchers and farmers for domestication in each region. For example, *Irvingia gabonensis*, *Dacryodes edulis*, *Chrysophyllum albidum*, *Garcinia kola*, *Garcinia livingstonei* and *Allanblackia floribunda* are the priority species in the humid regions of West Africa. *Adansonia digitata, Detarium microcarpum, Parkia biglobosa, Tamarindus indica*, *Vitellaria paradoxa* and *Ziziphus mauritiana* are the priority species in the Sahelian region of West Africa. Similarly, *Adansonia digitata*, *Balanites aegyptiaca, Carissa edulis*, *Cordeauxia edulis*, *Sclerocarya birrea*, *Tamarindas indica*, *Vitellaria paradoxa*, *Vitex doniana* and *Ziziphus mauritania* were ranked the top most species in sub-humid to arid areas of East Africa. In the southern African region, *Uapaca kirkiana*, *Strychnos cocculoides*, *Strychnos spinosa*, *Parinari curatellifolia*, *Vangueria infausta* and *Ziziphus mauritiana* were identified as priority species for domestication (see Sileshi et al. 2023a, b, c for details). Over the years, researchers have developed techniques for propagation and on-farm management of the selected tree species.

5.3.4 Timber and Fuel Wood Trees

Timber trees and other trees used for producing fuelwood are often planted in specific arrangements in woodlots, i.e., sole stands of trees planted on farms, communal lands or degraded lands to produce wood for fuel, construction and or pulp (Nyadzi et al. 2003). In the rotational woodlot system, food crops are intercropped with leguminous trees during the first 2–3 years. Then the trees are left to grow and harvested in about the fifth year, and food crops are replanted to benefit from the improved soil (Nyadzi et al. 2003). The tree species tested in rotational woodlots include *Acacia crassicarpa*, *Acacia polyacantha*, *Leucaena leucocephala* and eucalypts (*Eucalyptus* spp). In Tabora, *Acacia crassicarpa* produced 24–77 tons of wood per hectare in 4–5 years, while *Leucaena leucocephala* produced 71–89 tons per hectare during the same period (Nyadzi et al. 2003). In the early 2000s the concept of Taungya was developed into the rotational woodlot as an alternative to establishing fast growing Acacia species to overcome wood shortage. Taungya starts as an agroforestry system during the initial three years, and then evolves to a plantation when the trees form a closed canopy (Kalame et al. (2011). This optimises the use of the land, reduces tree establishment costs such as weeding, increases productivity and generates early income for farmers around forests. Kalame et al. (2011) reported a modified taungya system in Ghana as a win–win practice for forestry and adaptation to climate change. Taungya was introduced in Ghana, Nigeria and Kenya by colonial authorities and since then has been modified and practised in West Africa (Kalame et al. 2011). It remains one of the cheapest means of establishing plantations in the rainforest zones.

The various practices described above involve planting of trees in specific arrangements. More recently, farmer managed natural regeneration of trees (FMNR) has been promoted as an innovation in various parts of Africa. FMNR involves farmers nurturing natural woody regeneration on their fields, while keeping them under the primary function of agricultural production (Kuyah et al. 2023). Farmer management includes removal of undesired trees especially in active crop fields to avoid competition with crops (Lohbeck et al. 2020). It also involves fencing off a portion of land from human activities in order to allow the vegetation to recover. Initially, FMNR was initiated as a community-led approach to re-establish trees and shrubs on deforested and degraded land in Niger. The practice immediately spread to the neighbouring countries because of its benefits, and is now practised in more than 18 countries in Africa (Kuyah et al. 2023). Farmers used this adaptation to diversify production of food, fodder, fuel, wood, and improve system sustainability (Kuyah et al. 2023). FMNR holds a great potential for increasing tree cover in the agropastoral, pastoral and arid pastoral farming systems, where low soil moisture, poor access to seedlings and grazing limit establishment of trees through planting (Sileshi et al. 2023b).

5.4 Improved Productivity and Ecosystem Services from Multifunctional Landscapes

The application of various forms of agroforestry has been shown to play a significant role in raising productivity (Kuyah et al. 2019; Sileshi et al. 2008a, b, 2014, 2023a), carbon sequestration and mitigation of greenhouse gas emissions (Corbeels et al. 2019; Muchane et al. 2020; Luedeling et al. 2011; Shi et al. 2018) and biodiversity conservation on farmland (Abebe and Asfaw 2023; Chirwa et al. 2008; De Beenhouwer et al. 2013). The productivity, climate change adaptation and mitigation benefits of the various agroforestry practices have been documented in significant detail in Sileshi et al. (2023a). Therefore, we will only briefly highlight key findings. Meta-analyses of studies conducted across Africa have demonstrated that fertilizer trees can significantly increase crop productivity (Sileshi et al. 2008a, b, 2014). Demographic health surveys across Africa have also revealed that agroforestry can improve household nutrition and dietary diversity (Ickowitz et al. 2014). The climate change adaptation benefits of agroforestry practices include provision of soil cover, maintenance of better microclimate conditions for crops, increase in water infiltration, reduction of runoff and soil erosion, increase in the water holding capacity of the soil, increase in rain use efficiency, and reducing risks of crop failure and yield variability (Sileshi et al. 2023a). With carbon sequestration rates of 1.2–5.8 tons per hectare per year (Luedeling et al. 2011), trees in agroforestry practices can also provide significant climate change mitigation benefits (Muchane et al. 2020).

Trees in the various farming systems are also sources of income and employment for many people in Africa. For example, agricultural populations in a number of farming systems derive their livelihoods from tree crops such as coffee, cocoa, rubber, oil palm, fruit trees and timber trees managed on cropland. As a result, African agriculture is a significant contributor to the global supply of tree commodities and a source of revenue for many countries (Table 5.5). Tree commodities are also tied to the identities of many African countries as they represent the main source of foreign exchange (Minang et al. 2021; Piabuo et al. 2021). For example, cocoa accounts for 40% and 25% of the export revenues in Cote d'Ivoire and Ghana (Piabuo et al. 2021). Cocoa alone contributes 15% and 7% to the GDP of Cote d'Ivoire and Ghana, respectively. Similarly, coffee contributes to 34% of total exports and 5% of Ethiopia's GDP (Piabuo et al. 2021). Côte d'Ivoire and Ghana alone account for over 70% of the global cocoa supply (Jagoret et al. 2020), while Nigeria and Cameroon contribute a substantial portion.

In terms of area, cocoa occupies the largest annual area of production in Africa, followed by oil palm, olives, cashew and coffee (Table 5.5). Cocoa also stood first in terms of annual export value (7.4 billion US$) followed by coffee, cashew and oranges (Table 5.5). Much of the production come from monoculture plantations, which are unsustainable. For example, the current cocoa production system remains productive only for 20–30 years, while forest areas available for setting up new orchards are dwindling (Jagoret et al. 2020). In a recent study, Kouassi et al. (2023) showed the failure of plantation programs in cocoa fields and suggested promotion of natural tree regeneration to ensure sustainability. If scaled up in the right context, agroforestry practices involving tree crops such as cocoa can help maintain significant biodiversity in agricultural landscapes, support local and global ecosystem services and play a role in maintaining landscape connectivity (Asare et al. 2014; Sassen et al. 2022). Species diversity is often very high in traditional agroforestry practices such as home gardens. For example, according to various studies in Ethiopia, home gardens contain up to 200 plant species in Dwaro zone (Agize et al. 2013), 120 woody species in Sidama, 78 species in Wolayita and Sidama zones, 59 species in West Oromia Region, and 69 species in west Gojam Zone (Abede et al.

Table 5.5 The annual areal extent (in million ha), production (in million tonnes), gross production value (in billion US$ in 2021) and export value (in billion US$ in 2021) of tree crops across Africa

Tree crops	Area (CV%)[a]	Production (CV%)[a]	Gross production value	Export value	Top three countries in terms of area of production in 2021
Cocoa	7.5 (10.5)	3.5 (9.6)	4.3	7.4	Côte d'Ivoire, Ghana, Nigeria
Oil palm	5.2 (9.0)	21.0 (8.0)	1.4	0.1	Nigeria, Côte d'Ivoire, Ghana
Olives	3.6 (23.6)	4.2 (19.8)	1.2	0.0001	Tunisia, Morocco, Algeria
Cashew	3.6 (14.7)	1.7 (12.0)	0.3	1.8	Côte d'Ivoire, Tanzania, Benin
Coffee	3.0 (15.4)	1.2 (11.8)	0.6	2.4	Côte d'Ivoire, Uganda, Ethiopia
Rubber	0.9 (20.6)	0.9 (28.9)	1.8	0.0002	Côte d'Ivoire, Nigeria, Liberia
Mangoes and guavas	0.9 (15.9)	7.7 (15.1)	2.6	0.2	Côte d'Ivoire, Egypt, Nigeria
Coconuts	0.6 (26.3)	2.1 (3.5)	0.3	0.02	Tanzania, Ghana, Mozambique
Karite nuts (shea nuts)	0.6 (13)	0.7 (16)	0.1	0.1	Nigeria, Burkina Faso, Mali
Oranges	0.5 (7.2)	9.5 (6.8)	4.7	1.6	Egypt, Morocco, Algeria
Kola nuts	0.5 (3.7)	0.3 (5.1)	0.1	0.01	Nigeria, Cameroon, Ghana
Dates	0.4 (4.8)	3.7 (9.3)	4.4	0.5	Algeria, Tunisia, Morocco
Tea	0.4 (12.4)	2.9 (10.7)	0.5	1.5	Kenya, Mozambique, Uganda
Apples	0.2 (6.1)	2.9 (11.9)	2.5	0.5	Morocco, Egypt, Algeria
Figs	0.2 (2.1)	0.5 (12.7)	0.2	0.001	Morocco, Egypt, Algeria
Tangerines/ mandarins	0.2 (11.4)	2.9 (15.3)	1.5	1.1	Morocco, Egypt, South Africa
Peaches and nectarines	0.1 (8.1)	1.0 (8.6)	0.4	0.1	Algeria, Morocco, Tunisia
Avocados	0.1 (18.7)	0.8 (23.4)	0.5	0.4	Ethiopia, Kenya, Malawi
Lemons and limes	0.1 (16.5)	1.5 (17.6)	1.0	0.4	Sudan, South Africa, Egypt
Pears	0.05 (8.1)	0.8 (3.4)	0.6	0.21	Algeria, South Africa, Egypt
Pistachios	0.04 (14.3)	0.01 (69.0)	–	0.002	Tunisia, Madagascar, Mauritius

[a]Means and coefficients of variation (CV%) for 10 years (2012–2021) calculated from FAOSTAT (2023)
NA estimates not available

2010; Abebe and Asfaw 2023). In Uganda, Negawo and Beyene (2017) recorded a total of 63 tree and shrub species in coffee based agroforestry practices. Diversity indexes were significantly higher in coffee based agroforestry than nearby forest reserves when exotic tree species were accounted for Negawo and Beyene (2017).

A growing body of literature suggests that conversion of forests into monoculture plantations reduces biodiversity, soil organic carbon (SOC) and nutrients stocks and overall ecosystem health (Liu et al. 2023). For example, according to a recent global meta-analysis conversion of forest to rubber plantations significantly reduced SOC, microbial carbon, total nitrogen and total magnesium pools by 24, 35, 13 and 94%, respectively (Liu et al. 2023). On the other hand, intercropping rubber with different crops in the early stages of plantation establishment (Taungya) increased SOC and microbial biomass carbon pools by 6% and 24%, respectively (Liu et al. 2023). Agroforestry practices can also help in converting simplified production systems such as plantations into multi-functional landscapes under the right combination of technological innovations, policy, and market incentives. The multifunctional use of landscapes entails the sharing of land to produce food and resources and to conserve ecosystems and their components (Fischer et al. 2017; Schümann et al. 2022). In multifunctional agricultural landscapes, increased connectivity between habitats can promote spill-over of ecosystem services (Schümann et al. 2022). However, the design of multifunctional landscapes requires a good understanding of the risks and trade-offs associated with each tree-based intervention. This will be briefly discussed in the following passages.

5.5 Trees-Crop Interplay: Risks and Trade-Offs in Multifunctional Landscapes

Tree-crop interactions may be competitive or complementary. Competition is said to occur when trees exploit the same resources as crops, and consequently exert negative effects on the crop. Complementarity occurs when trees acquire resources that crops cannot acquire, and in the process benefit the crop. Although trees compete with crops for resources, they can also improve the growing conditions of crops, for example by providing shade under drought conditions or nutrients in nutrient-limited areas (Sileshi 2016; Sileshi et al. 2020a; Temani et al. 2021). Consequently, the tree-crop association can promote higher crop yields or higher harvest quality (Temani et al. 2021). The beneficial effects of trees may depend on the seasonal pattern of rainfall, which determines the compensation between yield components (Temani et al. 2021). Shading by trees may penalizes crop growth under favourable conditions due to lower light interception and photosynthesis rates. The reverse holds true for shade loving plants, which may grow better under the shade. The shade also creates a beneficial microclimate including relatively lower temperature, wind, higher air humidity and reduced evapotranspiration which can reduce water and heat stresses especially in drylands. These may create more favourable conditions to certain microfauna and macrofauna (Barrios et al. 2023). This highlights the need for understanding the benefits and trade-offs of different agroforestry arrangements.

The presence of trees on cropland almost always entails competition for resources between the tree and associated crops depending on site conditions, tree species, tree management and crop species. This may lead to reduction in crop yields. In northern Rwanda, maize grain yields were about 45% and 19% less than in sole maize at 1 m and 3 m from alder trees, respectively (Ndoli et al. 2017). In a study by Muthuri et al. (2005), a reduction of 21% compared to the sole maize control was observed in the Naro Moru site in Kenya.

The most recognized advantage of agroforestry is the increase in resource-use efficiency, which results from complementarity or facilitative interaction between trees and crops (Temani et al. 2021). However, such complementarity may be either spatial or temporal and therefore context specific (Ong et al. 2004). Spatial complementarity occurs when trees exploit deep reserves of water and nutrients, which crops cannot access (Akinnifesi et al. 2004; Makumba et al. 2009). Temporal complementarity occurs when trees and crops make their primary demand for resources at different times, for example, if the tree's resource demand is higher during the dry season after crops have been harvested (Ong et al. 2004). This is observed in agroforestry parklands with *Faidherbia*. Farmers in the agropastoral farming system and parts of the maize mixed farming system have capitalized on the high degree of complementarity between *Faidherbia* and cereal crops. Recent analyses have demonstrated that increases in crop yields under the *Faidherbia* canopy is due to improvements in water and nutrient availability (Sileshi 2016). Hundreds of studies conducted across Africa have demonstrated that planting of fertilizer trees on cropland can increase nutrient supply to crops through various mechanisms (Sileshi et al. 2014, 2020a).

5.6 Chapter Highlights

This chapter synthesized the state of knowledge on the role of trees in African farming systems and highlighted the role that multifunctional landscapes with trees can play in increasing productivity, biodiversity conservation, and climate change adaptation and mitigation. It has also key farming systems in which indigenous and introduced agroforestry technologies have been adopted by farmers. It also presented key data on the contribution of trees on farmland to livelihoods of communities in Africa, through provision of tree products, supporting crop and livestock production, and income, which plays a key role in adaptation of local communities to climate change. We strongly recommend transformation of forest land converted to cropland or plantations of cash crops into agroforestry. This should be pursued through economic incentives such as payment for ecosystem services.

References

Abebe H, Asfaw Z (2023) Review on homegarden agroforestry and its contribution in Ethiopia. Int J Agric Econ 8:7–13. https://doi.org/10.11648/j.ijae.20230801.12

Abede T, Wiersum KF, Bongers F (2010) Spatial and temporal variation in crop diversity in agroforestry homegardens of Southern Ethiopia. Agrofor Syst 78:309–322

Adjei-Nsiah S, Asumugha G, Njukwe E, Akoroda M (2020) The root and tuber crop farming system: diversity, complexity and productivity potential. In: Dixon J et al (eds) Farming systems and food security in Africa: priorities for science and policy under global change. Routledge, London/New York, pp 182–213

Aerts R, Hundera K, Berecha G, Gijbels P, Baeten M, Van Mechelen M, Hermy M, Muys B, Honnay O (2011) Semi-forest coffee cultivation and the conservation of Ethiopian Afromontane rainforest fragments. For Ecol Manag 261:1034–1041. https://doi.org/10.1016/j.foreco.2010.12.025

Agize M, Demissew S, Asfaw Z (2013) Indigenous knowledge on management of home gardens and plants in Loma and Gena Bosa districts (Weredas) of dawro zone, Southern Ethiopia: plant biodiversity conservation, sustainable utilization and environmental protection. Int J Sci Basic Appl Res 10:63–99

Akinbisoye OS, Oke SO, Adebola SI, Mokwenye A (2014) Influence of Taungya agroforestry system on diversity of native woody species and soil physico-chemical properties in Nigeria. Int J Sci Res 4:1–15

Akinnifesi FK, Rowe EC, Livesley SJ, Kwesiga F, Vanlauwe B, Alegre J (2004) Tree root architecture. In: Ong C, van Noordwijk M, Cadisch G (eds) Belowground interactions in tropical agroecosystems: concepts and models with multiple plant components. CAB Publishing, Wallingford, pp 61–81

Akinnifesi FK, Leakey RB, Ajayi OC, Sileshi G, Tchoundjeu Z, Matakala P, Kwesiga FR (eds) (2008) Indigenous fruit trees in the tropics: domestication, utilization and commercialization. CABI, Wallingford

Akinnifesi FK, Ajayi OC, Sileshi G, Chirwa PW, Chianu J (2010) Fertiliser trees for sustainable food security in the maize-based production systems of East and Southern Africa. A review. Agron Sustain Dev 30:615–629

Alemie TC (2009) The effect of *Eucalyptus* on crop productivity, and soil properties in the Koga watershed, western Amhara region, Ethiopia. MSc. thesis, Cornell University, USA

Amede T, Lemenih M (2020) The highland mixed farming system of Africa: diversifying livelihoods in fragile ecosystems. In: Dixon J et al (eds) Farming systems and food security in Africa: priorities for science and policy under global change. Routledge, London and New York, pp 248–281

Asare R, Afari-Sefa V, Osei-Owusu Y, Pabi O (2014) Cocoa agroforestry for increasing forest connectivity in a fragmented landscape in Ghana. Agrofor Syst 88:1143–1156. https://doi.org/10.1007/s10457-014-9688-3

Bandi MM, Mahimba MB, Mpie PMM, M'vubu ARN, Khasa DP (2022) Adoption of agroforestry practices in and around the Luki Biosphere Reserve in The Democratic Republic of the Congo. Sustain For 14:9841. https://doi.org/10.3390/su14169841

Barrios E, Coe R, Place F, Sileshi GW, Sinclair F (2023) Nurturing soil life through agroforestry. In: Uphoff N, Thies J (eds) Biological approaches to regenerative soil systems. CRC Press, Boca Raton. https://doi.org/10.1201/9781003093718-27

Ben zineb A, Barkaoui K, Karray F, Mhiri N, Sayadi S, Mliki A, Gargouri M (2022) Olive agroforestry shapes rhizosphere microbiome networks associated with annual crops and impacts the biomass production under low-rainfed conditions. Front Microbiol 13:977797. https://doi.org/10.3389/fmicb.2022.977797

Betemariyam M, Kefalew T (2022) Carbon stock estimation of mixed-age date palm (*Phoenix dactylifera* L.) farms in northeastern Ethiopia. Heliyon 8(1):e08844. https://doi.org/10.1016/j.heliyon.2022.e08844

Birhane E, Treydte AC, Eshete A, Solomon N, Hailemariam M (2017) Can rangelands gain from bush encroachment? Carbon stocks of communal grazing lands invaded by *Prosopis juliflora*. J Arid Environ 141:60–67. https://doi.org/10.1016/j.jaridenv.2017.01.003

Blackie M, Dixon J, Mudhara M, Rusike J, Snapp S, Mekuria M (2020) Maize mixed farming system: an engine for rural growth and poverty reduction. In: Dixon J et al (eds) Farming systems and food security in Africa: priorities for science and policy under global change. Routledge, London and New York, pp 67–104

Boffa JM (1999) Agroforestry parklands in sub-Saharan Africa. FAO Conservation Guide 34. Food and Agriculture Organization, Rome

Boffa JM, Sanders J, Taonda SJB, Hiernaux P, Bagayoko M, Ncube S, Nyamangara J (2020) The agropastoral farming system: achieving adaptation and harnessing opportunities under duress. In: Dixon J et al (eds) Farming systems and food security in Africa: priorities for science and policy under global change. Routledge, London/New York, pp. 105–147

Branthomme A, Merle C, Kindgard A, Lourenço A, Ng W-T, D'Annunzio R, Shapiro A (2023) How much do large-scale and small-scale farming contribute to global deforestation? Results from a remote sensing pilot approach. FAO, Rome. https://doi.org/10.4060/cc5723en

Chaibou M, Bonnet B (2020) The arid pastoral and oasis farming system: key centres for the development of trans-Saharan economies. In: Dixon J et al (eds) Farming systems and food security in Africa: priorities for science and policy under global change. Routledge, London/New York, pp 450–481

Chakeredza S, Hove L, Akinnifesi FK, Franzel S, Ajayi O Sileshi G (2007) Managing fodder trees as a solution to human-livestock food conflicts and their contribution to income generation for smallholder farmers in Southern Africa. Nat Res For 31:286–291

Chirwa PW, Akinnifesi FK, Sileshi G, Syampungani S, Kalaba FK, Ajayi OC (2008) Opportunity for conserving and utilizing agro-biodiversity through agroforestry in Southern Africa. Biodiversity 9:45–48

Corbeels M, Cardinael R, Naudin K, Guibert H, Torquebiau E (2019) The 4 per 1000 goal and soil carbon storage under agroforestry and conservation agriculture systems in sub-Saharan Africa. Soil Till Res 188:16–26. https://doi.org/10.1016/j.still.2018.02.015

Cuni-Sanchez A, Sullivan MJ, Platts PJ, Lewis SL, Marchant R, Imani G, Hubau W, Abiem I, Adhikari H, Albrecht T, Altman J (2021) High aboveground carbon stock of African tropical montane forests. Nature 596:536–542

Curtis PG, Slay CM, Harris NL, Tyukavina A, Hansen MC (2018) Classifying drivers of global forest loss. Science 361:1108–1111. https://doi.org/10.1126/science.aau3445

Cyamweshi RA, Muthuri CW, Kuyah S, Mukuralinda A, Sileshi GW (2023) Pruning and green manure from *Alnus acuminata* improve soil health, and potato and wheat productivity in Rwanda. Agrofor Syst 98:269–282. https://doi.org/10.1007/s10457-023-00904-5

De Beenhouwer M, Aerts R, Honnay O (2013) A global meta-analysis of the biodiversity and ecosystem service benefits of coffee and cacao agroforestry. Agric Ecosys Environ 175:1–7. https://doi.org/10.1016/j.agee.2013.05.003

De Leeuw J, Osano P, Said M, Ayantunde A, Dube S, Neely C, Vrieling A, Thornton P, Ericksen P (2020) The pastoral farming system: balancing between tradition and transition. In: Dixon J et al (eds) Farming systems and food security in Africa: priorities for science and policy under global change. Routledge, London/New York, pp 318–353

De Pinto A, Robertson R, Begeladze S, Kumar C, Kwon HU, Thomas T, Cenacchi N, Koo J (2017) Cropland restoration as an essential component to the forest landscape restoration approach-global effects of wide-scale adoption. IFPRI discussion paper 01682

Dixon J, Garitty D, Boffa J-M, Williams TO, Amede T et al (eds) (2020) Farming systems and food security in Africa: priorities for science and policy under global change. Routledge, London/New York

Dubiez E, Freycon V, Marien JN, Peltier R, Harmand JM (2019) Long term impact of *Acacia auriculiformis* woodlots growing in rotation with cassava and maize on the carbon and nutrient contents of savannah sandy soils in the humid tropics (Democratic Republic of Congo). Agrofor Syst 93:1167–1178. https://doi.org/10.1007/s10457-018-0222-x

Feyisa TH, Aune JB (2003) Khat expansion in the Ethiopian highlands: effects on the farming system in Habro district. Mtn Res Dev 23:185–189

Fischer J, Meacham M, Queiroz C (2017) A plea for multifunctional landscapes. Front Ecol Environ 15:59. https://doi.org/10.1002/fee.1464

Geist JH (1999) Global assessment of global deforestation related to tobacco farming. Tob Control 8:18–28

Gibbs HK, Rueschb AS, Achard F, Clayton MK, Holmgren P, Ramankutty N, Foley JA (2010) Tropical forests were the primary sources of new agricultural land in the 1980s and 1990s. Proc Natl Acad Sci USA 107:16732–16737. https://doi.org/10.1073/pnas.0910275107

Gockowski J (2020) The tree crop farming system: stagnation, innovation and forest degradation. In: Dixon J et al (eds) Farming systems and food security in Africa: priorities for science and policy under global change. Routledge, London/New York, pp 282–317

Haddad NM et al (2015) Habitat fragmentation and its lasting impact on Earth's ecosystems. Sci Adv 1:e1500052

Hall JM, Gillespie TW, Mwangoka M (2011) Comparison of agroforests and protected forests in the East Usambara Mountains, Tanzania. Environ Manag 48:237–247. https://doi.org/10.1007/s00267-010-9579-y

Hamerlynck O, Nyingi WD, Paul JL, Duvail S (2020) The fish-based farming system: maintaining ecosystem health and flexible livelihood portfolios. In: Dixon J et al (eds) Farming systems and food security in Africa: priorities for science and policy under global change. Routledge, London and New York, pp 354–391

Hauser S, Norgrove L, Tollens E, Nolte C, Robiglio V, Gockowski J (2020) The forest-based farming system: highly diverse, annual and perennial systems under threat. In: Dixon J et al (eds) Farming systems and food security in Africa: priorities for science and policy under global change. Routledge, London/New York, pp 393–422

Hemp A, Hemp C (2009) Environment and worldview: the Chaga homegardens, part 1—ethnobotany and ethnozoology. In: Clack T (ed) Culture, history and identity: landscapes of inhabitation in the Mount Kilimanjaro Area. Archaeopress, Tanzania

Hervé BD, Albanie MS (2021) Evolutions and Innovations in the Cocoa Supply chain. In: Minang PA, Duguma LA, van Noordwijk M (eds) Tree commodities and resilient green economies in Africa. World Agroforestry (ICRAF), Nairobi

Ickowitz A, Powell B, Salim MA, Sunderland TCH (2014) Dietary quality and tree cover in Africa. Glob Environ Change 24:287–294. https://doi.org/10.1016/j.gloenvcha.2013.12.001

Jagoret P, Michel I, Ngnogué HT, Lachenaud P, Snoeck D, Malézieux E (2017) Structural characteristics determine productivity in complex cocoa agroforestry systems. Agron Sustain Dev 37:60. https://doi.org/10.1007/s13593-017-0468-0

Jagoret P, Saj S, Carimentrand A (2020) Cocoa agroforestry systems in Africa—the art of reconciling sustainable production and ecological services. CIRAD Perspective, Montpellier, 54. https://doi.org/10.19182/perspective/31916

Jew EKK, Dougill AJ, Sallu SM (2017) Tobacco cultivation as a driver of land use change and degradation in the miombo woodlands of south-west Tanzania. Land Degrad Dev 28:2636–2645. https://doi.org/10.1002/ldr.2827

Jimu L, Mataruse L, Musemwa L, Nyakudya NW (2017) The miombo ecoregion up in smoke: The effect of tobacco curing. World Dev Perspect 5:44–46. https://doi.org/10.1016/j.wdp.2017.03.007

Johnson JA, Runge CF, Senauer B, Foley J, Polasky S (2014) Global agriculture and carbon tradeoffs. Proc Natl Acad Sci USA 11:12342–12347. https://doi.org/10.1073/pnas.1412835111

Kachaka ÉY, Munson AD, Gélinas N, Khasa D (2020) Adoption of an improved fallow practice using *Acacia auriculiformis* on the Batéké Plateau in The Democratic Republic of the Congo. Agrofor Syst 94:1047–1058. https://doi.org/10.1007/s10457-019-00474-5

Kalame FB, Aidoo R, Nkem J, Ajayi OC, Kanninen M, Luukkanen O, Idinoba M (2011) Modified taungya system in Ghana: a win–win practice for forestry and adaptation to climate change? Environ Sci Pol 14:519–530. https://doi.org/10.1016/j.envsci.2011.03.011

Kalinganire A, Weber JC, Uwamariya A, Kone B (2008) Improving rural livelihoods through domestication of indigenous fruit trees in the parklands of the Sahel. In: Akinnifesi et al. (eds) Indigenous fruit trees in the tropics: domestication, utilization and commercialization, pp. 186–203, CABI

Kalischek N, Lang N, Renier C, Daudt RC, Addoah T, Thompson W, Blaser-Hart WJ, Garrett R, Schindler K, Wegner JD (2023) Cocoa plantations are associated with deforestation in Côte d'Ivoire and Ghana. Nature Food 4:384–393

Kang BT, Reynolds L, Atta-Krah K (1990) Alley farming. Adv Agron 43:316–359

Kassam A, Kueneman E, Lott R, Friedrich T, Lutaladio N, Norman D, Bwalya M, Poisot AS, Mkomwa S (2020) The cereal-root crop mixed farming system: a potential bread basket transitioning to sustainable intensification. In: Dixon J et al (eds) Farming systems and food security in Africa: priorities for science and policy under global change. Routledge, London/New York, pp 214–247

Katsoulis GI, Kimbaris AC, Anastasaki E, Damalas CA, Kyriazopoulos AP (2022) Chamomile and Anise cultivation in olive agroforestry systems. Forests 13:128. https://doi.org/10.3390/f13010128

Kengoum F, Pham TT, Moeliono M, Dwisatrio B, Sonwa DJ (2020) The context of REDD+ in the Democratic Republic of Congo: drivers, agents and institutions, 2nd edn. Occasional paper 207. CIFOR, Bogor

Kouassi AK, Zo-Bi IC, Aussenac A, Kouamé IK, Dago MR, N'guessan AE, Jagoret P, Hérault B (2023) The great mistake of plantation programs in cocoa agroforests—Let's bet on natural regeneration to sustainably provide timber wood. Trees For People 12: 100386. https://doi.org/10.1016/j.tfp.2023.100386

Kuntashula E, Mafongoya PL, Sileshi G, Lungu S (2004) Potential of biomass transfer technologies in sustaining vegetable production in the wetlands (*dambo*s) of eastern Zambia. Expl Agric 40:37–51

Kuyah S, Whitney CW, Jonsson M, Sileshi GW, Öborn I, Muthuri CW, Luedeling E (2019) Agroforestry delivers a win-win solution for ecosystem services in sub-Saharan Africa. Agron Sustain Dev 39:47. https://doi.org/10.1007/s13593-019-0589-8

Kuyah S, Buleti S, Dimobe K, Nkurunziza L, Moussa S, Muthuri C, Öborn I (2023) Farmer managed natural regeneration in Africa: Evidence for climate change mitigation and adaptation in drylands. In: Dagar JC et al (eds) Agroforestry for sustainable intensification of agriculture in Asia and Africa. Springer Nature Singapore Pte Ltd, pp 53–89. https://doi.org/10.1007/978-981-19-4602-8_3

Kwesiga F, Coe R (1994) The effect of short rotation *Sesbania sesban* planted fallows on maize yield. For Ecol Manag 64:199–209

Leakey RRB, Tientcheu ML, Awazi NP, Assogbadjo AE, Mabhaudhi T, Hendre PS, Degrande A, Hlahla S, Manda L (2022) The future of food: domestication and commercialization of indigenous food crops in Africa over the third decade (2012–2021). Sustain For 14:2355. https://doi.org/10.3390/su14042355

Leauthaud C, Ben Yahmed J, Husseini M, Rezgui F, Ameur F (2022) Adoption factors and structural characteristics of irrigated olive grove agroforestry systems in Central Tunisia. Agroecol Sustain Food Syst 46:1025–1046. https://doi.org/10.1080/21683565.2022.2085230

Lee-Smith D, Prain G, Cofie O, van Veenhuizen R, Karanja N (2020) Urban and peri-urban farming systems: feeding cities and enhancing resilience. In: Dixon J et al (eds) Farming systems and food security in Africa: priorities for science and policy under global change. Routledge, London/New York, pp 504–532

Liu Y, Shen J, Zhang C, Chen Z (2023) Impact of rubber-based land use changes on soil properties and carbon pools: a meta-analysis. Catena 227:107–121. https://doi.org/10.1016/j.catena.2023.107121

Lohbeck M, Albers P, Boels LE, Bongers F, Morel S, Sinclair F, Smith-Dumont E (2020) Drivers of farmer-managed natural regeneration in the Sahel. Lessons for restoration. Sci Rep 10(1):1–11. https://doi.org/10.1038/s41598-020-70746-z

Lovell S (2020) Urban agroforestry and its potential integration into city planning efforts. Urban Agric Reg Food Syst 5:e20000. https://doi.org/10.1002/uar2.20000

Luedeling E, Sileshi G, Beedy T, Dietz J (2011) Carbon sequestration potential by agroforestry in Africa. In: Kumar BM, Nair PKR (eds) Carbon sequestration in agroforestry systems. Springer, Dordrecht, pp 61–83

Lynam J (2020) The highland perennial farming system: sustainable intensification and the limits of farm size. In: Dixon J et al (eds) Farming systems and food security in Africa: priorities for science and policy under global change. Routledge, London/New York, pp 148–181

Ma J, Li J, Wu W, Liu J (2023) Global forest fragmentation change from 2000 to 2020. Nat Commun 14:3752. https://doi.org/10.1038/s41467-023-39221-x

Makumba W, Akinnifesi FK, Janssen BH (2009) Spatial rooting patterns of gliricidia, pigeon pea and maize intercrops and effect on profile soil N and P distribution in southern Malawi. Afr J Agric Res 4:278–288

Masolele RN, Marcos D, De Sy V, Abu IO, Verbesselt Y, Reiche J, Herold M (2024) Mapping the diversity of land uses following deforestation across Africa. SC Rep 14:1681. https://doi.org/10.1038/s41598-024-52138-9

Miller DC, Muñoz-Mora JC, Christiaensen L (2018) Do trees on farms matter in African agriculture? In: Christiaensen L, Demery L (eds) Agriculture in Africa: Telling Myths from Facts. World Bank, Washington, DC, pp 115–121

Minang PA, Kamwilu E, van Noordwijk M (2021) Pathways to achieving sustainable development goals through tree commodities in Africa. In: Minang PA, Duguma LA, van Noordwijk M (eds) Tree commodities and resilient green economies in Africa. World Agroforestry (ICRAF), Nairobi

Moat J, Jenny W, Baena S, Wilkinson T, Gole TW et al (2017) Resilience potential of the Ethiopian coffee sector under climate change. Nature Plants 3(17081):1–14. https://doi.org/10.1038/nplants.2017.81

Molinario G, Hansen MC, Potapov PV (2015) Forest cover dynamics of shifting cultivation in the Democratic Republic of Congo: a remote sensing-based assessment for 2000–2010. Environ Res Lett 10:094009. https://doi.org/10.1088/1748-9326/10/9/094009

Muchane MN, Sileshi GW, Gripenberg S, Jonsson M, Pumarino L, Barrios E (2020) Agroforestry boosts soil health in the humid and sub-humid tropics: a meta-analysis. Agric Ecosyst Environ 295:106899. https://doi.org/10.1016/j.agee.2020.106899

Mudhara M, Lombard J, Palmer A, Dixon J (2020) Perennial mixed and Island farming systems: exploiting synergies for maximum system productivity. In: Dixon J et al (eds) Farming systems and food security in Africa: priorities for science and policy under global change. Routledge, London/New York, pp 482–503

Muthuri CW, Ong CK, Black CR et al (2005) Tree and crop productivity in Grevillea, Alnus and Paulownia-based agroforestry systems in semi-arid Kenya. For Ecol Manag 212:23–39. https://doi.org/10.1016/j.foreco.2005.02.059

Nath AJ, Reang D, Sileshi GW (2022) The shifting cultivation juggernaut: an attribution problem. Glob Chall 6:2200051. https://doi.org/10.1002/gch2.202200051

National Academy of Sciences (2008) Lost Crops of Africa vol III. Fruits. National Academies Press, Washington, DC

Ndoli A, Baudron F, Schut AGT et al (2017) Disentangling the positive and negative effects of trees on maize performance in smallholdings of Northern Rwanda. Field Crop Res 213:1–11

Negash M, Kaseva J, Kahiluoto H (2022) Perennial monocropping of khat decreased soil carbon and nitrogen relative to multistrata agroforestry and natural forest in southeastern Ethiopia. Reg Environ Chang 22:38. https://doi.org/10.1007/s10113-022-01905-3

Negawo JW, Beyene DN (2017) The role of coffee based agroforestry system in tree diversity conservation in Eastern Uganda. J Landsc Ecol 10(2):5–18. https://doi.org/10.1515/jlecol-2017-0001

Nyadzi et al (2003) Rotational woodlot technology in northwestern Tanzania: tree species and crop performance. Agrofor Syst 59:253–263. https://doi.org/10.1023/B:AGFO.0000005226.62766.05

Ong C, Black CR, Wilson J, Muthuri C, Bayala J, Jackson NA (2004) Agroforestry: hydrological impacts. In: Van Noordwijk M, Cadisch G, Ong CK (eds) Below-ground interactions in tropical agroecosystems: concepts and models with multiple plant components. CAB International, Wallingford

Ordway EM, Asner GP, Lambin EF (2017) Deforestation risk due to commodity crop expansion in sub-Saharan Africa. Environ Res Lett 12:044015. https://doi.org/10.1088/1748-9326/aa6509

Piabuo SM, Tieguhong JC, Minang PA, Foundjemi-Tita D, Duguma LA (2021) Industrializing Africa through tree commodities. In: Minang PA, Duguma LA, van Noordwijk M (eds) Tree commodities and resilient green economies in Africa. World Agroforestry (ICRAF), Nairobi

Reyes T, Quiroz R, Luukkanen O, de Mendiburu F (2009) Spice crops agroforestry systems in the East Usambara Mountains, Tanzania: growth analysis. Agrofor Syst 76:513–523. https://doi.org/10.1007/s10457-009-9210-5

Rulli MC, Saviori A, D'Odorico P (2013) Global land and water grabbing. Proc Natl Acad Sci USA 110:892–897. https://doi.org/10.1073/pnas.1213163110

Saj S, Jagoret P, Etoa LE, Fonkeng E, Tarla JN, Essobo Nieboukaho J-D, Mvondo Sakouma K (2017) Lessons learned from the long-term analysis of cacao yield and stand structure in central Cameroonian agroforestry systems. Agric Syst 156:95–104. https://doi.org/10.1016/j.agsy.2017.06.002

Santoro A (2023) Traditional oases in Northern Africa as multifunctional agroforestry systems: a systematic literature review of the provided Ecosystem Services and of the main vulnerabilities. Agrofor Syst 97:81–96. https://doi.org/10.1007/s10457-022-00789-w

Sassen M, van Soesbergen A, Arnell AP, Scott E (2022) Patterns of (future) environmental risks from cocoa expansion and intensification in West Africa call for context specific responses. Land Use Policy 119:106142. https://doi.org/10.1016/j.landusepol.2022.106142

Schümann H, Knierim A, Bellingrath-Kimura SD, Kernecker M (2022) Working with or against multifunctional landscapes? A case study of land users' local knowledge of grassland–forest transition zones in northeastern Germany. Ecol Soc 27:38. https://doi.org/10.5751/ES-12970-270138

Shi L, Feng W, Xu J, Kuzyakov Y (2018) Agroforestry systems: meta-analysis of soil carbon stocks, sequestration processes, and future potentials. Land Degrad Dev 29:3886–3897

Sileshi GW (2016) The magnitude and spatial extent of *Faidherbia albida* influence on soil properties and primary productivity in drylands. J Arid Environ 132:1–14. https://doi.org/10.1016/j.jaridenv.2016.03.002

Sileshi GW, Nath AJ (2023) A framework for analysing spatial patterns and extent of influence by single trees on ecosystem properties in agroforestry, agroforestry for sustainable intensification of agriculture in Asia and Africa, sustainable agriculture and food security. Springer Nature Singapore Pte Ltd., pp 685–718. https://doi.org/10.1007/978-981-19-4602-8_21

Sileshi G, Akinnifesi FK, Ajayi OC, Chakeredza S, Mng'omba S, Nyoka BI (2008a) Towards sustainable management of soil biodiversity in agricultural landscapes in Africa. Biodiversity 9:64–67

Sileshi G, Akinnifesi FK, Ajayi OC, Place F (2008b) Meta-analysis of maize yield response to planted fallow and green manure legumes in sub-Saharan Africa. Plant Soil 307:1–19. https://doi.org/10.1007/s11104-008-9547-y

Sileshi GW, Akinnifesi FK, Gondwe F, Ajayi OC, Mng'omba SA, Mwafongo K (2011) Effect of organic fertilizer on the growth and fruit yield of six paprika (*Capsicum annum* L.) cultivars in Malawi. Agrofor Syst 83:361–372

Sileshi GW, Mafongoya PL, Akinnifesi FK, Phiri E, Chirwa P, Beedy T et al (2014) Fertilizer trees. Encyclopedia of agriculture and food systems, vol 1. San Diego: Elsevier, pp 222–234

Sileshi GW, Mafongoya PL, Nath AJ (2020a) Agroforestry systems for improving nutrient recycling and soil fertility on degraded lands. In: Dagar JC et al (eds) Agroforestry for degraded landscapes: recent advances and emerging challenges, vol 1. Springer Nature, Singapore, pp 225–254. https://doi.org/10.1007/978-981-15-4136-0_8

Sileshi GW, Akinnifesi FK, Mafongoya PL, Kuntashula E, Ajayi OC (2020b) Potential of Gliricidia-based agroforestry systems for resource limited agro-ecosystems. In: Dagar JC et al (eds) Agroforestry for degraded landscapes: recent advances and emerging challenges.vol 1. Springer Nature, Singapore, pp 255–281. https://doi.org/10.1007/978-981-15-4136-0_9

Sileshi GW, Dagar JC, Nath AJ, Kuntashula E (2023a) Agroforestry as a climate-smart agriculture: strategic interventions, current practices and policies, agroforestry for sustainable intensification of agriculture in Asia and Africa, Sustainable Agriculture and Food Security. Springer Nature Singapore Pvt Ltd, pp 589–640. https://doi.org/10.1007/978-981-19-4602-8_18

Sileshi GW, Dagar JC, Kuyah K, Datta A (2023b) The Great Green Wall initiatives and opportunities for integration of dryland agroforestry to mitigate desertification, Agroforestry for sustainable intensification of agriculture in Asia and Africa, sustainable agriculture and food security. Springer Nature Singapore Pte Ltd, pp 175–206. https://doi.org/10.1007/978-981-19-4602-8_6

Sileshi GW, Dagar JC, Akinnifesi FK, Mng'omba SA (2023c) Potentials of indigenous fruit trees in enhancing nutrition, income and biodiversity conservation in African agroforestry, Agroforestry for sustainable intensification of agriculture in Asia and Africa, sustainable agriculture and food security. Springer Nature Singapore Pte Ltd, pp 321–362. https://doi.org/10.1007/978-981-19-4602-8_11

Temani F, Bouaziz A, Daoui K, Wery J, Barkaoui K (2021) Olive agroforestry can improve land productivity even under low water availability in the South Mediterranean. Agric Ecosyst Environ 307:107234. https://doi.org/10.1016/j.agee.2020.107234

Termote C, Odongo NO, Dreyer BS, Guissou B, Parkouda C, Vinceti B (2022) Nutrient composition of *Parkia biglobosa* pulp, raw and fermented seeds: a systematic review. Criti Rev Food Sci Nutr 62:119–144. https://doi.org/10.1080/10408398.2020.1813072

Tian G, Kang BT, Kolawole GO, Idinoba P, Salako FK (2005) Long-term effects of fallow systems and lengths on crop production and soil fertility maintenance in West Africa. Nutr Cycl Agroecosyst 71:139–150

Wairore JN, Mureithi SM, Wasonga OV, Nyberg G (2016) Benefits derived from rehabilitating a degraded semi-arid rangeland in private enclosures in west Pokot County, Kenya. Land Degrad Dev 27:532–541

Williams TO, Faurès JM, Namara R, Snyder K (2020) Large-scale irrigated farming system: the potential and challenges to improve food security, livelihoods and ecosystem management. In: Dixon J et al (eds) Farming systems and food security in Africa: priorities for science and policy under global change. Routledge, London/New York, pp 423–449

Winter JC (2009) Environment and worldview: The Chaga homegardens-part II: the social history of the Chagga in outline with special reference to the evolution of the homegardens. In: Clack T (ed) Culture, history and identity: landscapes of inhabitation in the Mount Kilimanjaro Area, Tanzania. Archaeopress, Oxford

Zabel F, Delzeit R, Schneider JM, Seppelt R, Mauser W, Václavík T (2019) Global impacts of future cropland expansion and intensification on agricultural markets and biodiversity. Nat Commun 10:2844. https://doi.org/10.1038/s41467-019-10775-z

Chapter 6
Promoting Crop Productivity from Improved Soil Management in Smallholder Farming Systems: Policy Implications for Enhanced Soil Management in Africa

Agnes Kalibata, Paul Thangata, Vine Mutyasira, Sibusiso Nhlengethwa, and Paxie W. Chirwa

Abstract In sub-Saharan Africa (SSA), achieving sustainable agricultural intensification and meeting the growing food demands of a burgeoning population in the face of growing threats to food systems requires a rethink of soil health solutions for smallholder farming systems. During the recently ended Fertilizer and Soil Health Summit (2024), African leaders committed to practical measures to safeguard African soil health, recognizing how this is critical to achieving food security and environmental sustainability. The Summit also saw the launch of the Great Green Wall Initiative Strategy and Ten-Year Implementation Framework (2024–2034), whose aim is to strengthen ecosystem restoration, landscape management, soil health, and livelihood enhancement across the continent. Over the years, many African governments have implemented input subsidy programs, mainly focusing on promoting use of inorganic fertilizers to boost agricultural productivity. However, due to several challenges, including issues around distributional efficiencies and the economic costs associated with importation of fertilizer, input subsidies have had mixed results across the continent. Additionally, fertilizer subsidy programs have often failed to achieve their food security objectives, partly because soil health is in decline due to lack of complementary soil ameliorative measures. This chapter submits that agroecological principles are key in improving soil health and nutrient-use efficiency and are critical to sustaining crop productivity and safeguarding ecological services in the long term. The chapter calls for a rethink of existing input support strategies and a broader focus on complementary approaches to enhancing soil

---

A. Kalibata · P. Thangata (✉) · V. Mutyasira · S. Nhlengethwa
AGRA, West End Towers, Nairobi, Kenya
e-mail: pthangata@agra.org

P. W. Chirwa
Department of Plant and Soil Sciences, Faculty of Natural and Agricultural Sciences, University of Pretoria, Pretoria, South Africa

P. W. Chirwa et al. (eds.), *Trees in a Sub-Saharan Multi-functional Landscape*,
https://doi.org/10.1007/978-3-031-69812-5_6

health. We argue that for Sub-Saharan Africa, with an average fertilizer application rate of 22 kilograms per hectare against a global average of 146 kilograms per hectare, fertilizers remain critical but should be complemented with agroecological practices. To achieve this, adequate policy support is needed. Reforming policies and subsidies related to chemical fertilizers can significantly support the adoption of organic fertilizers, which are more sustainable for soil health, water quality, and overall environmental impact. The chapter discusses the need to repurpose policies to enhance soil health and incentivize the uptake of sustainable soil management practices for improved farmer incomes. The chapter discusses some of the policy options for supporting transition to more sustainable production practices.

Keywords Fertiliser use efficiency · Integrated soil health · Smart fertilizer subsidy · Organic fertilizers · Integrated soil fertility management (ISFM)

6.1 Introduction

The fragility of African food systems is evident in the face of global shocks and climate variability, hence urgent attention is needed to support technological, and management measures and practices that foster climate-smart, biodiversity, and ecosystem resilience across the entire food system. Recent studies have shown soil degradation and lack of investments in rehabilitation practices as significant factors in the declining yield response to fertilizers across Africa's smallholder farmers (Burke et al. 2022). The need for regenerative agriculture has, therefore, become more urgent than ever. Among the various approaches, agroforestry stands out as a promising and scalable solution (Sarveswaran et al. 2023). Africa's smallholder farmers, whose livelihoods depend on rain-fed agriculture and who are especially vulnerable to climate change, can benefit from agroforestry through improved food security and enhanced resilience. Besides promoting soil health, biodiversity, and climate resilience, agroforestry provides a range of social and economic benefits to smallholder farmers (Bayala et al. 2014), including enhancing productivity and diversifying income sources (Ekpo and Asuquo 2012). Studies have shown that fertilizer trees (leguminous) can improve soil fertility by incorporating biomass and fixing nitrogen, which could potentially increase crop yields by 20–200%, depending on the species, density, and management of the trees, and the type and quality of the soil (Sileshi et al. 2014). This represents an enormous and sustainable solution to the challenge of nutrient depletion, amidst the high cost of inorganic fertilizer (Ajayi et al. 2008; Franzel 1999).

However, despite the known benefits of agroforestry and its role in regenerative agriculture,[1] the adoption rates among smallholder farmers, particularly in Sub-

[1] Regenerative Agriculture aims to increase levels of productivity while also increasing the soil health with built in economic and biological stability thereby maintaining the environmental integrity of the land unit (Giller et al. 2021).

Saharan Africa, remain disappointingly low (Ajayi et al. 2008). This poses an obstacle to realizing the numerous benefits of agroforestry and regenerative agricultural systems. The widespread adoption of agroforestry has mainly been hampered by unsupportive policies. The existing policy and institutional architecture do not provide the necessary incentives to get farmers, where land is not a constraint, to bring more hectares under agroforestry and regenerative agriculture. For instance, given that agroforestry systems involve permanent and long-term investments, it is crucial to have appropriate property rights and land tenure arrangements that provide not only the security needed to encourage investments but also facilitate de-risking of the transitioning process practices (Mutyasira 2023). Appropriate and effective policy measures are needed to promote widespread agroforestry adoption and enhance its contribution to regenerative agriculture.

Review of several research works presented in several chapters of this book show that the solutions to soil health decline are well known, and include various combinations of fallowing, crop diversification, intercropping and crop rotations (especially legume rotations), soil organic matter additions, liming and applications of the appropriate inorganic fertilizers. However, many of these come with significant socio-economic and technical challenges to farmers.

While chemical fertilizers can quickly provide plants with the nutrients they require to support high crop yields, certain chemical fertilizers can bind to the soil and cause acidification (Zhang et al. 2019) or are lost via leaching, runoff, and erosion, which reach water bodies and contribute to eutrophication. Additionally, nutrient release by some chemical fertilizers can be delayed and thereby, uptake by the crop is dependent on the availability of other micronutrients that may not be available in the inorganic fertilizer. Phosphorus (P)-induced modification of microbial communities alters both microbial activities and carbon (C) cycling rates, which leads to a reduction in the Soil Organic Matter (SOM) content and soil fertility. This inhibits plant growth, reduces fruit yields, and reduces their quality. Thus, it is necessary to develop prudent and sustainable management practices that can mitigate any deleterious environmental costs.

Many governments in Africa have introduced subsidies on chemical fertilizers as a way of enhancing agricultural productivity. The idea behind the subsidies is to make them available to smallholders at below-market costs as a way of incentivizing adoption, and hence increasing agricultural productivity and profitability, increasing food availability and access, and ultimately reducing poverty and stimulating economic growth (Hemming et al. 2018). However, as has been highlighted in several chapters of this book, even subsidized chemical fertilizers are financially unaffordable to many smallholders. In many cases, the subsidies are a drain on the countries' forex. On top of the subsidized chemical fertilizers, most smallholders require food support year in and year out. For instance, IFPRI noted that in Malawi, the high expenditures on fertilizer subsidies have not been associated with growth in national harvests or improvements in maize self-sufficiency (Benson et al. 2024).

Nonetheless, the integrated use of organic and inorganic fertilizers can reduce the consumption of costly chemical fertilizers (inorganic fertilizers). Excessive application of inorganic nitrogen fertilizers alone leads to soil degradation, while the application of organic materials restores and maintains soil microecological balance. Hence, the combined use of organic and chemical fertilizers, through an Integrated Soil Fertility Management (ISFM) framework, does not only improve soil quality, and increase crop productivity, but also improves soil nutrient levels and soil enzyme activities than the application of only chemical fertilizers alone. Soil rich in organic fertilizers tend to exhibit higher levels of N, P, K, Ca, and Mg. Using organic fertilizers is an efficient way to achieve sustainable agricultural development (Nygren et al. 2012).

This chapter reviews the importance of agroforestry and other organic soil ameliorating activities in promoting the transition to regenerative agriculture. The chapter also reviews how policies have been used to alter the structure of incentives, remove barriers, and encourage the adoption of agroforestry by smallholder farmers. The chapter proposes policy options for governments to support the integrated use of organic and inorganic fertilizers as part of smart input subsidies to reduce the importation of costly chemical fertilizers (inorganic fertilizers), thereby reducing forex expenditure and redirecting it to other equally important sectors like health and education. This chapter does not intend to present comprehensive literature on the role of organic fertilizers. Other reviews in this book have extensively covered this (see Chaps. 2, 3, and 5). However, we shall try where applicable, to provide background information to inform the goals of this chapter, thus, advocating for policy options that take cognizant of the role of ISFM as a smart subsidy option that reduces government expenditure on chemical fertilizers alone.

In the chapter, we use agroforestry, regenerative agriculture, multipurpose trees, and fertilizer trees to mean all organic enhancement soil fertility improvement paradigms in multifunctional landscapes (see Chaps. 2, 3, and 5). Where needed, and reviewing specific studies, we use organic ameliorating technologies highlighted in those studies. We use Integrated Soil Fertility Management (ISFM) to mean the combination of the use of chemical and organic fertilizers in smallholder farmers' fields.

6.2 The Rising Challenges of Inorganic Fertilizer Use in Sub-Saharan Africa

For many countries in Sub-Saharan Africa, importing chemical fertilizers is a major expense due to high costs, logistic constraints, and environmental impacts. Countries spend not just through direct imports, but also through direct and indirect subsidies to marketers and farmers. The high costs are exacerbated by the volatility of international prices, limited foreign exchange reserves, poor infrastructure, and inefficient distribution systems. Many countries have attempted to reduce dependence on imported fertilizers by promoting local production and blending. However, local producers face challenges such as high production costs, lack of raw materials, and competition from subsidized imports (AGRA 2022). For instance, in Ethiopia where the total cost of importing fertilizers is equivalent to 4.4% of the country's agricultural GDP (Abay et al. 2023), the average cost of importing a metric ton of urea fertilizer was \$522 in 2018, 42% higher than the world average price of \$368. Kenya, with an annual fertilizer demand of about 600,000 metric tons, spends an equivalent of 2.7% of the country's agricultural GDP on inorganic fertilizer (AGRA 2022). The country relies on imports to meet about 80% of its fertilizer needs, as domestic production is limited and insufficient. However, importing fertilizers in Kenya is also costly and challenging, due to factors such as high tariffs, taxes, transport costs, and regulatory barriers. Consequently, the average cost of importing a metric ton of diammonium phosphate (DAP) fertilizer in Kenya was \$689 in 2017, 58% higher than the world average price of \$436.

Nigeria is the largest importer of chemical fertilizers in Africa, with an annual demand of about four million metric tons. The country imports about 90% of its fertilizer needs, as domestic production is inadequate and unreliable. However, importing fertilizers in Nigeria is also expensive and difficult, due to factors such as high exchange rate volatility, port congestion, smuggling, and corruption. According to a study by the African Fertilizer and Agribusiness Partnership (AFAP), the average cost of importing a metric ton of urea fertilizer into Nigeria was \$494 in 2019, which was 34% higher than the world average price of \$369. The study also estimated that the total cost of importing fertilizers in Nigeria was equivalent to 1.9% of the country's agricultural GDP (AGRA 2019).

As with other African countries, Zambia relies heavily on imported fertilizers and has experienced increased expenditure due to rising global fertilizer prices. The Zambian Kwacha has depreciated against the dollar by 18 percent from 18.5 ZMW/USD to 21.9 ZMW/USD between July 2020 and July 2021. Bulk shipping rates have more than doubled in the same period (Grain and IATP 2022). Due to its landlocked status and delays, congestion, inefficiencies, corruption, and high costs of import routes, it takes about 45 days and costs about \$200 per ton to transport fertilizer from Durban to Lusaka, and up to 15 days and \$80 per ton from Dar es Salaam. The import value of fertilizers into Zambia amounted to an approximate value of \$410 million in 2020, an increase from \$336 million in 2019.

In Malawi, fertilizer prices have skyrocketed due to a combination of factors including global trends, exchange rate depreciation, and supply chain inefficiencies. Global USD-denominated fertilizer prices increased on average by 82 percent in the 12 months ending in July 2021 due to a combination of strong demand and high input costs. In the same period, the Kwacha depreciated against the dollar by 9 percent from 756 MWK/USD to 821 MWK/USD. In Kwacha terms, global fertilizer prices, therefore, increased by 98 percent, which is even sharper than retail prices in Malawi. Duchoslav and Rusike (2021) reported that bulk shipping rates of fertilizers more than doubled in the 12 months ending July 2021, further increasing the landed cost of fertilizer in Malawi.

To promote the use of fertilizers as an avenue to boost productivity, many governments in Africa implement some form of agricultural input subsidies to reduce the cost of inputs. These subsidies exert a heavy fiscal burden on country budgets and distort markets and therefore their results are mixed (Nhlengethwa et al. 2022).

Some of the most well-known examples are Malawi's Farm Input Subsidy Programme (FISP), which started in 2005 and expanded to the Affordable Inputs Program (AIP) in the 2020/21 agricultural season; Zambia's Farm Input Subsidy Programme, Ethiopia's Agricultural Growth Programme (AGP), which has provided subsidised inputs and extension services to smallholders since 2010; and Nigeria's Growth Enhancement Support Scheme (GESS 2012).

In Malawi, the AIP accounts for close to 80 percent of the agriculture budget. The Malawian government spends over 160 billion MK on the program that allows Malawian subsistence farmers to purchase farm inputs at a subsidized cost with the government paying over 70% of the cost. In the 2021–22 agricultural season, for example, the government spent about MWK 40,000 per participating household to subsidize 2 bags of fertilizer. Similar to Malawi, the Zambian Government allocates about $200 million annually for FISP, which accounts for about 30 percent of the total agricultural budget.

6.3 Integrated Soil Fertility Management (ISFM): Transition to Regenerative Agriculture

ISFM is a means to increase crop productivity in a profitable and environmentally friendly way (Vanlauwe et al. 2015). It is one means to regenerative agriculture and reverses natural resource degradation in SSA. A review by Vanlauwe et al. 2015 has shown that the adoption of ISFM through the combined use of organic manure and mineral fertilizers benefits smallholder farming systems by increasing crop productivity, enhancing other ecosystem services and resilience by diversifying farming systems and increasing the availability of organic resources within farms, mainly as crop residues and/or farmyard manure. The integration of chemical fertilizers and organic resources is a universally applicable principle because removing crops requires nutrients to be replenished and applied organic inputs mineralize their carbon over time (Vanlauwe et al. 2015). Delving deeper into regenerative agriculture entails looking at the notable influence of agroforestry and other organic fertilizers to ascertain their significance.

6.3.1 The Potential Impacts of Agroforestry and Other Organic Fertilizer Options

Agroforestry involves the integration of woody perennials on the same land management unit as crops and/or animals, in some form of spatial arrangement or temporal sequence to harness both ecological and economic benefits from the interactions between the different components (ICRAF 2011). Agroforestry farms and landscapes are a major part of Africa's rural landscapes (Mbow et al. 2014), and the social-ecological interactions are systemic. It has widely been promoted as a land-use strategy to address global climate change and provide other environmental, economic, and social benefits (Jose et al. 2012; Waldron et al. 2017). The enormous potential for carbon sequestration, soil erosion and runoff control, and improved nutrient and water cycling, as well as the socio-economic benefits and greater agricultural productivity, is also widely acknowledged in literature (Udawatta et al. 2012, Wolz et al. 2018, Dollinger and Jose 2018, Holzmueller et al. 2012).

The utilization of multipurpose trees might improve the profitability of agroforestry as they can serve various functions such as alternative incomes, and sources of fodder or food (i.e., wild edible fruits) during deficit periods among rural communities (Gebru et al. 2019). For instance, the adoption of coffee agroforestry in Wey-Besay Watershed, Lampung, Indonesia contributed to more than 50% of household income compared to only 12% from the conventional agriculture system (non-agroforestry system) (Suyanto et al. 2007). In addition, agroforestry reduces the risk of losing crops due to high solar radiation by lowering the risks associated with prolonged dry spells and droughts (Musa et al. 2018).

Across the continent, several studies have shown the positive impacts of fertilizer trees on crop productivity. For instance, in Malawi, Zambia and Nigeria, studies have shown that maize yields increased by 0.7–2.5 Mg/ha and were more stable in fields intercropped with *Leucaena leucocephala* (*leucaena*) and *Gliricidia sepium* (*gliricidia*) than in fully fertilized sole maize (Isaac et al. 2007). There were also prominent yield increases for millet and sorghum when planted alongside fertilizer trees. These crops are staples in many countries hence their increase in yield through intercropping with fertilizer trees is critical in enhancing food security. Vegetable productivity has also been notably increased through the use of biomass transfer of fertilizer tree prunings (Nygren et al. 2012). Cash crop productivity has also been seen to increase when intercropped with fertilizer trees. This improves the financial earnings of farmers hence improving their ability to purchase more food. In Malawi, both male and female smallholders who plant improved fallows, and are able to sell seed from the woody species, their households stopped taking credit for chemical fertilizers. They produced most of their maize without chemical fertilizers (Thangata et al. 2002). Similar results were found in Zambia and Zimbabwe where seed selling of agroforestry tree species as an incentive enhanced adoption through augmenting household income and benefited farmers by increasing funds available for discretionary use. In Zimbabwe there was a greater increase in discretionary cash for draft animal owners than non-owners. It was concluded that in Southern Africa, improved fallows are a viable alternative to chemical fertilizer use for small farmers (Thangata et al. 2007).

Remarkably, these outcomes bolster resilience to shocks, as well as boost overall household income and food security (Dollinger and Jose 2018). Figure 6.1 illustrates the agroforestry impact pathways.

Overall, fertilizer trees are a promising option for sustainable intensification of agriculture in developing countries. However, there are still some challenges and constraints that limit their adoption and impact. These include lack of awareness and knowledge among farmers and extension agents, inadequate availability and

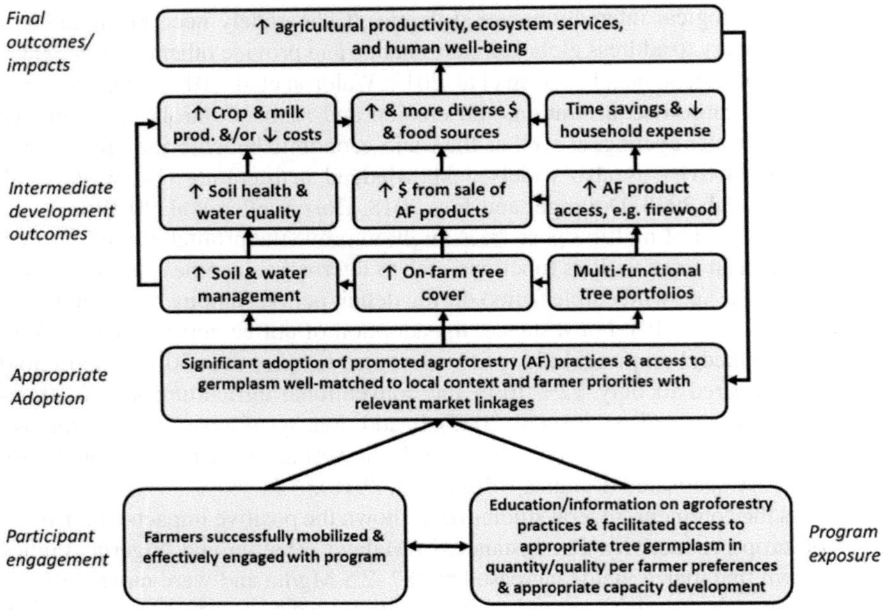

Fig. 6.1 Illustrative theory of change for an agroforestry (AF) intervention. (Source: Brown et al. 2018)

quality of tree seeds and seedlings, competition and allelopathy between trees and crops, pest and disease problems, land tenure and policy issues, and market access and value addition for tree products. Hence, more research and development efforts are needed to address these challenges and to promote the scaling up of fertilizer tree technologies (Isaac et al. 2007).

6.4 Rethinking Soil Health Solutions for Smallholder Farmers in SSA

In sub-Saharan Africa (SSA), achieving sustainable agricultural intensification and meeting the growing food demands of a burgeoning population in the face of growing threats to food systems requires a rethink of soil health solutions for smallholder farming systems. SSA is amongst the regions with the lowest cereal self-sufficiency ratio, with projections that the sub-continent's population will grow by 40% over the next 30 years, from the current 648 million people to 909 million in 2050 (United Nations 2022). The scenario requires a significant jump in production and productivity. However, the million-dollar question is how this could be achieved without excessive use of inputs and expansion of agricultural areas, which are often associated with deleterious environmental impacts and biodiversity loss.

Generally, fertilizers are important in increasing the yield of maize. In a review by Széles et al. (2019), nitrogen fertilizer is the key element that influences the biomass and grain yield due to the development and durability of the leaf area. The review noted that nitrogen uptake is the lowest at the time of maize germination and it becomes intensive from the 6–7-leaf stage reaching its peak during silking. Nitrogen uptake and incorporation are significant also during the grain-filling process where 60% of the total nitrogen uptake is incorporated into the grain. The review, however, notes that the uptake and accessibility of nitrogen are greatly influenced by climatic conditions. Traditionally, inorganic fertilizer use has been a significant factor in production intensification and the agricultural boom of the 1960s and 1970s. However, several challenges within Africa's food systems warrant a rethink of fertilizer use and options for soil health management. Firstly, the cost of inorganic fertilizer is usually prohibitively high for smallholder farmers. Recently, disruptions in global supply chains, rising energy prices and export restrictions, have seen a doubling in the weighted average price of fertilizer between June 2021 and April 2022, further putting a strain on farmers across the continent. The average application rate of fertilizer in sub-Saharan Africa averages 13 kg N h^{-1} $year^{-1}$ (FAOSTAT 2023), which is at least 10 times lower than corresponding figures in South Asia, EU, South America, and East Asia. Secondly, widespread soil degradation and limited investments in rehabilitation practices are contributing to the declining yield response to inorganic fertilizers in smallholder farming systems.

There is a need for complementary approaches to addressing Africa's soil health challenges. A study from Malawi examining yield response to nitrogen on farmer-managed fields over time, between 2014 and 2018 found a decline in productivity over some period of increasing fertilizer use (Burke et al. 2022). Severely degraded soils tend to be poorly responsive to mineral fertilizer additions. The implication is that increased fertilizer use alone is not sufficient to achieve sustainable agricultural intensification and complementary investments in land management practices. While ecological approaches have been suggested as alternatives to inorganic fertilizers, it is important to stress that in most of SSA's agroecosystems, inorganic fertilizers remain essential as soils are highly degraded, low in organic matter and inherently nutrient-poor (Campbell et al. 2023). Completely substituting for inorganic fertilizers could hamper productivity gains and contribute indirectly to agricultural expansion and deforestation. For instance, a recent review of agroecological principles observed that although legumes can contribute substantially to nitrogen input to cropping systems, it is unlikely that they alone can sustain agricultural productivity at the levels required for food security (Falconnier et al. 2023).

A systematic review of studies identifies several crucial points that could contribute to the call for complementary approaches to soil health management for smallholder farmers:

- Legume trees and shrubs could contribute as much as 100 kg N h^{-1} $year^{-1}$ through N fixation. However, only mature trees aged 20–40 years fix substantial amounts of nitrogen, which makes it a long-term investment for farmers (Falconnier et al. 2023). Despite the nitrogen fixation benefits, trees often compete strongly with the crops for light and water, especially during dry years.

- Green manures have a potential nitrogen fixation contribution to cropland. If green manures are rotated with cereals, they could bring a substantial yield benefit to the cereal. However, this comes at the cost of foregoing a part of the cereal yield. It is observed that the increase in cereal yield from the green manure often does not sufficiently compensate for the yield penalty (Falconnier et al. 2023).
- The labour demand is generally high. For instance, incorporating the biomass from green manure into the soil can take as much as 80 persons per day per hectare for a prolific Mucuna (velvet beans) in Malawi, which is roughly 40 times more than what is required to apply mineral nitrogen fertilizer (Ojiem et al. 2014).
- Organic fertilizers can increase the effective utilization rate of chemical fertilizers by improving the soil structure, enhancing the soil water-holding capacity, stimulating microbial activity, increasing the cation exchange capacity, and buffering the soil pH (Jose et al. 2012; Waldron et al. 2017).
- Using organic and chemical fertilizers in a balanced and complementary way is the most effective way to achieve optimal crop yield and quality, as well as the sustainable development of agriculture (Isaac et al. 2007; Nygren et al. 2012).
- The adoption of agroforestry technologies in the tropics stores high biomass for soil fertility replenishment and offers potential for carbon (C) storage. Research from Malawi has shown that agroforestry can increase C sequestration under smallholder farm management. The results indicated smallholders accumulating to annual mean C amounts of up to 393 Mg, translating to 4.17 Mg C ha^{-1}, of which agroforestry's contribution was 39% (Thangata and Hildebrand 2012).

6.5 Towards Holistic Soil Health Solutions for Africa's Smallholder Farmers: The Role of Policy

6.5.1 Support Government Efforts to Repurpose Subsidy Policies

The agricultural sector is a vital source of income, food security and employment for many developing countries. However, it also faces many challenges such as climate change, low productivity, market distortions and environmental degradation. To address these challenges, governments often resort to subsidies that aim to support farmers and promote agricultural development. Yet, not all subsidies are equally effective or efficient. Some subsidies may have negative impacts on the environment, trade, fiscal balance, and social equity. Therefore, it is important to design and implement subsidies that are based on sound evidence and aligned with the goals of sustainable development (Holden and Lunduka 2010).

Evidence-based subsidies are those that are informed by rigorous research and evaluation, that target the most relevant problems and beneficiaries, minimize market distortions and negative externalities, and are regularly monitored and adjusted to ensure their effectiveness and efficiency. Evidence-based subsidies can help save foreign exchange (forex) by reducing the dependence on imported inputs, increasing the competitiveness of domestic products, diversifying the export basket, and enhancing the resilience of the agricultural sector. Evidence-based subsidies can also contribute to sustainable development by improving the livelihoods of farmers, enhancing food security and nutrition, reducing greenhouse gas emissions, and preserving natural resources.

Some examples of evidence-based subsidies are (UNEP 2020):

- Input subsidies that are smart, targeted, and time-bound, such as vouchers or coupons that enable farmers to access improved seeds, fertilizers, pesticides, and irrigation equipment at subsidized prices.
- Output subsidies that are performance-based and quality-oriented, such as price premiums or bonuses that reward farmers for producing high-quality products that meet certain standards or certifications.
- Infrastructure subsidies that are public goods-oriented and demand-driven, such as grants or loans that support the construction or maintenance of roads, storage facilities, processing plants and marketplaces.
- Innovation subsidies that are knowledge-intensive and capacity-building, such as scholarships or fellowships that support the education and training of farmers, extension agents, researchers, and policymakers.
- Insurance subsidies that are risk-sharing and resilience-enhancing, such as premium subsidies or payouts that protect farmers from weather shocks, pests and diseases, price volatility and other uncertainties.

6.5.2 Policies for Supporting Agroecological Transitions in Africa

Policies and incentives can play a key role in supporting the use of organic fertilizers by addressing the barriers and creating enabling conditions. One possible policy option is to reform existing subsidies for chemical fertilizers, which can have negative effects on soil health, water quality, climate change, and human health. Chemical fertilizer subsidies are often costly, inefficient, and distortionary, and can undermine the competitiveness of organic fertilizers. For example, in India, fertilizer subsidies accounted for the second-largest government transfer at over USD 10 billion in 2018/19. Reforming these subsidies can free up public resources that can be used for different purposes, such as supporting more sustainable practices in the sector (UNEP 2020).

Policies and incentives in support of organic fertilizers can help to address the challenges of low food production caused by poor soil and high food demand due to harsh weather and rapid population growth. However, such policies and incentives should be carefully designed and implemented to avoid unintended consequences and ensure efficiency, equity, and sustainability (UNEP 2020).

One policy consideration by governments is to consider new green strategies, standards and guidelines for organic farming and waste management, and integration of organic fertilizers into national and international agendas on sustainable development (Yang et al. 2020). More critically, governments should look at the evidence science and policy on organic fertilizers. As has been shown in this chapter, nitrogen fertilizer uptake and incorporation are significant during the grain-filling process where 60% of the total nitrogen uptake is incorporated into the grain (Széles et al. 2019). Therefore, when ISFM options are used, basal nitrogen dressing might not be needed. This is because some of the active substances of N applied before or just after sowing may evaporate or leach into groundwater. The reason for this phenomenon is that in the early stages of plant development, poorly developed root systems are less able to access reserves in the soil (Alley et al. 2009). Inorganic fertilizers are important but must be part of a package of public support measures. Governments can reduce their agricultural budget by reducing the amounts of basal fertilizers needed as part of the input subsidies. Instead, smallholder farmers should be encouraged to use ISFM options.

This requires strengthening the collaboration among researchers, policymakers, practitioners and other stakeholders across different sectors, scales, and regions. It also requires improving the data collection, analysis, and dissemination on the performance, impacts and trade-offs of organic fertilizers. Moreover, it requires enhancing capacity building, communication and advocacy on the benefits and challenges of organic fertilizers for various audiences (Huang et al. 2016).

Another possible policy option is to provide evidence-based subsidies for organic fertilizers, which can be designed to target specific groups of farmers, crops, or regions that can benefit the most from them. Evidence-based subsidies are based on rigorous research and evaluation of the impacts and effectiveness of different types of subsidies. For example, a study in Nigeria found that the adoption of organic fertilizers positively and significantly impacts the welfare of farmers, particularly female household heads, agricultural households that had access to credit, and farm household residents in the southern region of Nigeria. Evidence-based subsidies can also consider the long-term effects of organic fertilizers on soil quality and climate resilience (Gregory et al. 2000; Debrah and Breman 2002). Governments can also consider these evidence-based organic fertilizer subsidies as part of climate adaptation plans.

6.6 Chapter Highlights

There is a compelling case for promoting agroecological practices that help address the soil fertility challenges facing Africa's food systems. Reviews of literature across the continent show the promise of agroecology in soil health management. However, given the depleted nature of Africa's soils, radical measures that seek to reduce use of mineral fertilizer may jeopardize productive gains, especially in

Sub-Saharan Africa, where average fertilizer application rates are 22 kilograms per hectare against a global average of 146 kilograms per hectare.

However, soil health solutions must be crafted within the context of the smallholder farmer. Holistic approaches are needed, with a view to forge complementarity among different options, including crop diversification, intercropping and legume rotations, soil organic matter additions, liming and applications of the appropriate inorganic fertilizers, among others. These measures should be integrated alongside sustainable land management practices, focusing on conserving soil and water by reducing erosion and surface run-off.

Achieving healthier soil fertility outcomes will depend greatly on the actions of smallholder farmers and how policies can craft an appropriate incentive structure that ensures that farmers can easily access the range of options in a cost-efficient way. While agroecological practices remain a plausible option for promoting soil fertility management, reclamation of degraded soils and hence sustaining crop productivity in the long run, more integrated approaches must be adopted that acknowledge the continued importance of inorganic fertilizers in Africa's largely degraded soils. Inorganic fertilizer will need to be complemented with other context-specific agronomic management practices to ensure that maximum benefits are derived from each component practice.

References

Abay K, Chamberlin J, Chivenge P, Hebebrand C, Spielman D (2023) Fertilizer policies amid global supply and price shocks. Africa south of the Sahara. https://ssa.foodsecurityportal.org/node/2644

AGRA (2019) Feeding Africa's soils: fertilizers to support Africa's agricultural transformation. Alliance for a Green Revolution in Africa (AGRA), Nairobi. https://hdl.handle.net/10855/43479

AGRA (2022) Price and policy bi-weekly monitor. Policy and Price Monitor, 6th edn. Alliance for a Green Revolution in Africa, 15p. https://agra.org/wp-content/uploads/2022/10/Price-and-Policy-Monitor-September-Edition-6.pdf

Ajayi O, Akinnifesi F, Mullila-Mitti J, Dewolf J, Matakala P, Kwesiga F (2008) Adoption, profitability, impacts and scaling-up of agroforestry technologies in southern African countries. In: Batish D, Kohli R, Jose S, Singh H (eds) Ecological basis of agroforestry. Boca Raton, Taylor and Francis Group

Alley MM, Martz ME, Davis PH, Hammons JL (2009) Nitrogen and phosphorus fertilization of corn. Virginia cooperative extension, publication 424-027. Virginia Polytechnic Institute and State University. https://www.pubs.ext.vt.edu/424/424-027/424-027.html

Bayala J, Sanou J, Teklehaimanot Z, Kalinganire A, Oue'draogo SJ (2014) Parklands for buffering climate risk and sustaining agricultural production in the Sahel of West Africa. Curr Opin Environ Sustain 6:28–34

Benson T, De Weerdt J, Duchoslav J, Masanjala W (2024) Fertilizer subsidies in Malawi: from past to present. International food policy research institute (IFPRI). Working paper 44, 23p. https://massp.ifpri.info/files/2024/01/Working-Paper-44-Fertilizer-subsidies-in-Malawi-from-past-to-present.pdf

Brown SE, Miller DC, Ordonez PJ, Baylis K (2018) Evidence for the impacts of agroforestry on agricultural productivity, ecosystem services, and human well-being in high-income countries: a systematic map protocol. Environ Evid 7:24. https://doi.org/10.1186/s13750-018-0136-0

Burke W, Snapp S, Brad G, Peter B, Jayne T (2022) Sustainable intensification in jeopardy: transdisciplinary evidence from Malawi. MwAPATA Institute working paper, No. 21/0, 155758

Campbell B, Thornton P, Loboguerrero A, Dinesh D, Nowak A (eds) (2023) Transforming food systems under climate change through innovation. Cambridge University Press, Cambridge. https://doi.org/10.1017/9781009227216

Debrah SK, Breman H (2002) To subsidize or not to subsidize: the subsidy issue revision. Draft IFDC

Dollinger J, Jose S (2018) Agroforestry for soil health. Agrofor Syst 92:213–219

Duchoslav J, Rusike J (2021) Why are fertilizer prices in Malawi high? And what can be done?. MaSSP policy notes 42,. Intl Food Policy Res Inst (IFPRI)

Ekpo FE, Asuquo ME (2012) Agroforestry practice as adaptation tools to climate change hazards in Itu LGA, Akwa Ibom state. Nigeria Glob J Hum Soc Sci Geogr Environ GeoSci 12(11):27–36

Falconnier GN, Cardinael R, Corbeels M, Baudron F, Chivenge P, Couëdel A, Ripoche A, Affholder F, Naudin K, Benaillon E, Rusinamhodzi L (2023) The input reduction principle of agroecology is wrong when it comes to mineral fertilizer use in sub-Saharan Africa. Outlook Agricult 52(3):311–326

FAOSTAT (2023) World food and agriculture—statistical yearbook. Rome https://doi.org/10.4060/cc8166e

Franzel S (1999) Socioeconomic factors affecting the adoption potential of improved tree fallows in Africa. Agrofor Syst 47:305–321. https://doi.org/10.1023/A:1006292119954

Gebru KM, Leung M, Rammelt C, Zoomers A, van Westen G (2019) Vegetable business and smallholders' food security: empirical findings from northern Ethiopia. Sustain For 11(3):743

Giller KE, Hijbeek R, Andersson JA, Sumberg J (2021) Regenerative Agriculture: An agronomic perspective. Outlook Agricult 50(1):13–25. https://doi.org/10.1177/0030727021998063

Gregory DI, Roy AH, Bumb BL (2000) Agriculture subsidies in developing countries with a particular focus fertilizer. Paper presented at the Fertilizer Association of India (FAI) Annual Conference, New Delhi, India, December 11–13

Hemming DJ, Chirwa EW, Ruffhead HJ, Hill R, Osborn J, Langer L, Harman L, Coffey C, Dorward A, Phillips D (2018) Agricultural input subsidies for improving productivity, farm income, consumer welfare and wider growth in low- and middle-income countries: a systematic review. 3ie systematic review 41. International Initiative for Impact Evaluation (3ie), London. https://doi.org/10.23846/SR51062

Holden S, Lunduka (2010) Too poor to be efficient? Impacts of the targeted fertilizer subsidy programme in Malawi on farm plot level input use, crop choice and land productivity. NORAGRIC report no. 55. Department of International Environment and Development Studies, Norwegian University of Life Sciences

Huang ZH, Zhong YQ, Wang XL (2016) Study on the impacts of government policy on farmers' pesticide application behaviour. China Popul Resour Environ 8:148–155

Isaac ME, Timmer VR, Quashie-Sam SJ (2007) Shade tree effects in an 8-year-old cocoa agroforestry system: biomass and nutrient diagnosis of Theobroma cacao by vector analysis. Nutr Cycl Agroecosyst 78:155–165. https://doi.org/10.1007/s10705-006-9081-3

Jose S, Gold MA, Garrett HE (2012) The future of temperate agroforestry in the United States. In: Nair PKR, Garrity D (eds) Agroforestry—the future of global land use, Advances in agroforestry, vol 9. Springer, Dordrecht, pp 217–245. https://doi.org/10.1007/978-94-007-4676-3_14

Mbow C, Van Noordwijk M, Luedeling E, Neufeldt H, Minang P, Kowero G (2014) Agroforestry solutions to address food security and climate change challenges in Africa. Environ Sustain 6:61–67

Musa F, Kamoto J, Jumbe C, Zulu L (2018) Adoption and the role of fertilizer trees and shrubs as a climate smart agriculture practice: the case of Salima District in Malawi. Environments 5(11):122. https://doi.org/10.3390/environments5110122

Mutyasira V (2023) Transforming Africa's food systems: a smallholder farmers' perspective. Global Social Challeng J 1(aop):1–13

Nhlengethwa S, Thangata P, Muthini D, Djido A, Njiwa D, Nwafor A (2022) Review of agricultural subsidy programmes in sub-Saharan Africa: the impact of the Russia—Ukraine War. AGRA HAPA Policy Brief, 14p

Nygren P, Fernández MP, Harmand J-M, Leblanc HA (2012) Symbiotic dinitrogen fixation by trees: an underestimated resource in agroforestry systems? Nutr Cycl Agroecosyst 94:123–160. https://doi.org/10.1007/s10705-012-9542-9

Ojiem JO, Franke AC, Vanlauwe B, de Ridder N, Giller KE (2014) Benefits of legume–maize rotations: assessing the impact of diversity on the productivity of smallholders in Western Kenya. Field Crop Res 68:75–85. https://doi.org/10.1016/j.fcr.2014.08.004

Sarveswaran S, Johar V, Singh V, Raghunanyhan C (2023) Agroforestry: a way forward for sustainable development. Ecol Environ Conserv 29:S300–S309

Sileshi GW, Mafongoya PL, Akinnifesi FK, Phiri E, Chirwa P, Beedy T, Makumba W, Nyamadzawo G, Njoloma J, Wuta M, Nyamugafata P, Jiri O (2014) Agroforestry: fertilizer trees. In: van Alfen N (ed) Encyclopedia of agriculture and food systems, vol 1. Elsevier, San Diego, pp 222–234

Suyanto S, Khususiyah N, Leimona B (2007) Poverty and environmental services: case study in way Besai watershed, Lampung Province. Indonesia Ecol Soc 12(2):13. http://www.ecologyandsociety.org/vol12/iss2/art13/

Széles A, Kovács K, Ferencsik S (2019) The effect of crop years, nitrogen basal, and top dressing on the yield of different maize genotypes and marginal revenue. IDŐJÁRÁS Quart J Hungarian Meteorol Ser 123(3):265–278

Thangata PH, Hildebrand PE (2012) Carbon stock and sequestration potential of agroforestry systems in smallholder agroecosystems of sub-Saharan Africa: mechanisms for 'reducing emissions from deforestation and forest degradation' (REDD+). Agric Ecosyst Environ 158:172–183

Thangata PH, Hildebrand PE, Gladwin CH (2002) Modeling agroforestry adoption and household decision making in Malawi. Afr Stud Quart 6(1 & 2):271–293. http://www.africa.ufl.edu/asq/v6/v6i1-2a11.pdf

Thangata PH, Mudhara M, Grier C, Hildebrand PE (2007) Potential for agroforestry adoption in southern Africa: a comparative study of improved fallow and green manure adoption in Malawi, Zambia and Zimbabwe. Ethnobot Res Appl 5:067–075

Vanlauwe BK, Descheemaeker KE, Giller J, Huising R, Merckx G, Nziguheba J, Wendt Z (2015) Integrated soil fertility management in sub-Saharan Africa: unravelling local adaptation. Soil 1:491–508. https://soil.copernicus.org/articles/1/491/2015/

Waldron A, Garrity D, Malhi Y, Girardin C, Miller DC, Seddon N (2017) Agroforestry can enhance food security while meeting other sustainable development goals. Trop Conserv Sci 10:1–6

Wolz KJ, Lovell ST, Branham BE, Eddy WC, Keeley K, Revord RS, Wander MM, Yang WH, DeLucia EH (2018) Frontiers in alley cropping: transformative solutions for temperate agriculture. Glob Chang Biol 24:883–894

Yang Y, Li Z, Zhang Y (2020) Incentives or restrictions: policy choices in farmers' chemical fertilizer reduction and substitution behaviours. Intl J Low-Carbon Technol 16:351–360

Zhang D, Wang C, Li X-L, Yang X-S, Zhao L-B, Xia S-J (2019) Correlation of production constraints with the yield gap of apple cropping systems in Luochuan County. China J Integr Agric 18:1714–1725

Chapter 7
Land Use and Land Cover Change in Africa: Implications on Landscape Multifunctionality

Jonathan M. Kamwi, Paxie W. Chirwa, Stephen Syampungani, Josephine M. Kamelo, Theo N. Tsuaneng, and Abel Ramoelo

Abstract Despite a favourable natural resource base, Africa continues to suffer from high rates of poverty and slow rates of economic growth. Two-thirds of the population in Africa depends on agriculture for their livelihoods. However, a major portion of Africa's agricultural productivity is static or declining due to soil degradation and climate change. Degradation of forests and woodlands, for instance, has intensified due to rapid growth of human populations that rely on fuelwood for cooking and heating. A few other spatially explicit factors that influence land use and land cover are population density, agricultural expansion, elevation, rainfall, and the distance to roads and settlements. The interactive effects of these drivers that influence land use and land cover change should be considered during decision-making to reduce degradation of land resources used to sustain livelihoods. In this study, Land Use and Land Cover (LULC) detection of Africa over 2017–2022 was analysed using the ESRI Global Land Cover dataset. The results show significant change during the period. Trees, crops, rangeland and built-up areas showed an

J. M. Kamwi (✉)
Department of Natural Resource Sciences, Namibia University of Science and Technology, Windhoek, Namibia
e-mail: jkamwi@nust.na

P. W. Chirwa
Department of Plant and Soil Sciences, Faculty of Natural and Agricultural Sciences, University of Pretoria, Pretoria, South Africa

S. Syampungani
Department of Plant and Soil Science, Faculty of Natural and Agricultural Sciences, University of Pretoria, Pretoria, South Africa

Chair of Environment and Development, Oliver R. Tambo Africa Research Chairs Initiative (ORTARChI), Copperbelt University, Kitwe, Zambia

J. M. Kamelo
Integrated Rural Development and Nature Conservation, Windhoek, Namibia

T. N. Tsuaneng · A. Ramoelo
Centre for Environmental Studies, Department of Geography, Geoinformatics and Meteorology, University of Pretoria, Pretoria, South Africa

P. W. Chirwa et al. (eds.), *Trees in a Sub-Saharan Multi-functional Landscape*,
https://doi.org/10.1007/978-3-031-69812-5_7

increasing trend of 0.62%, 0.36%, 0.27 and 0.15%, respectively, while bare ground showed a decreasing trend of 1.26%. For the vast number of countries in Africa whose economies rely on agriculture, net gain changes in LULC associated with crops and rangelands are inevitable in achieving food security. Sustainable land management practices and their upscaling in sub-Saharan Africa are essential for a variety of reasons, but the most basic is to sustain and improve livelihoods while protecting the land's resources and ecosystem functions. There are some successes in implementing sustainable land management in Africa including agroforestry, community-based natural resource management and the great Greenwall Initiative. Sustainable management of African savanna calls for developing silvicultural systems in the context of understanding the key attributes of African savanna plant species, as this is the basis for addressing the current unsustainable harvesting practices across African savanna landscapes.

Keywords Land use land cover · Agricultural expansion · Settlement · Forest or woodland loss · Fuelwood · Biodiversity loss · Climate change · Drought · Floods · Fire

7.1 Land Use and Practices in Africa

7.1.1 Land Use Practices and Population Dynamics

Many obstacles that include population growth, dwindling land availability, degrading land resources, and changes in global trade markets of land resources must be overcomed in Africa for the land's ability to support household needs, and to generate income on a national scale to maintain the growth of national economies. Traditional techniques for raising crops and cattle, land tenure structures, the roles of women, and strategies for using forest resources have important connections. When human population densities were low, the old systems and practices were well suited to people's needs for survival. Populations slowly expanded, and these practices changed as a result. However, with the acceleration of population expansion in the 1950s, these conventional methods became increasingly overburdened. Rapid population increase in conjunction with a high reliance on wood fuels has accelerated the degradation of forests and woodlands, particularly around major urban areas. Despite being increasingly scarce, efficient markets for fuelwood have not yet emerged as it is typically seen as a free good and is mostly harvested on freely accessible land. Only 10 to 20% of Sub-Saharan Africa's forest degradation is directly attributable to commercial logging' However, this triggers a second round of devastation as logging roads give access to settlers who carry on the deforestation initiated by loggers. Ultimately, the deterioration and eradication of forests and woods has a negative impact on local and regional hydrological regimes, local and regional climates, and agriculture.

In response, some governments have nationalised land, allowed customary laws to regulate the use of some land, and distributed other land to private investors and public enterprises. Some countries, like Kenya, have established individual land titles without considering the pre-existing nature of traditional tenure systems. Commonly, this has led to unintended outcomes. The well-connected have frequently been successful in stealing land from its long-time owners and occupants. Government interference has commonly decreased land security, sped up the demise of traditional land management practices in forest and range areas and encouraged the emergence of de facto open-access conditions that are harmful to private investment in resource preservation or land improvement. For instance, the nationalisation of the majority of the land in Côte d'Ivoire, along with partial land titling, de facto acknowledgement of traditional tenure patterns, and significant Sahelian immigration, have resulted in an open access (UNHCR 2016).

7.1.2 Land Use Practices and Climate Change

The nexus between global change and land-use dynamics in Africa is characterised by unprecedented population growth in most African countries. The most common tree farm-based mitigation and adaptation activities in sub-Saharan African countries include climate-smart agriculture, establishment of woodlot for energy and poles, agroforestry systems and technologies and restoration. In Zambia, for example, the main farm tree-based activities identified include tree planting, reforestation, assisted natural regeneration in the miombo woodlands, agroforestry/ climate-smart agriculture. In Mozambique, the activities include promoting multipurpose planted forests, agroforestry systems, and sustainable charcoal production. In Uganda, many players contribute to climate change forest-based mitigation and adaptation activities. The main activities include tree planting or natural regeneration in afforestation, reforestation and agroforestry. On the other hand, in Ghana, most of the tree farm-based mitigation and adaptation activities include promoting a sustainable cocoa and agroforestry landscape off-reserve that is productive, climate-smart and environmentally responsible: (i) integration of trees into the cocoa landscape, and (ii) managing naturally occurring trees in food and other tree-crop farming systems. Another relevant component is the support for community restoration of degraded forest and agricultural landscapes. In southern Africa (South Africa), farm-based management of natural resources that address mitigation and adaptation interventions include afforestation, and reforestation activities. All these land use activities addressing sustainable land management, restoration, and rehabilitation aimed at improved food security and wood energy supplies are prevalent in sub-Saharan Africa (see Maisharou et al. 2015; Chirwa et al. 2015a, b).

However, in a study conducted in Ghana, Mozambique and Zambia, the main factors influencing successful implementation of tree farm- based land use activities for climate-smart adaptation and mitigation seem to be influenced by the desire to increase productivity on farms. Thus, land rehabilitation, agroforestry practices, and

Table 7.1 Determinants/factors that have influenced the successful implementation of tree farm-based adaptation and mitigation projects/activities

Factors	Zambia	Mozambique	Ghana	Average
Soil improvement	2.8	2.0	2.0	2.3
Land rehabilitation	2.8	2.5	3.0	2.8
Biodiversity conservation	2.0	2.0	3.0	2.3
Climatic change - carbon sequestration, Clean Development Mechanisms, REDD+, voluntary carbon markets etc.	2.0	1.0	2.0	1.7
Socioeconomic benefits (balance gender roles security of land and tree tenure availability of inputs -labour, planting materials, accessibility to information and markets - demand, supply, prices)	2.0	2.5	1.0	1.8
Potential of the agroforestry system (affordable costs, livelihoods and income benefits)	2.3	3.0	2.0	2.4
Financial motivation to invest (access to capital/credit, incentives -cash or noncash)	1.8	2.0	1.0	1.6

Key: rating scale: *1 = not successful, 2 = partly successful, 3 = very successful*

soil improvement were found to be the main drivers (Table 7.1). This is expected since most farming systems in Africa primarily involve food crops, and the main constraints to food production in Africa are declining soil fertility, and expensive inputs (McKenzie and Williams 2015, see also Chap. 6). Agroforestry has been reported to improve soil fertility and has even been recommended by the IPCC for mitigating climate change (Tschora and Cherubini 2020). Other practices such as climate-smart agriculture and conservation agriculture have also been promoted for rehabilitating degraded soils. In terms of tree-based interventions, agroforestry with multiple tree species has been promoted widely in many parts of Africa. These include intercropping with multipurpose trees such as *Faidherbia albida*, *Gliricidia*, *Sesbania* (Garrity et al. 2010, see also Chap. 5). Other notable interventions are farmer managed natural tree regeneration. This has been very successful, especially in the Sahel and is the basis of the Great Green Wall in West Africa (Chirwa and Larwanou 2017; Maisharou et al. 2015).

7.1.3 Land-Use Patterns

African land use techniques vary depending on the prevalent agroecological potential. For instance, intensive farming is practised in humid and sub-humid areas due to more rainfall and fertile soils (Fig. 7.1). Land use in semi-arid regions is characterised by the conservation of wildlife and cattle grazing systems. In humid and sub-humid regions, agriculture is the predominant land use, whereas agropastoral land-use systems are becoming more common in semi-arid ecological zones due to the blending of agriculture and grazing (Foley et al. 2011).

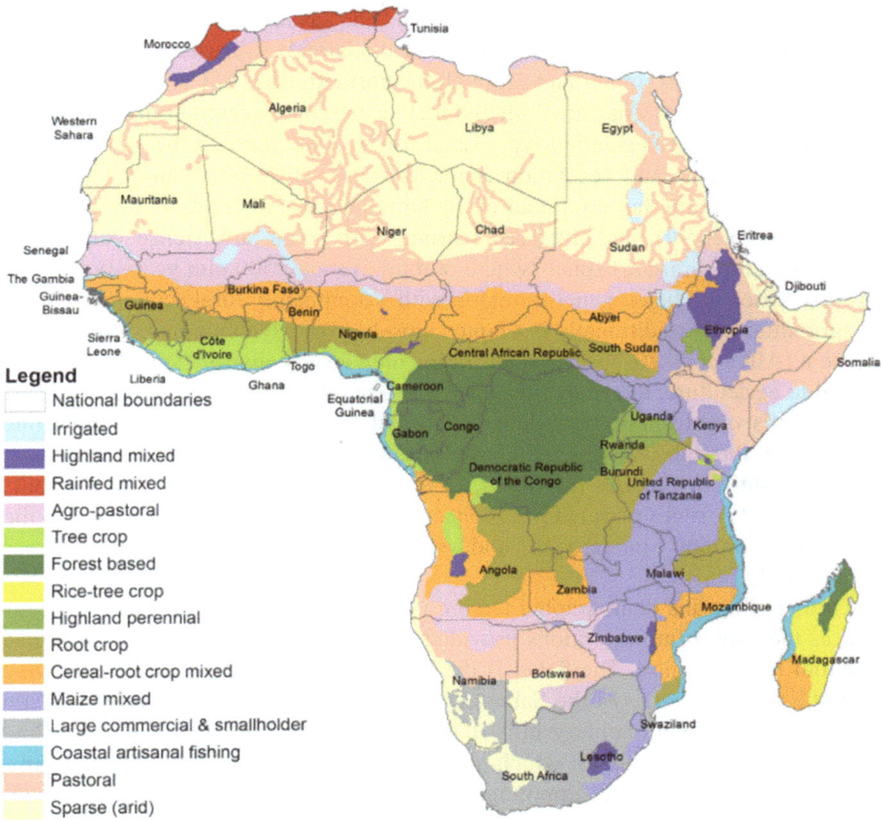

Fig. 7.1 Farming systems in Africa. (Dixon et al. 2001)

With a total land area of 6.9 million km^2, southern Africa has a land cover that is roughly 33% forest, 21% desert, and primarily savannahs and grasslands. Annual rainfall in the sub-region ranges from 50 mm in the arid deserts of Botswana, Namibia, and South Africa to more than 1000 mm in the equatorial forests of northern Zambia, Malawi, Angola, and Mozambique. Rainfall is typically seasonal in most areas, occurring over a short period and frequently taking the form of violent thunderstorms or showers. In some countries, like Malawi, where tree cover has reduced, soil erosion can reach 50 t ha^{-1} year^{-1} (Asfaw et al. 2020). High levels of rainfall variability, as well as frequent or protracted drought and flooding seasons, are experienced by most of the sub-region (Shackleton et al. 2009). Grazing lands make up 49% of the total area, mainly in savannahs and grasslands, particularly in arid countries with low forest cover. Permanent crops and arable lands cover around 6% of the earth's surface and most of them are rain-fed, except for South Africa, where the irrigation system is extremely advanced.

Central Africa is overwhelmingly timberland and savannah (Sayer and Campbell 2001). The coastal sticky belt, with tall and moderately consistent precipitation, bolsters thick tropical timberlands, though the northern parts of Cameroon, Central African Republic, and Chad are drier and include variable precipitation, with savannah as the dominant vegetation type. Land use in some parts of the sub-region is sensitive to climatic and vegetation characteristics, with forestry and commercial plantation agriculture largely found in the humid zones (where rainfall reaches up to 4000 mm year$-^{1}$), and livestock rearing, with some subsistence cultivation, occurring in the semi-arid zones (where rainfall averages 500 mm year$-^{1}$).

West Africa is a heterogeneous region with a dominantly rural economy centred on agriculture that traditionally is adapted closely to the dynamic biophysical environment (Allen et al. 2018). The status of the biophysical environment is, therefore, crucial to agriculture, the economies, and the general development process in West Africa. Soils vary from nutrient-deficient coastal sands, plateau laterites, and saline desert soils, through waterlogged coastal swampy soils, dry savannah, and desert soils, including potentially rich loess, to fertile alluvium and humic forest soils. More of West Africa's population is concentrated in the rainforest zone, and in the adjoining swampy mangroves and sandy coconut-growing coastal areas (Olukoshi 2001).

Furthermore, this demographic feature is dependent on the forest resources, as well as the export-import options provided by the sea, making the rainforest zone essential for protection. The Sahel is the largest ecosystem in the area, extending from the southern edges of the Sahara Desert northwards of the savannah zone on an elevated plain with solitary hills. It includes resistant stunted trees, including Acacia sp., and short grasses, which provide the foundation of the cattle economy, which is fundamentally all-important, and it receives 200–500 mm of very unimodal rainfall. The actual desert is located in the far north, beyond this region. In Western Africa, climate has a significant role in determining both land use and cover, with rainfall and vegetation cover varying dramatically from north to south. Savannahs are the dominant ecosystems in the southern boundary of the Sahel (Mali, Mauritania, Niger, and northern Senegal), where the average annual rainfall is 350–850 mm year$-^{1}$. Droughts are frequent and frequently severe in this region, where climate variability is greatest.

The primary agricultural activity in the Sahel is pastoral livestock husbandry due to little room for cultivation. For instance, only 4% of the land is used for agriculture in Mali and Niger. In comparison, 25% of the land in Mali is used for permanent grazing. Greater inter-annual and intra-annual rainfall predictability, as well as sporadic flooding, are all characteristics of the equatorial and coastal zone's higher annual rainfall rates, which range from 1000 to 4000 mm. The forest, which made up approximately 12% of the land area in 2000, is severely fragmented, under threat from slash-and-burn farming, commercial logging, plantations, charcoal burning, wood gathering for fuel, and other activities. Currently, rain-fed agriculture, covering around 11% of western Africa's total land area is practised, primarily in the equatorial zone. The biggest proportions of cultivable land are in Togo and Nigeria (42% and 33%, respectively), followed by Côte d'Ivoire and Ghana (23% each).

Mountains and hills, primarily found in Burundi, Rwanda, Uganda, Kenya, and Ethiopia, as well as the semiarid or arid (drier) regions, primarily found in Ethiopia, Kenya, and Somalia, are what define the eastern African sub-region. In Ethiopia, the highlands (above 1500 m asl) make up about 45% of the country's total land area, are home to 80% of the population and 75% of the country's livestock and support the majority of the sub-region's population (with densities of more than 200 people per km^2) (EPA and MEDC 1997). Higher and more reliable rainfall encourages more widespread farming in Burundi, Rwanda, and Uganda (42%, 35%, and 45%, respectively). However, the dryland ecosystems receive little rainfall and are highly susceptible to drought.

7.2 Drivers of Land Cover Change in Africa

Land use and land cover changes (LULCC) affect multiple systems such as global climate, greenhouse gas emissions and mitigation, livelihoods and biodiversity (Lambin et al. 2000, Schneider and Pontius 2001, IPCC 2000; IPBES 2018). Changes in LULC are the result of complex interactions between people, the environment, and politics that occur across space and time. Some studies have been done from a Malthusian perspective (Allen and Barnes 1985). According to the perspective, population growth probably causes changes in LULC. In this case, population growth in Africa implies increasing food demand and thus more pressure on land resources to sustain the demand (Bongaarts and O'Neill 2018). Proponents of the tragedy of the commons equation argue that the common resources of a nation are used without regard to the sustainability of these resources (Hardin 1968; Agrawal 2001). In addition, economists believe that market access and distance affect LULC changes (Von Thunen 1960). Furthermore, others identified indirect causes of LULC change such as demographic, economic, technical, political, institutional and cultural factors (Geist and Lambin 2002). Among direct and indirect causes, factors such as environmental and biophysical factors, and social triggering events have been identified as intermediates in LULC change (Fig. 7.2).

Over the past few decades, significant biophysical and socioeconomic changes have occurred in Africa. Concerns over the rate and scope of land cover change are a top priority for climate change research, especially in light of the growing understanding of the significance of land cover for the climate system and rural livelihoods. Population increase and technological advancement are typically the driving forces behind the changes, which have transcended natural change to the point where the vast majority of terrestrial biomes now bear the spatial and temporal imprint of humanity. These variables affect LULC in a negative way by causing loss of vegetation cover, soil erosion and decreased agricultural production in particular regions.

According to studies (López-Carr 2021; Saint-Macary et al. 2013), changes in forest cover may be related to the population's level of poverty, the need for arable

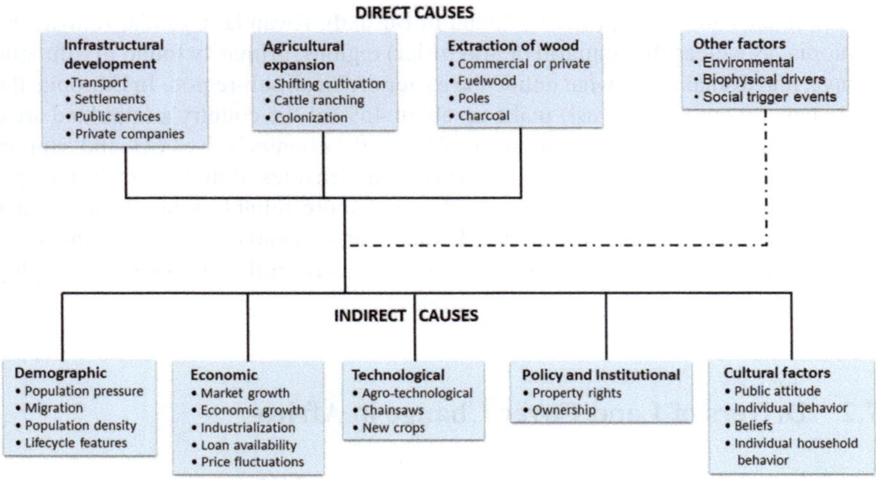

Fig. 7.2 Direct and indirect drivers of land use and land cover change. (Turner and Meyer 1991; Meyer and Turner 1992; Geist and Lambin 2002)

land, and the adoption of a market-oriented agricultural system. The cyclical nature of changes between closed deciduous woodland, grassy woodland, wooded grassland, and grassland as well as between primary and secondary tropical moist forests is indicated by field observations and recent studies in various parts of Africa (Syampungani et al. 2016; Mala et al. 2020). According to Gonzalez et al. (2012) and Portela and Rademacher (2001), changes in land use and land cover (LULC) may have an impact on ecosystem services that support human well-being, including food production, climate management, biodiversity preservation, and erosion control.

Drivers of LULC are temporally and spatially explicit. These include distance to roads, distance to settlements, elevation, rainfall and population density. The interactive effects of these drivers influence LULC change and should be considered during decision-making to reduce the fragilities and environmental problems, especially on land resources which are used to sustain livelihoods. Most of the spatial drivers are linked to distances with critical infrastructure such as settlements and roads. Studies have demonstrated that areas which are closer to settlements and roads have higher probabilities of conversion from forest land to other land use types since communities rely on forest resources closer to their settlements for sustenance (Kamwi et al. 2018).

The increased demand for land resources as a result of the growing population has resulted in both subtle changes, such as an increase in trees outside the forest and a decrease in woody vegetation in savannas, as well as more drastic changes, such as the conversion of land cover from savanna or forest to agriculture (Fig. 7.3). Studies at the local level suggest that the extension of farmland, particularly in semi-arid regions of the continent, is strongly influenced by rural population increase (Brink and Eva 2009). Figure 7.3 shows the two periods of LULC for Africa.

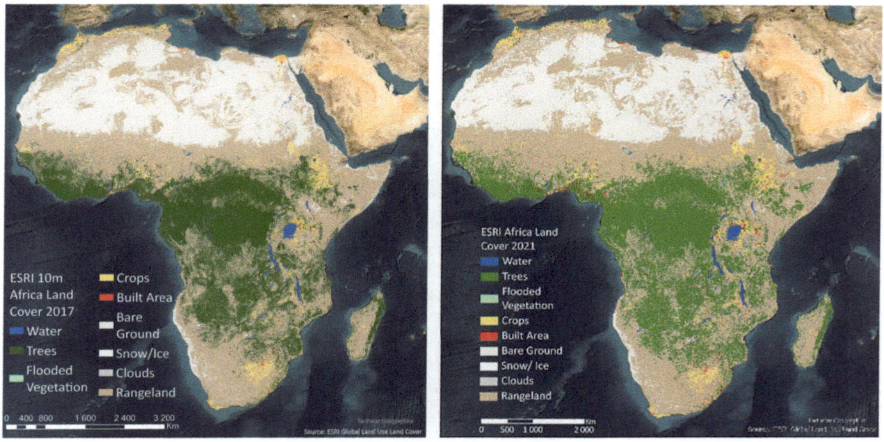

(a) (b)

Fig. 7.3 Maps of ESRI Global 250 m resolution land cover maps in (**a**) 2017 and (**b**) 2021

The multiple global LULC datasets have been developed to facilitate the prompt availability of accurate information on land cover at global, national, and local levels. These include European Space Agency's (ESA's) World Cover, Esri's Land Cover (ESRI) products, and Google's Dynamic World (Venter et al. 2022). Although these datasets possess fine detailed information with acceptable accuracy, they cannot track changes over several decades or tailored land cover typologies necessary in national and regional applications. Information on LULC helps to assess progress towards global initiatives, including, but not limited to the United Nations Framework Convention on Climate Change, the Convention on Biological Diversity, Ramsar Convention on Wetlands, United Nations Convention to Combat Desertification and the United Nations Sustainable Development Goals. Despite this background, Fig. 7.3a and b illustrate 2017 and 2021 maps of the ESRI land cover maps for Africa which were found to have high overall accuracies compared to Google's Dynamic World and ESA's world cover (Venter et al. 2022). Very little change can be visibly seen in the five-year period between Fig. 7.3a, b.

However, there has been an overall net gain in the number of trees of 0.6235% (Fig. 7.5), which is akin to a dense growth of woodland vegetation. This can be noticed in the northern region of southern and east Africa (Fig. 7.4). Trees experienced a transition of 1,449,157.82 km^2 of its area in 2021.

A large area covered by trees (1,307,235 km^2) was turned to rangelands between 2017 and 2021 (Table 7.2). Similarly, rangelands and croplands experienced a net gain of 0.2723% and 0.3566% respectively. In terms of unchanged LULC classes, it is interesting to note that rangelands (10,729,623 km^2) remain unchanged between 2017 and 2021. This was followed by bare ground 6,431,479 km^2 and trees (6,284,479 km^2). Obviously, the landscape in this part of Africa remained stable between 2017 and 2021. Unchanged regions are crucial for maintaining

Fig. 7.4 Land use land cover change map derived from land cover maps for the period between 2017 and 2021

biodiversity, supplying wood and non-timber products, and storing carbon in Africa. This is especially true of forest and shrubland. Furthermore, unchanged areas boost the connection between forest patches, sustain the amount of forest habitat, and buffer the effects of edges (Koizumi et al. 2010).

Whereas more cropping activity was prevalent in the north-east regions of east Africa, bare ground experienced a net loss of −1.5634% over the period, localized across the entire north of west and east Africa (Fig. 7.5).

Several studies conducted in Africa indicate that changes in land use have resulted in a reduction in the extent of woody vegetation (Assede et al. 2023). From 2000 to 2020, urban areas in all sub-Saharan African regions underwent significant growth (Yin et al. 2021) and conversion of grasslands, shrublands, and wetlands into agricultural land (Reid et al. 2000; Creutzig et al. 2019) was identified. In local studies, Namibia had two distinct stages of LULC change. The first phase, from

Table 7.2 Land use and land cover change matrix between 2017 and 2021 (km^2)

LULC class	Water	Trees	Flooded vegetation	Crops	Built Area	Bare Ground	Snow/ Ice	Clouds	Rangeland	Total 2017
Water	235,849	21,055	6091	5090	2403	4854	1	63	23,986	**299,393**
Trees	23,040	6,284,479	16,622	63,642	26,459	3546	16	8597	1,307,236	**7,733,637**
Flooded vegetation	5678	7116	26,692	1622	442	905	1	89	14,735	**57,281**
Crops	5706	45,060	2509	617,070	41,566	7923	2	180	237,879	**957,893**
Built area	2051	17,912	438	31,002	93,435	1997	1	212	36,194	**183,241**
Bare ground	9917	4070	1227	23,633	4053	6,431,480	44	132	939,449	**7,414,005**
Snow/ ice	0	13	1	3	2	33	3	2	59	**116**
Clouds	52	10,125	30	339	510	80	1	1369	2562	**15,067**
Rangeland	35,977	1,530,057	36,347	322,015	58,890	496,176	37	1256	10,729,623	**13,210,380**
Total 2021	**318,270**	**7,919,887**	**89,957**	**1,064,417**	**227,761**	**6,946,992**	**105**	**11,898**	**13,291,724**	**29,871,011**

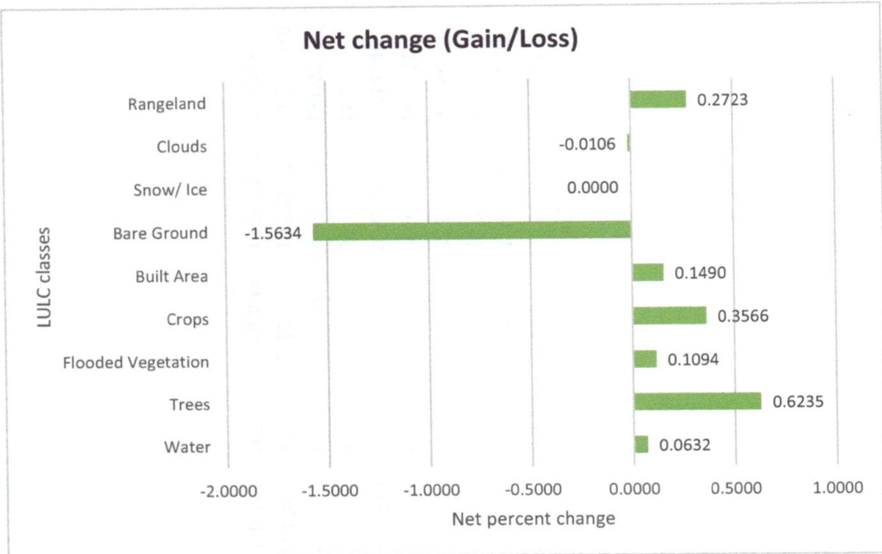

Fig. 7.5 Net change for each land use and land cover class

1984 to 1991, was characterised by significant deforestation and a progressive rise in shrubland, followed by a second phase, from 1991 to 2000 and from 2000 to 2010, which saw a decrease in deforestation and a slower development of agro-pastoral activities (Kamwi et al. 2018). Meanwhile in South Africa, it was reported that woodlands increased by 46.94% from 1990 to 2018. Conversely thickets and dense forests declined by 38.79% (Musetsho et al. 2021). These local patterns of LULC are replicated in all regions of Africa. In southern Africa, the focus is on growing eucalyptus, pine, and acacia plantations, while in east and west Africa, the emphasis is on teak plantations (Assede et al. 2023). A consistent trend of transforming natural vegetation into croplands is observable resulting in a decrease in woody vegetation cover and the subsequent loss of biodiversity. A study conducted in the south-west region of Côte d'Ivoire uncovered extensive deforestation occurring in the cocoa belt and forest reserves, driven by the expansion of croplands (Kouassi et al. 2021). The result was an annual reduction in the extent of dense forest cover by 1.44%, accompanied by an increase in croplands by 3.39%, rubber plantations by 6.98%, and settlements by 7%. Burkina Faso experienced significant growth in its agricultural area, with a remarkable 91% rise, reaching a total of 116,900 km^2 during the period from 2001 to 2014 (Knauer et al. 2017). Similarly, a review of land cover in Ghana revealed a substantial conversion of dense to open forest and non-forest land cover types between 1986 and 2014 (Baffour-Ata et al. 2021). According to Assede et al. (2023), changes in east Africa are due to variations in land cover and water balances. In Ethiopia, agricultural and settlement land expanded by 7054.6 hectares, whilst the grass and bush lands have declined by 4846.5 and 3376 hectares, respectively. Wetlands decreased from 580.2 to 68.3

hectares during the 38-year period (Sewnet and Abebe 2017). Substantial LULC changes took place from 1985 to 2015, and it is anticipated that these changes will accelerate by 2045 (Gashaw et al. 2017).

7.3 Implications of Land Use and Land Cover Change on Biodiversity and Food Production Systems

Land use, also known as land occupation, is the reduction in the amount of land that is available for uses other than its current use. Agriculture uses land in a way that displaces ecosystems and habitats, which results in a decrease in biodiversity and a reduction in ecosystem services (De Groot et al. 2012). Changes in land cover, also known as land use change or land transformation, can impact climate system and ecosystem services. The number of natural ecosystems that are altered over time is a component of this influence. Land use change reduces habitats and ecosystems, which results in a decline in biodiversity and a reduction in ecosystem services.

Biodiversity is defined as the diversity of life on earth, including diversity within species, between species, and of ecosystems (UN 1992). As a result, it supports the existence, functioning, and evolution of the biosphere and, consequently, the well-being of humans. Land use has an impact on ecosystem health, biodiversity, and the services ecosystems offer (Koellner et al. 2013). As a result, land use is a significant factor in the loss of ecosystem services and biodiversity worldwide (De Baan et al. 2013). In addition, international agreements such as the International Convention on Biodiversity (Aichi Biodiversity Targets) and the Sustainable Development Goals (UN 2015) acknowledge the significance of biodiversity and ecosystems. Within the scope of the United Nations (UN), land cover is important because it influences food security, biodiversity conservation, climate change, and sustainable development. Because land cover affects both natural and human systems by influencing ecosystems, it is recognized by the UN as one of the 14 essential data themes. Achieving the Sustainable Development Goals (SDGs) depends on conserving terrestrial ecosystems and practising sustainable land management.

The loss of biodiversity in Africa is being identified as a result of land use changes (Osei-Wusu et al. 2014). The ecosystem's long-term survival and functioning, as well as its economic and environmental worth for human life, are all threatened by the loss of biodiversity brought on by changes in land use (Mori et al. 2017).

7.3.1 Bibliometric Review of Remote Sensing and Land Use Land Cover

Remote sensing offers an in-depth assessment of the overall land use and land cover (LULC) conditions across a vast geographical region at different spatial, spectral, and temporal resolutions. Table 7.3 lists the number of retrieved articles, books,

Table 7.3 A bibliometric search result and associated number of documents retrieved between 1980 and 2023

Search criteria (TITLE-ABS-KEY)	Scopus	Web of science core collection
"Land use land cover change" OR "Forest" OR "biodiversity loss" OR "climate change" OR "drought" OR "floods" OR "fire" AND "remote sensing"	668	378
"Land use land cover" OR "land use land cover change" OR "agricultural expansion" OR "built-up OR settlement expansion" OR "Forest OR woodland loss" OR "biodiversity loss" OR "climate change" OR "drought" OR "floods" OR "fire" AND "remote sensing"	498	205

book chapters, reviews, conference, and preceding papers from a bibliometric search using relevant keywords from two frequented scientific research databases, i.e., Scopus and Web of Science. The results indicate that an average of 523 papers involving the use of remote sensing for land use and land cover change were published. The studies generally involve applying remote sensing to monitor several facets of the environment, including but not limited to assessing vegetation cover changes, land surface temperatures in urban settings and variables related to climate change.

An average of 352 papers were retrieved when the keywords were further refined to include agricultural expansion, built-up or settlement expansion was added to the search. This shows the need to further explore human conversion of land cover. During the period 1980–2023, the publication count demonstrated gradual growth, culminating in a total of 1043 entries and an annual growth rate of 14.11%. The bibliometric analysis revealed that, for Africa, South Africa and Ethiopia, had the highest number of publications for research in remote sensing and LULC (Fig. 7.6), with a combined 690 outputs.

After the literature search, it was found that the cumulative number of published research papers integrating remote sensing data in land cover change of Africa grew exponentially from 1 in 1985 to 151 in 2023. Figure 7.6 Illustrates the total number of scientific publications during the study period concerning the use of remote sensing and its application to land use and land cover change dynamics. The surge in scientific outputs can be credited to the rapid expansion of earth observation satellites i.e., Landsat 8 in 2013 and Sentinel satellite in 2014, respectively.

American Society for Photography and Remote Sensing (ASPRS) / American Congress on Surveying and Mapping (ACSM) /RT 92: Technical paper is the most relevant publication source of information related to remote sensing and LULC. The three most common sources, as indicated in Table 7.4. are international publications originating from the United States of America and the United Kingdom. Figure 7.7 depicts the annual publications and total citations (1985–2023) in Africa published between 1985 and 2023. Figure 7.8 indicates most used keywords associated with this topic and remote sensing came out on top of the list. This serves as evidence for the need for significant use of remote sensing in facilitating land use and land cover (LULC) change, climate change, and environmental monitoring.

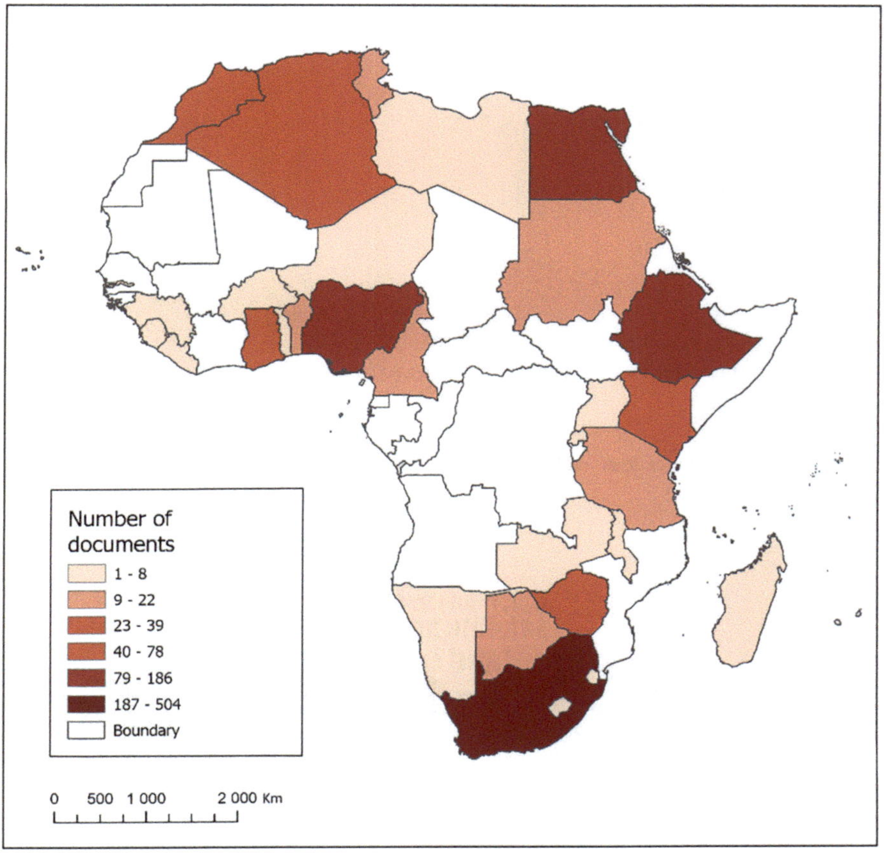

Fig. 7.6 Number of studies per country in Africa

Table 7.4 The top ten most relevant sources

Publication source	No. of documents	Percentage (%)
ASPRS/ACSM/RT 92: Technical papers, Vol 5: Resource technology 92	63	6
International journal of remote sensing	48	5
Geocarto international	37	4
Egyptian journal of remote sensing and space science	31	3
Environmental monitoring and assessment	30	3
Remote sensing applications: Society and environment	30	3
Remote sensing applications-society and environment	25	2
Journal of applied remote sensing	23	2
Physics and chemistry of the earth	19	2
South African journal of geomatics	18	2

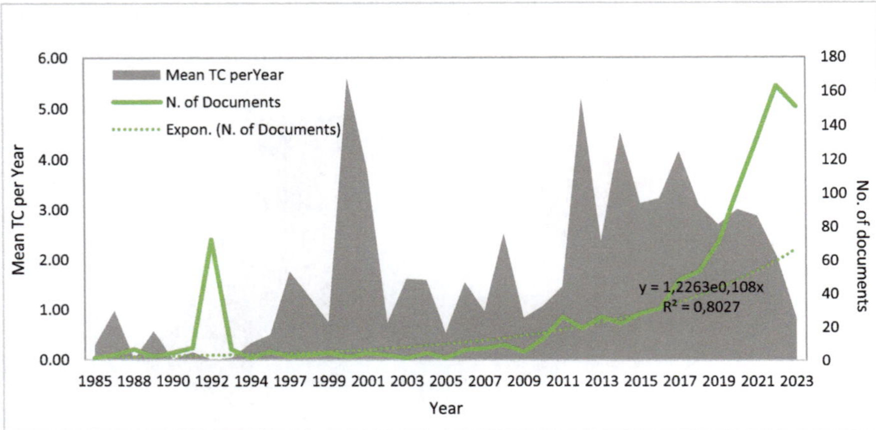

Fig. 7.7 Annual publications and total citations (1985–2023) in Africa published annually between 1985 and 2023

Fig. 7.8 World cloud illustrating the most used words in papers during the period from 1985 to 2023

7.4 Landscape Management Strategies in Africa (Drivers of Change and Response Strategies)

Increased pressure is being placed on land resources to concurrently deliver many benefits, and at high levels in order to meet the growing demand for ecosystem products and services in Africa. Significant changes in land use have been brought about by these shifting needs, and these changes have the potential to exacerbate stakeholder disputes by upsetting the balance of ecosystem services. Conflicts frequently arise because land-use modifications typically only support a certain number of ecosystem services, particularly those involved in provisioning (Tanentzap et al. 2015). In Africa, landscape management is receiving more attention as a framework for addressing intricate and connected agricultural and environmental problems. It provides cutting-edge methods to boost agricultural output, enhance rural livelihoods, and coordinate cross-sector planning. Landscape initiatives, however, encounter significant funding difficulties, much like many other endeavours.

Agroforestry systems are found in a wide range of configurations in Africa. Many of these systems consist of woody perennials in association with livestock or agricultural crops. In Africa, trees both inside and outside forests are important in ensuring food security for households in the face of climatic change and variability. Agroforestry also offers social and environmental benefits. In various regions of Africa, distinct ecological and socioeconomic circumstances have given rise to particular choices in agroforestry techniques. The lack of suitable methods for defining the area influenced by trees in a mixed stand of trees and crops is a major challenge in estimating the area covered by agroforestry in Africa and beyond. Agroforestry can be used to describe all simultaneous multi-strata systems, including home gardens, shaded perennial systems, and intense tree-intercropping scenarios. However, the majority of agroforestry systems, such as the parkland system and vast silvopastures, are relatively extensive, with the components, particularly trees, not planted at regular spacing or density. Although trees are planted at great distances between rows to create windbreaks and around agricultural or pastoral areas to create boundary planting, these techniques present a challenge.

Currently, agroforestry is seen as a system of trees incorporated into agricultural landscapes rather than a collection of methods. Agroforestry is just one of many ways that farmers can transform shifting cultivation into a more beneficial and productive cropping system that can be practised in complete harmony with the natural surroundings and the needs of their human inhabitants. Farmers are faced with the need to produce more food as well as an increasing scarcity of land and degrading soil. Any modifications should consider any physical or socioeconomic restrictions that may be present. More communication with farmers is required so that both foresters and agriculturalists can benefit from their vast local experience and any alternatives presented in specific locations can be perfectly matched to the demands of those areas. Agroforestry systems need to be recognized as valuable solutions for restoring damaged landscapes in restoration plans and strategies. Agroforestry systems are important for the implementation of international conventions such as the United Nations Convention to Combat Desertification (UNCCD), the United Nations Framework Convention on Climate Change (UNFCCC), and the United Nations Convention on Biological Diversity (UNCBD) because they provide a wide range of ecosystem services, including biodiversity conservation and carbon storage.

The term "Community-Based Natural Resource Management" (CBNRM) refers to the management of natural resources such as forests, land, wildlife, and water by collective, local institutions for local benefit. CBNRM is not a new phenomenon, it has been implemented with various degrees of success in most African countries. For millennia, local people have managed the land on which they live as well as the natural resources that surround them. Today, local groups of pastoralists, farmers, and hunter-gatherers throughout Africa maintain many traditional systems of collective natural resource management that help sustain the livelihoods and cultures of millions of people (Roe et al. 2009). The following are the key principles that emerge as crucial for success:

- In order to take management decisions, communities require rights over their land and resources, so that they can control access to resources and how they are used. They also need security of tenure (i.e. the knowledge that these rights will not be arbitrarily removed by the government and are secure over time).
- In order to take management decisions and manage the distribution of benefits, communities need strong, representative, and accountable institutions. These institutions require members who have solid financial and organisational management skills.
- Communities must be able to derive appropriate benefits from the resources that they manage. It is unlikely they will invest time, effort, and finances into managing a resource if the benefits of management do not exceed the costs.

CBNRM therefore rests on the following assumptions: If a resource is valuable and communities have the exclusive rights to use, benefit from and manage the resource, then sustainable management is likely to ensue. The benefits from management must exceed the perceived costs and must be secure over time.

Across the African region, CBNRM has led to the generation of income for local communities. Generally, communities are using their CBNRM income for social projects, community infrastructure such as classrooms and buildings for clinics and support to the aged and the needy. In some cases, communities pay cash dividends to households or individual members. Under CBNRM, jobs are created through trophy hunting and tourism and where community members develop successful enterprises.

CBNRM contributes to reducing poverty. CBNRM also helps to reduce poverty by empowering communities to manage their own affairs. Overall, CBNRM has the potential to support rural development through:

- Generating new and additional discretionary income at community level for social welfare or other purposes such as infrastructure development;
- Providing jobs and additional income for some residents;
- Increasing household assets (e.g., extra cash to buy a plough, a fridge or radio) and community assets (e.g., a vehicle for transport or community meeting hall);
- Providing land use diversification options in semi-arid and arid areas;
- Providing livelihood diversification options for some residents;
- Building skills and capacity;
- Empowering marginalised and disenfranchised rural people through devolved decision-making, fiscal devolution, improved advocacy, and institutional development;
- Promoting sustainable natural resource management; and
- Strengthening or building local institutions for common property resource management and driving local development.

The Great Greenwall, an extensive rural development program, began as an ambitious tree-planting campaign in 2007 and was initiated by the African Union. The objective has changed from building a literal wall of trees to improving millions of lives by assembling verdant, fruitful landscapes in 11 countries. Nearly 18 million hectares of degraded land covering more than 8000 km across the continent have been restored thus far. The main aim is to combat the effects of climate change and the loss of biodiversity.

7.5 Multifunctional Landscape as an Adaptive Response to the Current Trend of Land Use and Land Cover Change in Africa

Understanding socio-economic factors and land-use/land-cover (LULC) change dynamics is critical for sustainable land resource management in sub-Saharan Africa as the majority of the population depends on forest resources for their livelihood (Wood et al. 2004). Land-use and land-cover change has a significant impact on forest status, condition and management. Landscape fragmentation resulting from human-induced disturbances, such as logging, conversion of natural forests to agricultural lands (see Schwartz and Caro 2003; Assogbadjo et al. 2017) and charcoal production, may alter the ecological processes, thus resulting in changes in forest structure, floristic composition and plant assemblages depending on extent and type of disturbance. However, the distribution and extent of disturbance across spatial scales is not uniform and the impacts on forest diversity may not be avoidable (Ellis et al. 2013). Additionally, not all forests or plant species respond to such anthropogenic disturbance in the same way. In order to sustainably manage the African savanna landscape, it is important to consider management in the context of the landscape ecology and the major disturbances as this has direct implications for both species persistence and for people's livelihoods (Syampungani et al. 2013). The ecological dynamics of the savanna landscape is strongly influenced by the woody components, particularly by large trees that play a key role in ecosystem functions (Ribeiro et al. 2015). The woody component in turn is influenced by a combination of climate, disturbances (fires, drought and herbivory) and anthropogenic factors (e.g. shifting cultivation, charcoal production and timber harvesting) (Ribeiro et al. 2015).

Sustainable management of African Savanna aims at maintaining and enhancing the economic, social and environmental values from the landscape. It calls for developing silvicultural systems in the context of understanding the key attributes as this is the basis for addressing the current unsustainable harvesting practices across African landscapes. The key attributes that demonstrate their adaptation to the long dry season and other disturbance factors include the following:

- Sprouting from a persistent underground rootstock or the stem base: This allows them to persist on site through vegetative regrowth from underground rootstocks and cut stems to produce fast growing resprouts after disturbance;
- Most species (especially the tree canopy species) require light for their regeneration from either seed or sprouting. They cannot grow well under canopy of other trees; and

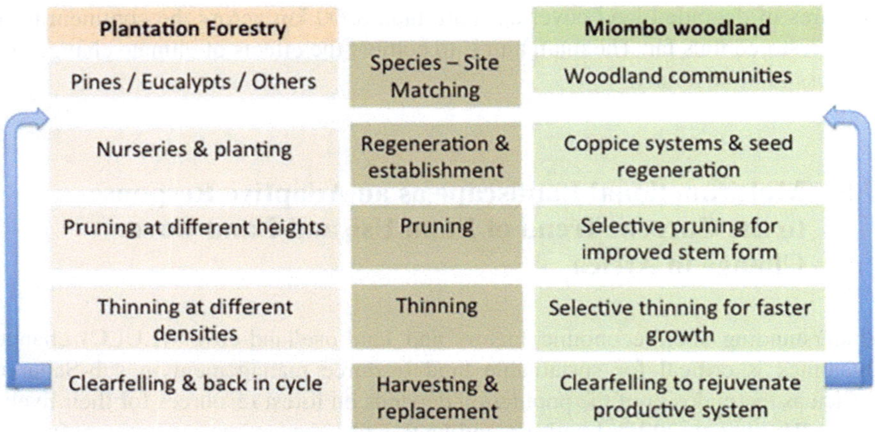

Fig. 7.9 A comparison of the silvicultural practices in plantation forestry and miombo woodland

- Most of the species are semi-deciduous or deciduous as a survival strategy to overcome the long dry season and seasonal changes.

The implications of these ecological attributes are that the African Savanna is in many ways similar to a plantation forestry system as demonstrated in their typical recovery after slash-and- burn agriculture or charcoal production. In some areas of the savanna landscape, the mosaic of tree regrowth stands of different development stages resembles a forestry plantation estate of different ages and represents stands that would produce all kinds of products (timber, poles, firewood etc) (Fig. 7.9).

The key attributes of savanna ecology and species have direct implications on developing a sustainable silvicultural system for the management of savanna ecosystems. A silvicultural system that integrates utilisation aspects that opens up the woodland can be developed (Mutanga et al. 2012). In developing silvicultural systems, it would be important to consider the mosaic of regrowth stands at different development stages, with different products important for meeting the daily livelihood needs of village households and ecosystem services. This means that sustainable management of savanna woodland and related coppicing systems need to be managed as a clear-felling system in the mature stage, with selective thinning and pruning in stand development stages towards the mature.

7.6 Chapter Highlights

- Climate change represents a major threat to Africa achieving the Sustainable Development Goals.
- The most popular tree farm-based mitigation and adaptation strategies against climate change across all countries are climate-smart agriculture, the establish-

ment of woodlots for energy and poles, various agroforestry systems and technologies, and replanting and restoration.
- The most common tree farm-based mitigation and adaptation activities in all the African countries include climate-smart agriculture, woodlot establishment for energy and poles, different agroforestry systems and technologies, reforestation and restoration.
- Land use and land cover changes are rampant in Africa driven by population growth and technological development. A few other spatially explicit factors that influence land use and land cover are population density, elevation, rainfall, and the distance to roads and settlements.
- Land Use and Land Cover (LULC) detection in Africa over 2017–2022 showed that significant change detection was observed during the period. Forests, crops, rangelands and settlement areas showed an increasing trend of 0.62%, 0.36%, 0.27 and 0.15%, respectively.
- For the vast number of countries in Africa whose economies rely on agriculture, net gain in LULC associated with crops and rangelands are inevitable in achieving food security.
- There are some successes in implementing sustainable land management in Africa including agroforestry, community based natural resource management and the great Greenwall initiative.
- Sustainable management of African savanna calls for developing silvicultural systems in the context of understanding the key attributes of African Savanna plant species as this is the basis for addressing the current unsustainable harvesting practices across African landscapes.
- The key attributes of savanna ecology and species have direct implications on developing a sustainable silvicultural system for the management of the savanna ecosystems.

References

Agrawal A (2001) Common property institutions and sustainable governance of resources. World Dev 29:1649–1672

Allen JC, Barnes DF (1985) The causes of deforestation in developing countries. Ann Assoc Am Geogr 75:163–184

Allen T, Heinrigs P, Heo I (2018) Agriculture, food and jobs in West Africa. West African Papers, N°14, OECD Publishing, Paris

Asfaw S, Pallante G, Palma A (2020) Distributional impacts of soil erosion on agricultural productivity and welfare in Malawi. Ecol Econ Vol 177, November 2020, 106764

Assede ESP, Orou H, Biaou SSH, Geldenhuys CJ, Ahononga FC, Chirwa PW (2023) Understanding drivers of land use and land cover change in Africa: a review. Curr Landsc Ecol Rep 8:62–72

Assogbadjo AE, Mensah S, Kakaï RG (2017) The relative importance of climatic gradient versus human disturbance in determining population structure of Afzelia africana in the Republic of Benin. Southern Forests: A Journal of Forest Science 79:(2). Forests, people and the environment. https://doi.org/10.2989/20702620.2016.1255406

Baffour-Ata F, Antwi-Agyei P, Nkiaka E (2021) Climate variability, land cover changes and livelihoods of communities on the fringes of Bobiri Forest reserve, Ghana Forests:12

Bongaarts J, O'Neill BC (2018) Global warming policy: is population left out in the cold? Science 361(6403):650–652. https://doi.org/10.1126/science.aat8680

Brink AB, Eva HD (2009) Monitoring 25 years of land cover change dynamics in Africa: a sample based remote sensing approach. Appl Geogr 29:501–512

Chirwa PW, Larwanou M (2017) Overview of restoration and management practices in the degraded landscapes of the Sahelian and dryland forests and woodlands of east and southern Africa. South For J For Sci 79(2):87–94. https://doi.org/10.2989/20702620.2016.1255419

Chirwa PW, Larwanou M, Syampungani, Babalola FD (2015a) Management and restoration practices in degraded landscapes of eastern Africa and requirements for up-scaling. Int For Rev 17(S3):20–30

Chirwa PW, Larwanou M, Syampungani S, Babalola FD (2015b) Management and restoration practices in degraded landscapes of southern Africa and requirements for up-scaling. Int For Rev 17(S3):31–42

Creutzig F, Bren D'amour C, Weddige U, Fuss S, Beringer T, Gläser A, Kalkuhl M, Steckel JC, Radebach A, Edenhofer O (2019) Assessing human and environmental pressures of global land-use change 2000–2010. Global Sustain 2(e1):1–17. https://doi.org/10.1017/sus.2018.15

De Baan L, Alkemade R, Koellner T (2013) Land use impacts on biodiversity in LCA: a global approach. Int J Life Cycle Assess 18(6):1216–1230. https://doi.org/10.1007/s11367-012-0412-0

de Groot R, Brander L, van der Ploeg S, Costanza R, Bernard F, Braat L, Christie M, Crossman N, Ghermandi A, Hein L, Hussain S, Kumar P, McVittie A, Portela R, Rodriguez LC, ten Brink P, van Beukering P (2012) Global estimates of the value of ecosystems and their services in monetary units. Ecosyst Serv 1(1):50–61. https://doi.org/10.1016/J.ECOSER.2012.07.005

Dixon J, Gulliver A, Gibbon D (2001) Farming systems and poverty: improving farmer's livelihoods in a changing world. FAO & World Bank, Rome/Washington, DC

Ellis EC, Fuller DQ, Kaplan JO, Lutters WG (2013) Dating the Anthropocene: towards an empirical global history of human transformation of the terrestrial biosphere. Elementa Sci Anthropocene 1:000018. https://doi.org/10.12952/journal.elementa.000018

EPA, MEDC (1997) The conservation strategy of Ethiopia: the Federal Policy on the environment, vol II. EPA/MEDC, Addis Ababa

Foley JA, Ramankutty N, Brauman KA, Cassidy ES, Gerber JS, Johnston M, Mueller ND, O'Connell C, Ray DK, West PC, Balzer C, Bennett EM, Carpenter SR, Hill J, Monfreda C, Polasky S, Rockstrom J, Sheehan J, Siebert S, Tilman D, Zaks D (2011) Solutions for a cultivated planet. Nature 478:337–342

Garrity DP, Akinnifesi FK, Ajayi OC, Sileshi GW, Mowo JG, Kalinganire A, Larwanou M, Bayala J (2010) Evergreen agriculture: a robust approach to sustainable food security in Africa. Food Secur 2:197–214. https://doi.org/10.1007/s12571-010-0070-7

Gashaw T, Tulu T, Argaw M, Worqlul AW (2017) Evaluation and prediction of land use/land cover changes in the Andassa watershed, Blue Nile Basin. Ethiopia Environ Syst Res 6:17. https://doi.org/10.1186/s40068-017-0094-5

Geist HJ, Lambin EF (2002) Proximate causes and underlying driving forces of tropical deforestation. Bioscience 52(2):143–150

González MG, Martínez M, Lithgow D, Pérez-Maqueo O, Simonin P (2012) Land use change and its effects on the value of ecosystem services along the coast of the Gulf of Mexico. Ecol Econ 82:23–32

Hardin G (1968) The tragedy of the commons. Science 162:3859–1248

IPBES (2018) Summary for policymakers of the regional assessment report on biodiversity and ecosystem services for Europe and Central Asia of the intergovernmental science-policy platform on biodiversity and ecosystem services. In Fischer M, Rounsevell M, Torre-Marin Rando A, Mader A, Church A, Elbakidze M, Elias V, Hahn T, Harrison PA, Hauck J, Martín-López B, Ring I, Sandström C, Sousa Pinto I, Visconti P, Zimmermann NE, Christie M (eds) IPBES secretariat, Bonn, Germany

IPCC (2000) The intergovernmental panel on climate change. Special report on land use, land use change and forestry. Cambridge University PressCambridge

Kamwi JM, Kätsch C, Cho M, Manda SO, Chirwa PWC, Graz FP (2018) Assessing the spatial drivers of land use and land cover change in the Zambezi region. Namibia Land 7(4):131. https://doi.org/10.3390/land7040131

Knauer K, Gessner U, Fensholt R, Forkuor G, Kuenzer C (2017) Monitoring agricultural expansion in Burkina Faso over 14 years with 30 m resolution time series: the role of population growth and implications for the environment. Remote Sens 9(2):132. https://doi.org/10.3390/rs9020132

Koellner T, De Baan L, Beck T, Brandão M, Civit B, Margni M, Milài Canals L, Saad R, Maia de Souza D, Müller-Wenk R (2013) UNEP-SETAC guideline on global land use impact assessment on biodiversity and ecosystem services in LCA. Int J Life Cycle Assess 18(6):1188–1202. https://doi.org/10.1007/s11367-013-0579-z

Koizumi T, Okabe, K, Thompson I, Sugimura K, Toma T, Fujita K (2010) The role of forest biodiversity in the sustainable use of ecosystem goods and services in agro-forestry, fisheries, and forestry. Forestry and Forest Products Research Institute, Ibaraki, 106 pp. ISBN 978-4-902606-68-3

Kouassi JL, Gyau A, Diby L, Bene Y, Kouamé C (2021) Assessing land use and land cover change and farmers' perceptions of deforestation and land degradation in south-west Côte d'Ivoire. West Africa, Land, p 10

Lambin EF, Rounsevell MD, Geist HJ (2000) Are agricultural land-use models able to predict changes in land use intensity? Agric Ecosyst Environ 82:321–331

López-Carr D (2021) A review of small Farmer land use and deforestation in tropical forest frontiers: implications for conservation and sustainable livelihoods. Land 10(11):113. https://doi.org/10.3390/land10111113

Maisharou A, Chirwa PW, Larwanou M, Babalola FD, Ofoegbu C (2015) Sustainable land management practices in the Sahel: review of practises, techniques and technologies for land restoration and strategy for up-scaling. Int For Rev 17(S3):1–19

Mala WA, Geldenhuys CJ, Prabhu R, Manga EF (2020) Forest-agriculture in the Centre-south region of Cameroon: how does traditional knowledge inform-integrated management approaches? In: Gasparatos A et al (eds) Sustainability challenges in sub-Saharan Africa I. Springer, pp 317–353. https://doi.org/10.1007/978-981-15-4458-3_10

McKenzie FC, Williams J (2015) Sustainable food production: constraints, challenges and choices errors. Food Secur 7(2):221–233. https://doi.org/10.1007/s12571-015-0441-1

Meyer WB, Turner BL (1992) Human population growth and global land-use/cover change. Annu Rev Ecol Evol Syst 23:39–61

Mori AS, Kenneth PL, Gustafsson L (2017) Biodiversity and ecosystem services in forest ecosystems: a research agenda for applied forest ecology, forest biodiversity and ecosystem service. J Appl Ecol 54. https://doi.org/10.1111/1365-2664.1266912-27

Musetsho KD, Chitakira M, Nel W (2021) Mapping land-use/land-cover change in a critical biodiversity area of South Africa. Int J Environ Res Public Health 18

Mutanga O, Adam E, Cho MA (2012) High density biomass estimation for wetland vegetation using WorldView-2 imagery and random forest regression algorithm. Int J Appl Earth Obs Geoinf 18:399–406

Olukoshi A (2001) West Africa's political economy in the next millennium: retrospect and Prospect. Monograph series 2/2001. Council for the Development of Social Science Research in Africa. Avenue Cheikh Anta Diop Angle Canal IV, BP: 3304, Dakar, Senegal

Osei-Wusu PA, Buor, D, Addrah P (2014) Geo-spatial analysis of land use and land cover changes in the Lake Bosomtwe Basin of Ghana. Ghana J Geogr:6

Portela R, Rademacher I (2001) A dynamic model of patterns of deforestation and their effect on the ability of the Brazilian Amazonia to provide ecosystem services. Ecol Model 143:115–146

Reid RS, Kruska RL, Muthui N, Taye A, Wotton S, Wilson CJ, Mulatu W (2000) Land-use and land-cover dynamics in response to changes in climatic, biological and socio-political forces: the case of southwestern Ethiopia. Landsc Ecol 15:339–355

Ribeiro EMS, Arroyo-Rodriguez V, Santos BS, Tabarelli M, Leal IR (2015) Chronic anthropogenic disturbance drives the biological impoverishment of the Brazilian Caatinga vegetation. J Appl Ecol 52(3):611–620

Roe D, Nelson F, Sandbrook C (2009) Community management of natural resources in Africa: Impacts, experiences and future directions (No. 18). iied

Saint-Macary C, Keil A, Nielsen T, Birkenberg A, Van LTA, Van DTT, Ufer S, Pham TMD, Franz Heidhues PTMDF, Zeller M (2013) Linkages between agriculture, poverty and natural resource use in mountainous regions of southeast Asia sustainable land use and rural development in Southeast Asia: innovations and policies for Mountainous Areas, 2013. ISBN: 978-3-642-33376-7

Sayer JA, Campbell B (2001) Research to integrate productivity enhancement, environmental protection, and human development. Conserv Ecol 5:32

Schneider LC, Pontius RG (2001) Modelling land-use change in the Ipswich watershed, Massachusetts, USA. Agricult Ecosyst Environ 85:83–94

Schwartz MW, Caro TM (2003) Effect of selective logging on tree and understory regeneration in miombo woodland in western Tanzania. Afr J Ecol 41:75–82

Sewnet A, Abebe G (2017) Land use and land cover change and implication to watershed degradation by using GIS and remote sensing in the Koga watershed, North Western Ethiopia. Earth Sci Inf 11:99–108

Shackleton CM, Cundill G, Knight AT (2009) Beyond just research: experiences from southern Africa in developing social learning partnerships for resource conservation initiatives. Biotropica 41:563–570

Syampungani S, Geldenhuys CJ, Chirwa PW (2013) Developing silvicultural systems based on disturbance-recovery knowledge of the southern African Miombo woodlands. In: Manzanares L (ed) Woodlands: structure, species diversity and sustainable management. NOVA Publishers, New York, pp 139–152

Syampungani S, Geldenhuys CJ, Chirwa PW (2016) Regeneration dynamics of miombo woodland in response to different anthropogenic disturbances: forest characterization for sustainable management. Agrofor Syst 90:563–576. https://doi.org/10.1007/s10457-015-9841-7

Tanentzap AJ, Lamb A, Walker S, Farmer A (2015) Resolving conflicts between agriculture and the natural environment. PLoS Biol 13(9):e1002242

Tschora H, Cherubini F (2020) Co-benefits and trade-offs of agroforestry for climate change mitigation and other sustainability goals in West Africa. Global Ecology and Conservation. Retrieved from: https://doi.org/10.1016/j.gecco.2020.e00919

Turner B, Meyer WB (1991) Land use and land cover in global environmental change. Int Soc Sci J 43:669–679

UNHCR (2016) Statelessness and nationality in Côte d'Ivoire. P. O. Box 2500, 1211, Geneva 2

United Nations (UN) (1992) Convention on biological diversity. Retrieved from: https://www.cbd.int/doc/legal/cbd-en.pdf

United Nations (UN) General Assembly (2015) Sustainable development goals. Retrieved from: https://sustainabledevelopment.un.org/

Venter ZS, Barton DN, Chakraborty T, Simensen T, Singh G (2022) Global 10 m land use land cover datasets: a comparison of dynamic world, world cover and Esri land cover. Remote Sens:14

Von Thunen JH (1960) Der Isolierte Staat in Beziehung auf Landwirtschaft und Nationalökonomie. Loyola University Press, Chikago

Wood EC, Tappan CG, Hadj A (2004) Understanding the drivers of agricultural land use change in Central Senegal. J Arid Environ 59:565–582

Yin Z, Kuang W, Bao Y, Dou Y, Chi W, Ochege FU, Pan T (2021) Evaluating the dynamic changes of urban land and its fractional covers in Africa from 2000–2020 using time series of remotely sensed images on the big data platform. Remote Sens 13

Chapter 8
The Role of Non-Timber Forest Products for Sustainable Livelihoods in African Multifunctional Landscapes

Tajudeen O. Amusa, Marie L. Avana-Tientcheu, Nyong Princely Awazi, and Paxie W. Chirwa

Abstract Within the African context, Non-Timber Forest Products (NTFPs) represent a rich array of biological resources drawn from forests, woodlands, savannas, as well as modified or managed tree-based landscapes. They constitute an essential part of the daily lives of numerous communities, providing livelihood support and opportunities for environmental conservation. Over 59 NTFPs have been identified as prime candidates for domestication in the over three decades of experimentation on the continent. In this chapter, we seek to underscore the paramount importance of NTFPs in shaping a sustainable and harmonious future for both human societies and the natural environments on the African continent. The chapter highlights the economic benefits of NTFPs and the importance of indigenous knowledge and practices on their sustainable harvesting, which has contributed to improved management of forest and tree-based ecosystems in different subregions of Africa. We paid particular attention to two important but often neglected and underutilized categories of NTFPs; edible insects and mushrooms. These categories of NTFPs are gaining recognition for their potential to alleviate food insecurity and reduce the environmental footprint of food production. Their ecological management within

T. O. Amusa (✉)
Department of Forest Resources Management, Faculty of Agriculture, University of Ilorin, Ilorin, Nigeria
e-mail: amusa.to@unilorin.edu.ng

M. L. Avana-Tientcheu
Département de Foresterie, Faculté d'Agronomie et des Sciences Agricoles, Université de Dschang, Dschang, Cameroun

African Forest Forum, United Nations, Nairobi, Kenya

N. P. Awazi
Department of Forestry and Wildlife Technology, College of Technology, The University of Bamenda, Bambili, Cameroon

P. W. Chirwa
Department of Plant and Soil Sciences, Faculty of Natural and Agricultural Sciences, University of Pretoria, Pretoria, South Africa

P. W. Chirwa et al. (eds.), *Trees in a Sub-Saharan Multi-functional Landscape*,
https://doi.org/10.1007/978-3-031-69812-5_8

the multifunctional landscape of Africa holds significant promise. We advocate a multifunctional landscape approach for NTFPs use and management to enhance the multiplicity of ecosystem services they provide including as source of food, energy, medicine, income, fibres and building materials among others, as well as the role of agroecological practices, in creating landscapes that are conducive to both NTFP diversity and agricultural production in many African subregions. By adopting a multifunctional landscape approach that values biodiversity conservation, agroecological practices, and ecosystem restoration, Africa can harness the nutritional and environmental benefits of NTFPs for sustainable livelihoods.

Keywords Sustainable harvesting · Environmental conservation · Edible insects · Mushrooms · Domestication

8.1 Introduction

In the intricate tapestry of Africa's diverse ecosystems, the interplay between humans and nature has historically forged a delicate balance that sustains both livelihoods and the environment. Central to this harmonious coexistence are Non-Timber Forest Products (NTFPs), a rich array of biological resources drawn from forests, woodlands, savannas, as well as modified or managed tree-based landscapes. These resources encompass a wide spectrum of natural goods, from plant-based products such as fruits, nuts, medicinal plants, firewood, resins and fodder-grasses, to animals, including fish, game and bushmeat, honey and other edible insects and mushrooms. Their significance in Africa's socio-economic landscape and pivotal role in fostering sustainable livelihoods and forest resources management cannot be overemphasized (Miller et al. 2021).

It is noteworthy that the utilization of NTFPs has been a cornerstone of subsistence and trade across the globe since time immemorial (Sills et al. 2011). They constitute an essential part of the daily lives of numerous communities, providing sustenance, medicine, income, and cultural significance. For many marginalized and remote populations in Africa, NTFPs represent a vital source of income, offering opportunities for entrepreneurship and market integration. Recent estimates by Shackleton and Alta de Vos (2021) indicate that nearly six billion people use NTFPs globally, including urban populations in the South and those located in the global North. This use involves collection by households as safety-nets to purchasing in the market, and for income generation through trade. A review study carried out in the Horn of Africa showed that NTFPs' overall revenue averaged 24.41% of the rural household income based on actual sale, significantly enhancing rural livelihood in the six nations of the sub-region (Derebe and Alemu 2023). In West Africa, the economic contribution of NTFPs to total household income varies from 12% in South-west Burkina Faso to 35% and 39% in Southern Ghana and Northern Benin respectively, and 45% in Sudanian Savanna of Burkina Faso (Leßmeister et al. 2018). The revenue generated from NTFPs often serves as a buffer against economic shocks and has significantly contributed to poverty alleviation and improved livelihoods. NTFPs are also reported to significantly contribute to African nations' economies as captured in selected countries' strategies for NTFPs development (AFF-FAO 2020).

NTFPs exploitation and use as a socio-economic incentive for forest conservation has been promoted in situations where regulation and enforcement are minimal (see Marshall et al. 2006; Gaoue and Ticktin 2009). While conventional forest management practices often emphasize timber extraction, the utilization of NTFPs offers an alternative approach that promotes biodiversity conservation and ecosystem resilience, specifically, for species favouring low-intensity and non-deleterious use (Harbi et al. 2018). Through the sustainable harvesting of NTFPs, communities have an incentive to preserve the forests and woodlands as well as trees that serve as the source of these valuable resources (Shackleton et al. 2018). This dynamic, including creating alternative supply sources through the domestication and cultivation of these high-value species, has the potential to alleviate the pressure on Africa's forests and curb deforestation rates, ultimately safeguarding critical ecosystems and the services they provide (Leakey et al. 2022). The potential of NTFPs to contribute to poverty alleviation is considered an opportunity to foster a win-win outcome in reconciling forest exploitation and livelihood improvement, with gains in environmental protection (Sunderlin et al. 2005). Moreover, NTFPs play a pivotal role in ecological management. A good understanding of species' level ecological attributes, local ecological knowledge and participatory approaches for natural resources management are documented as appropriate tools supporting sustainable exploitation of NTFPs while promoting ecological diversity and maintaining the health of ecosystems (Chirwa et al. 2008; Ticktin 2015; Sardeshpande and Shackleton 2019). Sustainable NTFP collection below the prescribed optimal harvest level can be a powerful tool for ecological restoration, helping to combat habitat degradation and invasive species proliferation (Regato and Berrahmouni 2005).

The multifunctionality of landscapes is another dimension through which NTFPs wield their influence (Regato and Berrahmouni 2005; Newton et al. 2016). African forested landscapes constitute a diversity of land use types, each purposely managed to provide multiple functions to the society and the global environment. These land uses contribute a diversity of products, enabling communities to extract value from their environments in multiple ways. Forests, woodlands and other tree-based land use often serve as a mosaic of resources, supporting the provision of ecosystem services through agriculture, agroforestry, and other land management practices (Sunderland et al. 2011). The multifunctionality of NTFP-based landscapes not only bolsters local resilience but also enhances the adaptability of landscapes in the face of environmental shocks. By maintaining healthy ecosystems and fostering biodiversity, sustainable management of NTFP-based landscapes helps to regulate climate, purify water, and support pollinators, thus enhancing the overall environmental quality (Sardeshpande and Shackleton 2019).

In the following sections, we explore further each of the above aspects of NTFPs. The chapter seeks to underscore the paramount importance of NTFPs in shaping a sustainable and harmonious future for both human societies and the natural environments. As a case study, special attention were paid to insects and mushrooms, representing lesser known, poorly documented, and largely underutilized groups of NTFPs, although widely used in all African subregions. Over the last two decades, insects and mushrooms have been increasingly profiled for their nutritional, medicinal and ecological values as key providers for provisional, regulatory and supporting services in the multifunctional landscape.

8.2 Defining Non-Timber Forest Products in the Context of Africa

Among the many definitions found in the literature for the term Non-Timber Forest Products (NTFPs), the one by Shackleton et al. (2011) was assessed as being more inclusive. The latter define NTFPs as largely wild (i.e. most populations exist and reproduce without human intervention), native or non-native biological organisms (thousands of plant, animal and fungal species) and materials, other than high-value timber, collected from any landscape and habitat (including human transformed, managed and dominated) for direct or indirect (e.g. for cultural purposes) human use (food, medicine, energy, construction, decoration, utensils, spirituality) and trade for local benefit. This definition encompasses most of the interests and uses for NTFPs on the one hand, while providing answers to questions related to people's objectives for conserving, managing, producing and marketing NTFPs, hence facilitating its incorporation into forest and development policies (Delgado et al. 2023). Most of the definitions found in NTFPs related studies conducted in the African context could fit within this broad but inclusive definition, clearing the confusion in the semantics usually observed in NTFPs literature (Ahenkan and Boon 2011). Therefore, this review that focuses on NTFPs in the African multifunctional landscape will adopt the above definition.

A key component of the definition of NTFPs is that they exclude timber and that the products, benefits or services derived should come from forests, other wooded lands, or trees outside forests. The central part of the concept is that the product of interest is of use to human society. This acknowledges their importance in subsistence economies and the informal sector, especially in rural Africa. It emphasizes their role in poverty alleviation and livelihood diversification. As such, any part of plant, animal, insects and other organisms harvested for use from the forest and/or other land areas can be described as NTFPs. Defining and understanding NTFPs in the African context is essential for sustainable forest management, conservation, and rural development on the continent. Meanwhile, the legal framework governing forest management also influences the classification of products as NTFPs in some African countries. For example, fuelwood and its derivatives are classified as NTFPs in several African countries including Gabon, Central African Republic and Cameroon (Ingram et al. 2013). In this regard, it could be stated that the concept of NTFPs remains an evolving process, shaped by economic, ecological, and cultural factors. Thus, as Africa continues to grapple with the challenges of sustainable forest management, NTFPs will undoubtedly remain at the centre of the discourse, emphasizing their importance in the continent's socio-economic and environmental landscape (Box 8.1).

Box 8.1: Defining Non-Timber Forest Products

NTFPs are largely wild, native or non-native biological organisms (thousands of plant, animal and fungal species) and materials, other than high-value timber, originating from forested or 'treed' landscapes and or collected from any landscape and habitat (including human transformed, managed and dominated) for direct or indirect (e.g. for cultural purposes) human use (food, medicine, energy, construction, decoration, utensils, spirituality) and trade for local benefit (Shackleton et al. 2011). The exclusion of timber from the definition of NTFPs is based on a conceptual distinction between products derived from trees that are not primarily harvested for their wood and those that are harvested for timber or wood products. This distinction arises from the unique characteristics, uses, and management considerations associated with timber compared to other forest products. NTFPs play a crucial role in the livelihoods of many African communities, serving as essential sources of food, medicine, cultural identity and income. Sustainable management and utilization of NTFPs are integral to the well-being of many local populations, promoting both economic resilience and environmental conservation in the diversity of ecosystems found across the African continent.

8.3 The Contribution of Non-Timber Forest Products Extraction to Forest Conservation

Non-Timber Forest Products extraction plays a significant role in the overall framework of the conservation of African forests. The diverse range of NTFPs, their economic benefits for local communities, indigenous knowledge and practices on sustainable harvesting, and collaborative efforts have collectively contributed to the sustainable management of forest ecosystems in many regions of Africa. Therefore, recognizing the value of NTFPs in forest conservation is essential for the sustainable management of Africa's rich and diverse forests, ensuring their continued existence for future generations.

The reliance of local communities on NTFPs for their livelihoods reduces interest in engaging in unsustainable timber logging (Melese 2016). This also reduces pressure on forest resources and helps protect mature trees from overexploitation. In a study in the coastal forest area of South Africa, Shackleton et al. (2007) investigated the sustainability of harvesting of *Ischyrolepis eleocharis*. This is a local Cape reed (locally called *umtshayelo*) which has been harvested for decades by residents of the local village of Tharfield to make hand brooms. These brooms are used by the villagers in their homesteads, are sold in a nearby town, have a cultural significance and are commonly presented as wedding gifts. The study revealed that the total annual demand for Ischyrolepis was approximately 2.7% of the standing crop. The bulk of the annual demand was for small-scale trade, the income from which was a primary source of income for the few harvesters. Very little evidence could be found indicating that harvesting was damaging the resource or its habitat, and local knowledge even suggested that the abundance of the species was stimulated by harvesting. This is consistent with its rhizomatous growth habit, which is well regarded as being tolerant of heavy harvesting or grazing (van den Wyngaert et al. 2003).

Additionally, the extraction of NTFPs provides incentives for forest stewardship among local communities. This is because sustainable harvesting of NTFPs often requires communities to manage and protect the forest ecosystems that produce these resources. This fosters a sense of ownership and stewardship, leading to better forest conservation practices. Besides, harvesting NTFPs often requires a deep knowledge of local ecosystems and their sustainable management. For instance, Gaoue and Ticktin (2009) showed that Fulani herders in West Africa have detailed ecological knowledge of their impacts on the resources they depend on, and this is finely tuned to local ecological conditions. This knowledge is also widely spread across different sectors of Fulani communities and is highly congruent with scientific findings. Traditional Fulani practices such as the *sopoodu* provide a basis for sustainable management of Fulani-owned *Khaya senegalensis* plantations. This study illustrates how local ecological knowledge and practices can play a key role in the design of culturally-appropriate conservation plans. Similarly, Cruz (2002) reported indigenous knowledge among local communities in Mozambique on sustainable harvesting of *Pterocarpus angolensis* and *Swartzia madagascariensis* through the practice of shifting cultivation and use of fire to encourage coppice regeneration of the two species that are exploited for varied purposes. This shows that local knowledge of NTFP utilization and management, passed down through generations, contributes to the preservation of traditional practices that are inherently conservation-oriented (Laird et al. 2010; Ruiz-Mallén et al. 2015). In the light of this, it appears that cultural and direct economic benefits gained through the use of NTFPs may be a contributory incentive for communities to conserve a species and/or habitats. Consequently, considerable local level contextual information and consultative approach are required in the sustainable management of NTFPs to meet both the objectives of commercialization and conservation.

Meanwhile, the sustainability of harvesting plant-based NTFPs depends on the plant parts collected. Studies have shown that harvesting bark and roots as NTFPs can negatively affect the health status of trees and could easily lead to unsustainable exploitation and overharvesting as observed on G*nidia glauca, Garcinia lucida* and *Prunus africana*, three species harvested for medicinal and fibre uses in central Africa (Avana-Tientcheu et al. 2018; Guedje et al. 2007; Momo et al. 2016). Although, in theory, bark harvesting can be sustainable through removing opposite quarters only, in practice it is not outside of certain restricted approaches, such as when there is a monopoly on harvesting. A free for all leads to removal of all bark. To this extent, plant domestication has been introduced for more than three decades as an integrated approach through which an alternative source of supply of NTFPs in the agricultural landscape is created to reconcile the increasing interest in collecting, using and trading NTFPs and the conservation of the natural resource base of selected high value species producing them. Evidence of cultivation supporting conservation could be found in the case of species such as *Irvingia gabonensis, Garcinia kola, Dacryodes edulis, Gnetum africana* and *Moringa oleifera* among others (Leakey et al. 2022). A recent review after third decade of domestication with an Africa focus has shown that over 59 priority indigenous food and non-food tree species (Table 8.1) have been identified based on high demand for their products at local, national, regional and international market as well as their conservation status hinged on their vulnerability in natural forested habitats (Leakey et al. 2022) (Box 8.2).

Table 8.1 African tree and other woody species producing edible products that have been identified as prime candidates for domestication in the literature over the third decade (species in bold have pan-African priority in the region)

Humid Lowlands of West and Central Africa	Sahelian Drylands of West and North Africa	Highlands and Dryland of East Africa	Savannah and Miombo Woodlands of Southern Africa
Allanblackia spp.	*Vitellaria paradoxa*	*Adansonia digitata*	*Sclerocarya birrea*
Prunus africana	*Adansonia digitata*	*Allanblackia stuhlmannii*	*Adansonia digitata*
Irvingia spp.	*Sclerocarya birrea*	*Zizyphus mauritiana*	*Azanza garckeana*
Dacryodes edulis	*Tamarindus indica*	*Vitellaria paradoxa*	*Uapaca kirkiana*
Garcinia kola	*Zizyphus mauritiana*	*Ficus natalensis*	*Strychnos spinosa*
Cola spp.	*Azanza garckeana*	*Balanites aegyptiaca*	*Strychnos cocculoides*
Gnetum spp.	*Moringa oleifera*	*Albizia coriaria*	*Dovyalis caffra*
Ricinodendron heudelotii	*Borassus aethiopum*	*Markhamia lutea*	*Vangueria infausta*
Chrysoplyllum albidum	*Carapa procera*	*Tamarindus indica*	*Parinaria curatellifolia*
Tetrapleura tetraptera	*Parkia biglobosa*	*Vitex doniana*	*Sizygium cordatum*
Canarium schweinfurthii	*Vitex doniana*	*Vitex mombassae*	*Vitex* spp.
Pentaclethera macrophylla	*Dalium guineense*	*Sclerocarya birrea*	*Flacourtia indica*
Baillonella toxisperma	*Garcinia kola*		*Zizyphus mauritiana*
Trichoscypha aciminata	*Xylopia aethiopica*		
Afrostyrax lepidophyllus	*Detarium microcarpum*		
Gnidia glauca	*Balanites aegyptiaca*		
Pausinystalia johimbe	*Prosopis africana*		
Tetracarpidium conophorum	*Argania spinosa*		
Garcinia lucida *Gambeya africanum*	*Irvingia gabonensis*		

Source: Leakey et al. (2022)

Box 8.2: Contributions of NTFPs to Forest Conservation

Biodiversity Preservation: NTFPs often involve the sustainable harvesting of a diverse range of plant and animal resources, promoting the conservation of biodiversity within forest ecosystems.

Ecosystem Resilience: The non-destructive nature of NTFP extraction helps maintain the overall health and resilience of forest ecosystems, as it minimizes the impact on the forest structure and composition.

Local Livelihoods: By providing alternative sources of income for forest-dependent communities, NTFPs reduce the pressure on forests for timber extraction and other potentially harmful activities, contributing to sustainable land use.

Cultural Preservation: Many NTFPs hold cultural significance for indigenous and local communities, fostering a sense of connection to the land. This connection can contribute to the preservation of traditional knowledge and practices that support forest conservation.

Reduced Deforestation: The economic value of NTFPs can incentivize communities to protect rather than exploit forests. This economic incentive helps combat deforestation by offering sustainable alternatives to more destructive land-use practices.

Carbon Sequestration: Intact forests, supported by sustainable NTFP management, play a crucial role in carbon sequestration, helping mitigate climate change by absorbing and storing atmospheric carbon dioxide.

Water Resource Management: Forests, sustained through NTFP conservation practices, contribute to water regulation, maintaining hydrological cycles that benefit both local communities and downstream ecosystems.

8.4 Rural Economy and Non-Timber Forest Product Trade in Africa: A Pathway to Improved Livelihoods

The global market value of NTFPs has been steadily rising, with estimates surpassing billions of dollars annually. A report by the Food and Agriculture Organization (FAO) suggests that NTFPs contribute approximately $21 billion annually to the global economy, though this figure is likely an underestimate due to the informal nature of many NTFP transactions (Lebedys and Li 2014). According to the World Bank (2002), NTFPs support the livelihoods of an estimated 1.2 billion people globally, contributing significantly to poverty reduction and rural development.

The trade of NTFPs in Africa is a dynamic and continuously evolving sector that connects rural producers to urban and global markets. NTFP trade on the continent represents a significant economic activity. Products such as shea-butter, locust bean seeds, baobab fruit, and various medicinal plants are exported to international markets, generating income for rural communities. Many NTFPs are traded within regional and local markets. These products are also sold in nearby towns and cities, providing income to rural producers. Honey, wild fruits, and craft materials are commonly traded in local markets, contributing to the rural economy.

The trade of NTFPs promotes sustainable forest management. Additionally, NTFP trade often provides opportunities for women to generate income and gain economic independence (Amusa and Abdul 2021). As a matter of fact, women are the primary collectors and processors of NTFPs in many African societies, making this trade crucial for gender equality and poverty reduction. NTFP trade offers rural households an additional income source beyond agriculture, which can be vital for coping with economic shocks and uncertainties. The revenue generated from NTFP trade can help lift rural communities out of poverty, improving access to education, healthcare, and better living conditions.

Iponga et al. (2018) showed that in two timber concession areas in Gabon, rural people's income from NTFPs was estimated at 9298 FCFA/18 USD/household/year around Bayonne and 20,725 FCFA/41 USD/household/year around Compagnie Equatoriale des Bois (CEB) while their consumption values were calculated as 72,630 FCFA/145 USD/household/year around Bayonne and 44,356 FCFA/89 USD/household/year around CEB. According to the United States Agency for International Development (USAID 2010), trade in shea-butter employs an estimated three million women across West Africa, generates between USD 90 million and USD 200 million a year from exports and promotes economic activity in rural communities. Adams et al. (2016) reported that about 93.7% of women in the Wa Municipality, Ghana are engaged in the shea value chain as a main source of livelihood strategy with the average annual income per person being around GH¢846.85. In Burkina Faso, shea sector contributes to women's economic well-being with an average of CFA 530000 (USD 892) per actor per year (Chen 2017).

Non-Timber Forest Products fulfill a wide variety of roles in improving rural livelihoods. These include food security and nutrition, energy, fibres and building materials, medicines, cultural value, and income, among others.

8.4.1 Food Security and Nutrition

As regards food security and nutrition, many NTFPs serve as staples for people living in/or near forest areas. They are a part of household subsistence strategies. Such products as wild fruits and edible roots, contribute to food security, ensuring that households have a diverse and reliable food supply. Many of them provide micro- and macro- nutrients, as well as proteins, carbohydrates and fats at various levels (Dounias et al. 2007; Johns and Sapitht 2014). Some NTFPs native to the African landscape including the African walnut (*Plukenetia conophora*), African pear or bush butter tree (*Dacryodes edulis*), Mobola plum (*Parinari curatellifolia*), white star apple (*Chrysophylum albidum*), and bush mango (*Irvingia gabonensis*) are particularly important for food security and nutrition. The African walnuts are reported to contain various minerals, including calcium, magnesium, phosphorus, potassium, and zinc as well as being rich in vitamin E. The fat contents are made up primarily

Fig. 8.1 *Plukenetia conophora* (African walnut) contributes significantly to food security and nutrition in southern part of Nigeria (© Amusa Tajudeen Okekunle)

of unsaturated fatty acids, including oleic acid (omega-9) and linoleic acid (omega-6), with traces of alpha-linolenic acid (omega-3) (Neji and Agwupuye 2019). This is similar to nutritional reports of minerals and vitamins found in the African pear (Ene-Obong et al. 2019). The white star apple is reputed to be rich in vitamin C (ascorbic acid), which acts as an antioxidant and supports immune function (Ibrahim et al. 2021).

The fruit pulp of shea tree is an important source of vitamins and minerals, especially for young children in rural areas. Shea-butter derived from the nut is used for cooking and serves as a crucial nutritional resource for millions of rural households within its range. In some cases, NTFPs provide exceptional 'treats' and may be important due to their cultural significance. For example, bushmeat in many urban areas of Africa is often consumed as a privileged food with cultural nostalgia and people are willing to pay high prices for them. Several authors have reported a significant correlation between tree cover and dietary diversity among children in several African countries (Ickowitz et al. 2014; Rasolofoson et al. 2018) (Fig. 8.1).

8.4.2 Energy

In terms of energy, fuelwood and charcoal as a form of NTFP contributes significantly to energy security in most rural areas. As an energy source, fuelwood is also collected and used for cooking and warmth in homes. Recent statistics indicate that 65% of charcoal production occurring in Africa is by smallholders who use unsustainable wood exploitation and carbonization systems (Mensah et al. 2020). Annual wood-energy consumption in Cameroon's urban areas is estimated at 2,203,496 tonnes for fuelwood, and 356,530 tonnes for charcoal, representing sales estimated at 188.33 billion FCFA per year. In rural areas, it is estimated that the local population consumes around four million tonnes of fuelwood a year, which they collect directly from the surrounding vegetation (Eba'a Atyi et al. 2016).

In DRC, the proportion of households involved in firewood extraction was substantial (80.70%) (Mendako et al. 2022). Makungwa et al. (2013) reported that over 97% of households in rural Malawi rely on firewood for cooking, making it the highest-value commodity rural households extract from forests. In Nigeria, more than 80% of domestic energy consumption is derived from woodfuel (Saad and Bugaje 2016; Suleiman et al. 2017). The charcoal business in Tanzania generates an estimated USD 650 million per year and employs over two million people (Nyamoga and Solberg 2019). The predominant energy source used by the rural population in the country is fuelwood, while charcoal is the most frequently used source of energy in many urban areas. Fuelwood and charcoal are sought after by both household and industrial sectors in all ecological zones of the country. Industrial uses include those by institutions, food and craft industries. They are also very important in local restaurants, bakeries, local breweries, pottery, blacksmith and burnt brick factories. Institutions such as hospitals, prisons and schools also demand fuelwood and charcoal for cooking and heating. The number of Africans who rely on charcoal is expected to rise to 823 million by 2030. In addition, charcoal is now an export commodity in Nigeria, with a large market in the EU, USA and Asia. The rising demand for charcoal and firewood in African countries coupled with unsustainability concerns associated with their production are stimulating interest in the formalization of the sector through enabling investments and promotion of carbon-neutral wood energy slogan (Mensah et al. 2020).

8.4.3 Fibres and Building Materials

Several NTFPs are also a source of fibre used in building, clothing, paper and ropes. These are derived from a variety of tree species, rattans and bamboos. Bamboos in particular are also greatly used as scaffolding, with large commercial markets in many countries catering to this use. Bamboo's rapid growth, renewability, and minimal environmental impact make it an eco-friendly alternative to traditional building materials like concrete and steel. In the same way, rattan-reinforced concrete can enhance the structural integrity of buildings, reducing the need for more resource-intensive materials. Jute, often referred to as the "golden fibre," is another NTFP with tremendous potential in the textile industry. Jute fibres are biodegradable and possess excellent insulation properties, making them suitable for various applications, from clothing to home furnishings. In Africa, although there is little cultivation of bamboo and little or no international trade in bamboo, many bamboo products are used domestically and can be very significant in both household and local economies (Bystriakova and Kapos 2006). In some parts of Africa, bamboo is also a source of food and drink. In Tanzania and Uganda young shoots and seeds of *Oxytenanthera abyssinica* are consumed as food. It is also used in the production of bamboo wine, also known as *ulanzi*.

8.4.4 Medicines

Similarly, many NTFPs are used in traditional systems of medicine. For instance, Ajaiyeoba and Fadare (2006) reported the leaves of *Plukenetia conophora* as being used for the treatment of dysentery and improve sperm- count in men. The bark is used in tea as a laxative, chewed for toothache, to prevent and control high blood pressure and as a male fertility agent. The root is used for the treatment of haemorrhoids, frostbite and varicose ulcers (Enitan et al. 2014). In the same vein, the medicinal uses of the African bush pepper (*Piper guineense*) have been reported by several workers. The leaves are considered appetitive (having a stimulating effect on the appetite), carminative (expelling gas from the body; relieving flatulence) and eupeptic (aiding good digestion; Okwute 1992). They are also used for the treatment of cough, bronchitis, intestinal diseases and rheumatism (Sumathykutty et al. 1999). There is a growing demand for the African bush pepper owing to its being used for a variety of purposes (Amusa et al. 2017). Amusa (2015) also related that *Annickia chlorantha* is a high value medicinal tree species in Nigeria whose stem bark is used for the treatment of malaria and other analgesic ailments of the human body. The intensely bitter stem bark is particularly used in traditional medicine in the West African sub-region to treat jaundice, fevers, tuberculosis, bloody vomit and urinary tract infections (Adesokan et al. 2007). The bark in powdered form is applied externally to treat sores, ulcers and wounds (Burkill 1985). Aqueous extract of the bark has been shown to have an effective antibacterial action, especially against gram-positive bacteria. The extract has also shown antiviral activity and has the potential in relieving pyrogen-induced fever. The alkaloids palmatine, coloumbamine and jatrorrhizine have been identified in the bark, and it is believed that these alkaloids are responsible for the medical activities in the product (Adesokan et al. 2007).

Traditional healing systems have been integral to African cultures for millennia, with local healers passing down knowledge of medicinal plants through oral traditions. This wealth of indigenous wisdom forms the foundation of traditional medicine, which continues to play a vital role in healthcare delivery, particularly in rural and underserved communities. The PROTA Medicinal Plants Database, a project initiated by Wageningen University in the Netherlands, has been instrumental in documenting the botanical diversity of tropical Africa. Its comprehensive database includes detailed information on over 7000 plant species, including their traditional uses, pharmacological properties, and distribution across the continent. Meanwhile, limited market access, inefficient value chains, and lack of market information hinder the economic potential of medicinal plants. Hence, strengthening market linkages, improving infrastructure, and providing training and technical assistance to smallholder farmers and entrepreneurs can unlock opportunities for income generation from NTFP-based medicinal plants (Amusa et al. 2010).

8.4.5 Cultural Value

Some NTFPs also have cultural significance. They are used in social activities including religious and various ceremonies. In Nigeria and many parts of West Africa, bitter kola (*Garcinia kola*) is used in the naming ceremony; wedding, hospitality, divination, traditional religious rites, spiritual invocations, initiation, oath taking, cursing, and semeinology, among others. NTFPs have also been instrumental in the creation of traditional crafts and artworks. For instance, in Africa, communities have long relied on palms, such as the doum palm, for weaving baskets, mats, and other artefacts. The intricate patterns and designs woven into these crafts often convey cultural narratives and stories, preserving the heritage of these communities (Cunningham 2001). The culinary traditions associated with some NTFPs such as bushmeat, Ackee apple/soapberry (*Blighia sapida*), locust bean seed, and shea butter among others are not only about sustenance but also about community, as the preparation and sharing of traditional dishes are central to many African cultures. Some NTFPs are sources of natural dyes which have been used in textile traditions for centuries. The rich colours extracted from plants, insects, and fungi have been used to create vibrant clothing and fabrics (Cardon 2007). All of these are a testament to the profound connection between humans and their natural environments through NTFPs. These products have not only sustained communities but have also enriched their cultures, traditions, and identities.

8.4.6 Income

Income from NTFPs contributes significantly to households' cash needs for meeting several necessities of harvesters and traders. They act as a safety net in times of crisis. (Awono et al. 2013) reported that 45 high-value NTFPs, including bushmeat, fuel wood, and various plants in Cameroon, are estimated to generate over $1 billion annually. Similarly, the number of people employed by the NTFP sectors in Cameroon and the Democratic Republic of Congo is double the number employed by the forestry industry (Awono et al. 2013). In DRC, the annual average income from forest-based activities was estimated at US$ 659, representing the second most crucial income portfolio, with the share of Bushmeat income being higher than other forest products income (Mendako et al. 2022). Trade in shea-butter is one of the most vibrant traditional businesses in many rural communities of West and Central Africa (Tweneboah-Kodua et al. 2018). Several studies have revealed the contributions of this trade to rural livelihoods in the regions. According to the United States Agency for International Development (USAID 2010), trade in shea-butter employs an estimated three million women across West Africa, generates between USD 90 million and USD 200 million a year from exports and promotes economic activity in rural communities. Adams et al. (2016) reported that about 93.7% of women in the Municipality, Ghana are engaged in the shea value chain as a main source of livelihood strategy with the average annual income per person being around GH¢846.85. In Burkina Faso, shea sector contributes to women's economic well-being with an average of CFA 530000 (USD 892) per actor per year (Chen 2017). There has been an increasing international interest in shea-butter trade owing to its numerous potentials (Nde et al. 2016).

8.5 Case Study 1: Ecological Management of Edible Insects in a Multifunctional Landscape

As a case study, an often neglected and underutilized group of NTFPs are edible insects. Edible insects are important NTFPs that have been a part of human diets in Africa for centuries, providing an essential source of nutrition and protein. As the world faces increasing pressure to find sustainable food sources to feed a growing population, edible insects are gaining recognition for their potential to alleviate food insecurity and reduce the environmental footprint of food production. Edible insects are known for their low environmental impact compared to traditional livestock. They require less land, water, and feed to produce the same amount of protein, and they emit fewer greenhouse gases. This ecological efficiency aligns with the need for more sustainable food production methods. In Africa, where biodiversity is rich and cultural acceptance of insect consumption is high, the ecological management of edible insects within a multifunctional landscape holds significant promise.

In the Central Africa country of Cameroon, caterpillars are among the most consumed, traded and economically valuable edible insects for many indigenous peoples. Ngute et al. (2019), through market surveys, semi-structured interviews and field observations, identified eleven species of caterpillars being traded in Cameroon Central region. Seven of the species accounted for a total of 69,482 kg and worth US$163,565 annually. Similarly, Hlongwane et al. (2021) assessed the socioeconomic benefits of insects in the Vhembe district of Limpopo province, South Africa and reported that edible insects are an important natural commodity in rural areas that is used for household consumption and to generate income through trade. It provides employment opportunities and improves the livelihoods of impoverished rural people. Five insect groups belonging to four insect orders are traded in informal markets of the studied district. Mopane worms (*Gonimbrasia belina*) were the most traded (42%) edible insects, while the provision of income (60%) and financial support (35%) were the primary benefits. In Zimbabwe, a study by Macheka et al. (2022) revealed that the consumption of edible insects and other NTFPs occurred in all provinces of the country irrespective of the agro-ecological region and only varied due to rainfall pattern and soil type. The study also showed that consumption of wild-harvested edible insects is not a coping strategy and in fact, households consuming wild harvested edible insects were associated with improved food and nutrition security outcomes as compared to non-consuming households. The authors submitted that consuming wild harvested edible insects may not be associated with poverty as indicated in most previous studies.

In East Africa, Okia et al. (2017) studied local perspectives of the catalogue, ecology, management, collection, processing and consumption of edible insects within the Lake Victoria Basin (LVB) in Uganda, Burundi and Rwanda using a household survey and focus group discussions. Results revealed that up to 20 insect species were eaten in Uganda, 13 in Burundi, and six in Rwanda. In Uganda, the most consumed insects were a katydid grasshopper (*Ruspolia differens*), palm weevil (*Rhynchophorus phoenicis*) larvae and termites (*Macrotermes spp*), while in Rwanda and Burundi, *Macrotermes spp* were the most consumed. In West Africa, Ebenebe et al. (2017) also assessed the consumption of edible insects in south east Nigeria using questionnaire surveys and oral interviews. The results showed that seventeen insect species belonging to six orders were consumed and harvesting is mainly during the rains, though some recorded year round availability. Restricted seasonal availability and high collecting costs are among the reported barriers to edible insect consumption in the region. Likewise, in north central Benue State of the country, Agbidye et al. (2009) documented ten species of edible insects that contribute significantly to the socio-economic well-being of the people in both rural and urban centres. They compete favourably with other meat products such as beef and mutton in the area.

In a synthesis study for Africa, Hlongwane et al. (2021) listed 212 edible insect species from eight orders that are consumed on the continent. These insects were reported to be rich in protein, fats, and fibre. The highest protein content was reported for Lepidoptera (range: 20–80%). Coleoptera had the highest carbohydrate content (7–54%), while Lepidoptera had the highest fat content (10–50%). Several defoliators of trees such as the pallid emperor moth (*Cirina forda*) and *Bunaea alcinoe* serve as a delicacy in many homes in Africa. Similarly, many termite and cricket species such as *Macrotemes natalensis* and *Brachytrupes membranaceus* feature prominently in traditional dishes on the continent. Several studies in different rural communities have shown that the volume of insects consumed as a percentage of total animal protein is quite appreciable. Scores of insect species are prominent items of commerce in markets within the villages and towns of Africa. Thus, considering the excellent source of nutrition, and potential socio-economic benefits from edible insects, it is safe to surmise that edible insects can contribute strongly to improved food security, and rural development in developing countries. In addition, edible insects can be used as a sustainable food source to combat food shortages in the future, for example, providing resilience during times of drought or other climate stressors.

Furthermore, van Itterbeeck and van Huis (2012) revealed that, throughout history, humans have manipulated their natural environment for increased predictability and availability of edible insects. The authors showed that palm weevil larvae collection in Africa is facilitated by manipulating host tree distribution and abundance by deliberately cutting palm trees at a chosen time and location. Accordingly, the larvae of *Rhynchophorus phoenicis* are harvested from trunks of palm trees that have been cut down to extract sap which is used for making palm wine. Adult weevils are attracted to the rotting trunk and subsequently invade it. It was suggested, therefore, that by felling palm trees beforehand the weevil larvae are a more predictable resource than by natural collection as search time is decreased. Similarly, the collection of caterpillars is facilitated by manipulating host tree distribution and abundance through shifting cultivation, fire regimes, host tree preservation, and manually introducing the insects to designated areas. Two methods of exploitation of palm weevils' grubs were observed in Cameroon, the traditional gathering with an average productivity of 35 ± 13.2 grubs per trunk and the semi-farming system with an average productivity of 50 ± 10.1 grubs per trunk. With a daily income from the trade of palm weevil grubs of 22,800 Central African XAF (around 44$US), it was estimated that professional grub collectors can reach a maximum monthly income of 456,000 XAF, contributing to 78% of the household revenue Nevertheless, both exploitation methods were destructive for the raphia swampy ecosystem where nearly 1200 raphia stems could be destroyed by a single collector per harvesting season (Ayemele et al. 2016). At any rate, in many landscapes, host trees of some edible insects are also timber species and the wood extraction regime may be a threat to the availability and distribution of edible insects.

Since insects can be a vital component of food security strategies, contributing to better nutrition and overall health, it is crucial to protect and manage their natural habitats. A multifunctional landscape approach recognizes that landscapes can provide multiple ecosystem services simultaneously, including food production, water regulation, carbon sequestration, and biodiversity conservation. To manage edible insects sustainably in African landscapes, agroecological practices, such as intercropping, crop rotation, and organic farming, can create landscapes that are conducive to both insect diversity and crop production. These practices reduce the need for chemical pesticides that can harm insect populations. Likewise, in areas where edible insect habitats have been degraded, restoration efforts should be employed to recreate suitable environments. This can include reforestation, wetland restoration, and the creation of insect-friendly corridors.

In sum, edible insects have the potential to play a significant role in addressing food security and promoting sustainability in Africa. However, their ecological management is critical to ensure their long-term availability without compromising other ecosystem services. By adopting a multifunctional landscape approach that values biodiversity conservation, agroecological practices, and ecosystem restoration, Africa can harness the nutritional and environmental benefits of edible insects while maintaining ecological balance (Box 8.3).

Box 8.3: Some Edible Insects in the Multifunctional Landscape of Africa

Edible insect	Culinary uses and contributions
Mopane Worms *Gonimbrasia belina or Imbrasia belina*	Caterpillars found in Southern Africa; are high in protein; dried or cooked, and used as a source of protein and income
Crickets Gryllus campestris	Insects with a high protein content; various species are consumed; fried, roasted, or ground into flour; rich in protein
Grasshoppers *Phymateus viridipes*	Commonly consumed in many African countries; rich in protein; fried, roasted, or ground; popular as a snack or in dishes
Mealworms *Tenebrio molitor*	Beetle larvae; high in protein and beneficial nutrients; used in baking, cooking, or as snacks; versatile in recipes
Silkworm Pupae *Bombyx mori*	Larvae of silk moths; consumed in some African regions; boiled, fried, or roasted; rich in protein and healthy fats
African palm weevil (*Rhynchophorus spp, phoenicis and ferrugineus*)	Caterpillars raised in raphia and other palm trunks. Highly nutritious with high fat, protein and mineral contents, consumed fried, roasted as snacks and side dishes in central and Northern Africa

Source: Authors compilation

8.6 Case Study 2: Improving the Multifunctionality of Landscapes with Edible and Medicinal Mushrooms: Prospects in Africa

Again, due to their underutilization, our second case study involves edible mushrooms. In recent years, there has been growing interest in harnessing the potential of edible and medicinal mushrooms to promote sustainability and biodiversity conservation. This is because edible mushrooms are a valuable source of nutrition and income in many parts of the world. In Africa, where food security remains a pressing concern, the cultivation and consumption of edible mushrooms, just like edible insects, can provide a sustainable source of protein and essential nutrients. More so, mushroom farming is relatively low-cost and can be practised in various settings, including rural communities and urban areas (Anchang 2014). Meanwhile, the medicinal potential of mushrooms in Africa remains largely untapped. Yet, many indigenous mushroom species in Africa have demonstrated significant therapeutic properties, including anti-inflammatory, antioxidant, and immune-boosting effects. Incorporating these mushrooms into traditional medicine practices can enhance healthcare accessibility and promote the sustainable use of natural resources (Bastos et al. 2023).

Mushroom, such as *Pleurotus ostreatus* has been reported for its traditional use as antimicrobial agent against *Escherichia coli*, *Staphylococcus aureus*, *Bacillus subtilis*, and *Candida albicans*. It is also used as an antioxidant (Stefanello et al. 2012). The mushroom, *Ganoderma lucidium* is reported to be used as an ointment to treat skin conditions in Cameroon after it has been dried and mixed with palm oil. It is also traditionally used to treat stomachache, haemorrhoids, rheumatism, arthritis, neoplasia, among others. *Daldinia concentrica*, is reputed to be used in Nigeria to treat hernia, stomachache, in the reduction of big navel of newborn babies and invigoration of kidneys activity, among other valences (Bastos et al. 2023).

The cultivation of edible and medicinal mushrooms can contribute to biodiversity conservation and the restoration of degraded landscapes. Mushroom mycelium helps improve soil health and structure, making it more suitable for plant growth. Furthermore, the integration of mushroom cultivation into agroforestry systems can enhance ecosystem services such as water purification and carbon sequestration (Oyetayo 2011). Additionally, mushroom cultivation and value-added product development offer economic opportunities for local communities, particularly in rural areas with limited access to formal employment. Small-scale mushroom farming can empower marginalized populations, reduce poverty, and enhance livelihoods. Moreover, the sale of mushrooms in local and international markets can generate income and boost economic resilience (Ariyo 2023).

In Africa, edible mushroom cultivation has found a niche among small scale farmers. The number of species consumed varies between different regions of Africa, with around 300 species being documented in the literature in total (Bastos et al. 2023). In Cameroon, Njouonkou et al. (2016) identified the various mushrooms exploited by the Bamoun people in the Noun Division. The authors showed that the Bamoun people use at least 40 species of mushrooms belonging to 8 genera: Auricularia, Cantharellus, Ganoderma, Pleurotus, Lactarius, Lactifluus, Russula, and Termitomyces. Species of genera Lactarius, Lactifluus, Russula, and Termitomyces are most often used for food, whereas Ganoderma spp. and Pleurotus tuber-regium are mainly exploited for medicinal purposes. In Nigeria, edible mushrooms such as *Auricularia auricular* Judae (Bull.) Quél, *Lentinus squarrosulus* Mont., *Pleurotus tuberregium* (Fr.) Singer and *Volvariella volvacea* (Bull.) Singer have been successfully cultivated on small-scale basis (Okhuoya et al. 2010). In Kenya, Waiganjo et al. (2009) reported that edible mushrooms were previously picked from the wild, but now, many farmers are growing mushrooms for their nutritive value as well as for industrial and medicinal purposes. The authors further averred that two main types of mushrooms are being commercialized in the country, the button (*Agaricus bisporus*) and oyster (*Pleurotus* species) with the former accounting for 95% (476 tonnes) of the 500 tonnes produced annually. Meanwhile, oyster mushroom, introduced in 2003, is the most popular among small-scale farmers, mainly because it can fruit over a wide range of temperatures. It offers lucrative business, requires no arable land for production, and provides diversification with benefits such as increased income, employment, and food and nutrition security. Furthermore, the abundant agricultural waste found countrywide offers an opportunity for production, which in turn provides a more economical and environmentally friendly disposal system. This is an indication of what could be achieved by building capacity in indigenous mushroom cultivation. The potential for mushroom production in Kenya is high. Demand outstrips supply as Kenya imports 150 t annually not to mention the feasible export market.

Edible and medicinal mushrooms have the potential to play a significant role in improving the multifunctionality of landscapes in Africa. From enhancing food security and traditional healing practices to promoting biodiversity conservation and creating economic opportunities, mushrooms offer a holistic solution to some of the continent's most pressing challenges. To harness this potential, it is crucial for governments, NGOs, and local communities to collaborate in developing sustainable mushroom cultivation practices and promoting their integration into African landscapes (Nyamoga 2023). The utilization of edible and medicinal mushrooms in Africa represents a sustainable and multifaceted approach to addressing pressing issues while conserving the continent's natural resources and cultural heritage. By recognizing and capitalizing on the potential of these fungi, Africa can move closer to achieving its goals of food security, biodiversity conservation, and economic development. Improving the multifunctionality of landscapes with edible and medicinal mushrooms would include different steps for their cultivation including spawn preparation, substrate preparation, fruiting body management, harvesting and storage with their own set of requirements and precautions for each step.

8.7 NTFPs as a Tool for Improved Environmental Services of Multifunctional Landscapes: Opportunities, Challenges, and Key Lessons

The utilization of NTFPs plays a pivotal role in improving the environmental services of many multifunctional landscapes. In particular, NTFPs contribute significantly to the preservation of biodiversity through their sustainable use. As pointed out earlier, indigenous and local communities who often rely on NTFPs for their livelihoods, have a vested interest in maintaining the health and diversity of various ecosystems where NTFPs are found. For this reason, sustainable harvesting of NTFPs can provide an economic incentive for the conservation of local flora and fauna, as it encourages responsible land management practices that ensure the long-term availability of these resources.

Several studies have indeed highlighted the positive relationship between NTFP extraction and biodiversity conservation. For example, Schreckenberg and Luttrell (2009) examining the role of forests in supporting livelihoods in multi-land-use landscapes of Sub-Saharan Africa showed that residents who engaged in NTFP harvesting tend to have a lower rate of deforestation compared to areas solely reliant on traditional agriculture or logging. As a consequence, NTFP collection can help create buffer zones around protected areas, reducing encroachment and poaching while helping to maintain ecological connectivity. Thus, multifunctional landscapes such as forests managed for timber and NTFPs retain more biodiversity and resilience than forest plantations or forests managed solely for industrial timber. They also protect ecosystem services such as hydrological functions and soil retention and act as a buffer against climate variability (Shanley et al. 2015).

Additionally, NTFP-bearing trees and plants in different landscapes play a crucial role in carbon sequestration. Trees, such as *Parkia biglobosa* (locust beans) and *Vitellaria paradoxa* (shea-butter trees) store substantial amounts of carbon in their biomass and canopies. By conserving and sustainably managing these NTFP-rich resources, the carbon storage potentials are further enhanced and contribute to global efforts to combat climate change. Besides, agroforestry practices, which involve planting NTFP-bearing trees alongside traditional crops, not only diversify income sources for local communities but also enhance soil fertility and sequester more carbon than conventional agriculture. This approach reduces the pressure on primary forests for agriculture and promotes sustainable land use practices.

Non-Timber Forest Products hold immense promise as tools for enhancing the environmental services of multifunctional landscapes. By promoting local livelihoods, biodiversity conservation, carbon sequestration, and landscape beauty, NTFPs offer a multifaceted approach to sustainable landscape management. Among opportunities offered by NTFPs to contribute to sustainable multifunctional landscapes are: the production of edible insects and mushrooms as well as the enhancement of rural livelihoods through use and trade of diverse functional categories of NTFPs.

However, several challenges associated with NTFPs collection including inadequate data for decision making, unsustainable harvesting, land use conflicts, lack of regulation, poor market access, and the problem of invasive species, among others must be addressed. Shackleton and Pandey (2014) suggested eight steps to facilitate the integration of NTFPs into the development agenda, for the benefit of local communities namely: (1) proper inventory of NTFP stocks, (2) research on NTFP ecology and sustainable harvest levels, (3) introduction of extension services for NTFPs, (4) inclusion of NTFPs in land management and trade-off decisions (Specifically measure links between commercialization and conservation and cultivation and conservation), (5) integration of NTFPs into sectoral policies, (6) ensuring NTFP commercialization is not at the expense of local livelihood needs, (7) promoting the security of access and use, and (8) examination of local contextual drivers of unsustainable use.

Similarly, the key lessons associated with the sustainable harvesting of NTFPs must be harnessed. These include:

1. **Community Involvement**

   Engaging local communities in NTFP management is essential. Their traditional knowledge and participation can lead to sustainable practices that benefit both people and the environment.
2. **Sustainable Harvesting**

   Implementing clear guidelines and regulations for NTFP harvesting is crucial. Sustainable harvesting practices, such as rotational harvesting and selective collection, can help prevent resource depletion.

3. **Market Development**
   Developing markets for NTFPs can enhance local economies and incentivize sustainable management. Value addition and market diversification can increase the income potential of NTFPs.
4. **Ecosystem-Based Approaches**
   Managing NTFPs within an ecosystem-based framework will promote the conservation of biodiversity and enhances the overall resilience of landscapes.
5. **Integration with Conservation**
   NTFP management should be integrated into broader conservation strategies, aligning with international agreements such as the Convention on Biological Diversity and the Sustainable Development Goals.

In light of the above, and given the increasing environmental challenges faced by humanity in contemporary times, recognizing the value of NTFPs in maintaining the delicate balance between human needs and environmental conservation will help achieve a more harmonious coexistence with nature while reaping the benefits of these diverse and valuable resources.

8.8 Chapter Highlights

The chapter explored the multifaceted role of NTFPs in Africa vis-à-vis sustainable livelihoods and forest conservation. Specifically, the chapter looked at:

- The value and significance of NTFPs in promoting sustainable livelihoods and forest conservation in Africa;
- NTFPs profound impact on the rural economy, including their economic benefits and the importance of indigenous knowledge and sustainable harvesting practices, which have enhanced the management of forest and tree-based ecosystems within the African landscape;
- Ecological management and potentials of edible insects and mushrooms, highlighting their potential to alleviate food insecurity and reduce the environmental impact of food production. These NTFPs are recognized for their ecological management potential within Africa's multifunctional landscapes;
- Multifunctional landscape approach, illustrating how NTFPs can provide a range of ecosystem services, including food, energy, medicine, income, fibers, and building materials. This approach also emphasizes the role of agroecological practices in creating landscapes that support both NTFP diversity and agricultural production.
- The opportunities, challenges, and key lessons learned from the utilization of NTFPs.

References

Adams AM, Abudulai I, Bashiru M (2016) The shea industry and rural livelihoods among women in the Wa municipality, Ghana. J Soc Sci Stud 3(2):40–56

Adesokan AA, Akanji MA, Yakub MT (2007) Antibacterial potentials of aqueous extract of *Enantia chlorantha* stem bark. Afr J Biotechnol 6(22):2502–2505. https://doi.org/10.5897/AJB2007.000-2397

African Forest Forum (AFF) & Food and Agriculture Organization of the United Nations (FAO) (2020) Value chain analysis of the non-timber forest products of the African Great Green Wall. Regional report covering Burkina Faso, Chad, Mali, Niger, Nigeria and Senegal, 47p

Agbidye FS, Ofuya T, Akindele SO (2009) Some edible insect species consumed by the people of Benue State, Nigeria. Pak J Nutr 8(7):946–950

Ahenkan A, Boon E (2011) Non-Timber Forest Products (NTFPs): Clearing the Confusion in Semantics. J Hum Ecol 33:1–9. https://doi.org/10.1080/09709274.2011.11906342

Ajaiyeoba EO, Fadare DA (2006) Antimicrobial potential of extracts and fractions of the African walnut–*Tetracarpidium conophorum*. Afr J Biotechnol 5(22):2322–2325

Amusa TO (2015) Stem bark harvesting and conservation status of *Annickia chlorantha* (Oliv.) Setten & Maas in Omo and Shasha Forest Reserves, South-western Nigeria. In Proceedings of the 5th annual biodiversity conference of Nigeria. Nigeria Tropical Biology Association (NTBA), Amusa TO, Babalola FD (eds), pp 6–13

Amusa TO, Abdul AM (2021) Shea butter production and marketing efficiency in Kwara State, Nigeria. Ilorin J Market 7(1):31–45

Amusa TO, Jimoh SO, Aridanzi P, Haruna M (2010) Ethnobotany and Conservation of Plant Resources of Kainji Lake National Park. Nigeria Ethnobot Res Appl 8(1):181–194

Amusa TO, Jimoh SO, Azeez IO (2017) Socio-economic factors influencing marketing of non-timber forest products in tropical lowland rainforests of south-western Nigeria. South For J For Sci 79(2):161–168

Anchang KY (2014) Current developments in mushroom biotechnology in Sub-Saharan Africa. World Soc Mushroom Biol Mushroom Prod (WSMBMP) Bull 11:4–13

Ariyo OO (2023) Edible Mushrooms: their Impact on Food Security. FUOYE J Pure Appl Sci (FJPAS) 8(2):57–68

Avana-Tientcheu ML, Solefack MCM, Kamga YB, Nguetsop VF (2018) Vulnerability assessment of *Gnidia glauca* (Thymelaeaceae) exploitation, traditional uses and domestication potential in the community forest of Kilum-Ijim, North Western Cameroon. J Ecol Nat Environ 10(8):182–191

Awono A, Ingram V, Schure J, Levang P (2013) Guide for small and medium enterprises in the sustainable non-timber forest product trade in Central Africa. CIFOR

Ayemele AG, Muafor FJ, Levang P (2016) Indigenous management of palm weevil grubs (*Rhynchophorus phoenicis*) for rural livelihoods in Cameroon. J Insects Food Feed 3(1):43–50. https://doi.org/10.3920/JIFF2016.0002

Bastos C, Liberal Â, Moldão M, Catarino L, Barros L (2023) Ethnomycological prospect of wild edible and medicinal mushrooms from Central and Southern Africa—a review. Food Front 4(2):549–575. https://doi.org/10.1002/fft2.215

Burkill HM (1985) The useful plants of west tropical Africa. Royal Botanical, Garden, Kew

Bystriakova N, Kapos V (2006) Bamboo diversity: the need for a Red List review. Biodiversity 6(4):12–16

Cardon D (2007) Natural dyes: Sources, tradition, technology and science. Stud Conserv 52(4):318–319. https://doi.org/10.1179/sic.2007.52.4.318Publications

Chen T (2017) The impact of the shea nut industry on women's empowerment in Burkina Faso: a multi-dimensional study focusing on the Central, Central-West and Hauts-Bassins regions. FAO- Social protection and forestry working paper 3, FAO, Rome, 60p. https://openknowledge.fao.org/server/api/core/bitstreams/4c577507-9ba7-4316-aa81-9064a62d0b1b/content

Chirwa PW, Syampungani S, Geldenhuys CJ (2008) The ecology and management of the Miombo woodlands for sustainable livelihoods in southern Africa: the case for non-timber forest products. South For J For Sci 70(3):237–245. https://doi.org/10.2989/SF.2008.70.3.7.668

Cruz M (2002) The impact of plant product harvesting on Derre miombo woodlands, Mozambique. Doctoral dissertation, MSc thesis. Department of Conservation Ecology, University of Stellenbosch, Stellenbosch)

Cunningham AB (2001) Applied ethnobotany: people, wild plant use and conservation. Earthscan Publications, London

Delgado TS, McCall MK, López-Binnqüist C (2023) Non-Timber forest products: small matters, big significance, and the complexity of reaching a workable definition for sustainability. Small-scale For 22:37–68. https://doi.org/10.1007/s11842-022-09517-9

Derebe B, Alemu A (2023) Non-timber forest product types and its income contribution to rural households in the Horn of Africa: a systematic review. For Sci Technol 19(3):210–220

Dounias E, Selzner A, Koizumi M, Levang P (2007) From sago to rice, from forest to town: The consequences of sedentarization for the nutritional ecology of Punan former hunter-gatherers of Borneo. Food Nutr Bull 28(2_suppl2):S294–S302

Eba'a Atyi R, Ngouhouo PJ, Mvondo AJP, Ngoungoure MA, Sufo KR (2016) Economic and social importance of fuelwood in Cameroon. Int For Rev 18(1):52–65

Ene-Obong H, Igile G, Ekpo A, Egbung E, Agbo M (2019) Variations in the nutrients and bioactive compounds of different accessions of the West African pear (*Dacryodes edulis*): Implications for dietary intake assessment and health. J Food Compos Anal 79:80–86

Enitan S, Olley M, Uhunmwangho SE, Akele YR, Ehiaghe AF, Enitan CB (2014) Antibacterial activity of methanolic leaf extract of *Plukenetia conophora* Mull. arg. against selected bacteria isolated from urinary tract infection. Intl J Microbiol Appl 1(1):1–10

Gaoue OG, Ticktin T (2009) Fulani knowledge of the ecological impacts of *Khaya senegalensis* (Meliaceae) foliage harvest in Benin and its implications for sustainable harvest. Econ Bot 63:256–270

Guedje NM, Zuidema PA, During H, Foahom B, Lejoly J (2007) Tree bark as a non-timber forest product: The effect of bark collection on population structure and dynamics of Garcinia lucida Vesque. For Ecol Manag 240(1–3):1–12

Harbi J, Erbaugh JT, Sidiq M, Haasler B, Nurrochmat DR (2018) Making a bridge between livelihoods and forest conservation: Lessons from non-timber forest products' utilization in South Sumatera, Indonesia. Forest Policy Econ 94:1–10

Hlongwane, Zabentungwa T, Rob S, Thinandavha C, Munyai (2021) The role of edible insects in rural livelihoods, and identified challenges in Vhembe District, Limpopo, South Africa. Resources 10(12):123. https://doi.org/10.3390/resources10120123

Ibrahim H, Haruna A, Abdullahi N (2021) Proximate and elemental analysis of african star Apple (*Chrysophyllum albidum*). J Appl Sci Environ Manag 25(2):253–256

Ickowitz A, Powell B, Salim MA, Sunderland TC (2014) Dietary quality and tree cover in Africa. Glob Environ Chang 24:287–294. https://doi.org/10.1016/j.forpol.2018.05.011

Iponga DM, Mikolo-Yobo C, Lescuyer G et al (2018) The contribution of NTFP-gathering to rural people's livelihoods around two timber concessions in Gabon. Agrofor Syst 92:157–168. https://doi.org/10.1007/s10457-016-0022-0

Johns T, Sapitht BR (2014) Biocultural diversity in the sustainability of developing country food systems. Food Nutr Bull 25(2):143–155

Laird SA, Wynberg R, Mclain RJ (2010) Wild product governance: finding policies that work for non-timber forest products. Earthscan, London

Leakey RR, Avana-Tientcheu ML, Awazi NP, Assogbadjo AE, Mabhaudhi T, Hendre PS, …, Manda L (2022) The future of food: Domestication and commercialization of indigenous food crops in Africa over the third decade (2012–2021). Sustain For 14(4): 2355

Lebedys A, Li T (2014) Contribution of the forest sector to national economies, 1999–2011. Forest Finance working paper FSFM/ACC/09. FAO, Rome

Leßmeister A, Katja H, Lykke AM, Thiombiano A, Wittig R, Hahn K (2018) Contribution of non-timber forest products (NTFPs) to rural household revenues in two villages in south-eastern Burkina Faso. Agrofor Syst 92:139–155

Macheka L, Manditsera FA, Ngadze RT, Mubaiwa J, Nyarugwe S, Bangira C, Kembo G (2022) Agro-ecological distribution and consumption of wild harvested edible insects, fruits, and vegetables in rural Zimbabwe. Future Foods 6:100187

Makungwa SD, Epulani F, Woodhouse IH (2013) Fuelwood supply: A missed essential component in a food security equation. J Food Secur 1(2):49–51

Marshall E, Schreckenberg K, Newton A (2006) Commercialization of non-timber forest products: factors influencing success: lessons learned from Mexico and Bolivia and policy implications for decision-makers. UNEP/World Conservation Monitoring Centre, Cambridg

Melese SM (2016) Importance of non-timber forest production in sustainable forest management, and its implication on carbon storage and biodiversity conservation in Ethiopia. Intl J Biodiver Conser 8(11):269–277

Mendako RK, Tian G, Ullah S, Sagali HL, Kipute DD (2022) Assessing the Economic Contribution of Forest Use to Rural Livelihoods in the Rubi-Tele Hunting Domain, DR Congo. Forests 13:130. https://doi.org/10.3390/f13010130

Mensah KE, Damnyag L, Kwabena NS (2020) Analysis of charcoal production with recent developments in Sub-Sahara Africa: a review. Afr Geograph Rev 41(1):35–55. https://doi.org/10.1080/19376812.2020.1846133

Momo SMC, Nguetsop VF, Dongmo N, Kenguem Kinjouo Ghislain J, Avana-Tientcheu ML (2016) The effects of some ecological and anthropic factors on the survival and bark recovery rate of *Prunus africana* (Rosaceae) in South-West Cameroon. Int J Curr Res Biosci Plant Biol 3:25–31

Nde DB, Boldor D, Astete C, Muley P, Xu Z (2016) Oil extraction from Sheanut (*Vitellaria Paradoxa* Gaertn CF) Kernels assisted by microwaves. J Food Sci Technol 53(3):1424–1434

Neji AP, Agwupuye A (2019) Proximate composition, mineral content, phytochemical screening and anti-nutritional constituents of walnut (*Tetracarpidium conophorum* OR *Plukenetia conophora*) seeds. Intl J Scient Res Publ (IJSRP) 9(7):91128

Newton P, Miller DC, Byenkya MAA, Agrawal A (2016) Who are forest-dependent people? A taxonomy to aid livelihood and land use decision-making in forested regions. Land Use Policy 57:388–395

Njouonkou AL, De Crop E, Mbenmoun AM, Kinge TR, Biyé EH, Verbeken A (2016) Diversity of edible and medicinal mushrooms used in the Noun Division of the West Region of Cameroon. Intl J Med Mushrooms 18(5):387–396. https://doi.org/10.1615/intjmedmushrooms.v18.i5.20

Nyamoga GZ (2023) Edible Miombo wild mushrooms and its potential for improving livelihoods of People in Songea and Tabora Districts. Tanzania Afr J Appl Res 9(1):20–37

Nyamoga GZ, Solberg B (2019) A review of studies related to charcoal production, consumption, and greenhouse gas emissions in Tanzania. In: Bamutaze Y, Kyamanywa S, Singh B, Nabanoga G, Lal R (eds) Agriculture and ecosystem resilience in sub Saharan Africa. Climate change management. Springer, Cham, pp 357–399. https://doi.org/10.1007/978-3-030-12974-3_17

Okhuoya J, Akpaja E, Omorefosa O, Abbot O, Ihayere C (2010) Nigerian mushrooms: underutilized non-wood forest resources. J Appl Sci Environ Manag 14(1):43–54. https://doi.org/10.4314/jasem.v14i1.56488

Okwute SK (1992) Plant derived pesticidal and antimicrobial agents for use in agriculture. A review of phytochemical and biological studies on some Nigerian plants. J Agric Sci Technol 2(1):62–70

Oyetayo OV (2011) Medicinal uses of mushrooms in Nigeria: towards full and sustainable exploitation. Afr J Tradit Complement Altern Med 8(3):275–283. https://doi.org/10.4314/ajtcam.v8i3.65289

Rasolofoson RA, Hanauer MM, Pappinen A, Fisher B, Ricketts TH (2018) Impacts of forests on children's diet in rural areas across 27 developing countries. Sci Adv 4(8):eaat2853

Regato P, Berrahmouni N (2005) Using non-timber forest products for restoring environmental, social, and economic functions. In: Forest restoration in landscapes. Springer, New York. https://doi.org/10.1007/0-387-29112-1_31

Ruiz-Mallén I, Schunk C, Corbera E, Rös M, Reyes-García V (2015) Meanings, drivers, and motivations for community-based conservation in Latin America. Ecol Soc 20(3), JSTOR. http://www.jstor.org/stable/26270245

Saad S, Bugaje IM (2016) Biomass consumption in Nigeria: trends and policy issues. J Agricult Sustain 9(2):127–157

Sardeshpande M, Shackleton C (2019) Wild edible fruits: a systematic review of an under-researched multifunctional NTFP (Non-timber forest product). Forests 10:467. https://doi.org/10.3390/f10060467

Schreckenberg K, Luttrell C (2009) Participatory forest management: a route to poverty reduction? Int For Rev 11(2):221–238

Shackleton CM, Pandey AK (2014) Positioning non-timber forest products on the development agenda. Forest Policy Econ 38:1–7

Shackleton CM, McGarry D, Fourie S, Gambiza J, Shackleton SE, Fabricius C (2007) Assessing the effects of invasive alien species on rural livelihoods: case examples and a framework from South Africa. Hum Ecol 35:113–127

Shackleton S, Delang CO, Angelsen A (2011) From subsistence to safety nets and cash income: exploring the diverse values of non-timber forest products for livelihoods and poverty alleviation. In: Shackleton S, Shackleton C, Shanley P (eds) Non-timber forest products in the global context, Tropical forestry, vol 7. Springer, Berlin/Heidelberg, pp 55–81. https://doi.org/10.1007/978-3-642-17983-9_3

Shackleton CM, Ticktin T, Cunningham AB (2018) Non-timber forest products as ecological and biocultural keystone species. Ecol Soc 23(4):22

Shanley P, Pierce AR, Laird SA, Binnqüist CL, Guariguata MR (2015) From lifelines to livelihoods: non-timber forest products into the twenty-first century. In: Pancel L, Köhl M (eds) Tropical forestry handbook. Springer, Berlin/Heidelberg, pp 1–50. https://doi.org/10.1007/978-3-642-41554-8_209-1

Sills E, Shanley P, Paumgarten F, de Beer J, Pierce A (2011) Evolving perspectives on non-timber forest products. In: Shackleton S, Shackleton C, Shanley P (eds) Non-timber forest products in the global context, Tropical Forestry, vol, vol 7. Springer, Berlin Heidelberg, pp 23–51

Stefanello FS, Cavalheiro CV, Dugatto JS, Kubota EH, Silva MS (2012) Avaliação da ativadade antioxidante in vitro do cogumelo shimeji (*Pleutorus ostreatus*). Simpósio de Segurança Alimentar 4, 29–31 May 2012

Suleiman MS, Wasonga VO, Mbau JS, Suleiman A, Elhadi YA (2017) Non-timber forest products and their contribution to households income around Falgore Game Reserve in Kano, Nigeria. Ecol Process 6:1–14

Sumathykutty MA, Rao JM, Padmakumari KP, Narayanan CS (1999) Essential oil constituents of some Piper species. Flavour Fragr J 14(5):279–282

Sunderland TC, Ndoye O, Harrison-Sanchez S (2011) Non-timber forest products and conservation: what prospects? In: Shackleton S, Shackleton C, Shanley P (eds) Non-timber forest products in the global context, Tropical Forestry, vol 7. Springer, Berlin/Heidelberg, pp 209–224

Sunderlin WD, Angelsen A, Belcher B, Burgers P, Nasi R, Santoso L, Wunder S (2005) Livelihoods, forests, and conservation in developing countries: an overview. World Dev 33(9):1383–1402

Ticktin T (2015) The ecological sustainability of NTFP harvest: principles and methods. In: Shackleton CM, Pandey AK, Ticktin T (eds) Ecological sustainability for non-timbe forest products: dynamics and case studies of harvesting. People and Plant International Conservation, Routelage, pp 31–53

Tweneboah-Kodua T, Ankama J, Addae M (2018) Assessing the profitability of small scale local shea butter processing: Empirical evidence from Kaleo in the Upper West region of Ghana. Cogent Food Agricult 4(1):1453318

USAID (2010) Investing in shea in West Africa: a US investor's perspective. West Africa Trade Hub technical report, Accra. http://pdf.usaid.gov/pdf_docs/Pnadu686.pdf

Van den Wyngaert IJ, Wienk LD, Sollie D, Bobbink R, Verhoeven JT (2003) Long-term effects of yearly grazing by moulting Greylag geese (*Anser anser*) on reed (*Phragmites australis*) growth and nutrient dynamics. Aquat Bot 75:229–248

Van Itterbeeck J, van Huis A (2012) Environmental manipulation for edible insect procurement: a historical perspective. J Ethnobiol Ethnomed 8:1–7

Waiganjo MW, Ngeli P, Gateri MW, Muriuki AW (2009) Cultivation and commercialization of edible mushrooms in Kenya: A review of prospects and challenges for smallholder production. Intl Symp Underutil Plants Food Sec, Nutrition, Income and Sustainable Development, pp 473–480

World Bank (2002) A revised forest strategy for World Bank Group. World Bank, Washington, DC

Chapter 9
Commercialization Potential of Africa's Indigenous Fruit Trees (IFTs) and Their Products for Enhancing Nutrition and Rural Livelihoods

Gillian Kabwe, Achille E. Assogbadjo, Festus K. Akinnifesi, Fifanou G. Vodouhe, Paxie W. Chirwa, and Tajudeen O. Amusa

Abstract Indigenous fruits have the potential to address food security in Africa through the provision of nutrition and maintenance of sustainable livelihoods. Fruit trees are not only an important food source but also provide valuable vitamins, fats, sugar, protein and fibres that human bodies require. Whereas the contribution of Indigenous Fruit Trees (IFTs) to household nutrition and rural livelihoods is evident, their potential for commercialization needs to be enhanced by improving the standards of value-added products for the international markets. The few IFT products that have already penetrated the international markets have shown potential for more growth owing to their nutritional, adaptative and market values. Commercialization can further be improved through domestication of important IFTs and the use of eco-labelling through the fair-trade. This chapter shows that

G. Kabwe (✉)
School of Natural Resources, Plant and Environmental Sciences Department, Copperbelt University, Kitwe, Zambia
e-mail: gillian.kabwe@cbu.ac.zm

A. E. Assogbadjo
NTFPs and Orphan Crops Unit/Lab of Applied Ecology, Faculty of Agronomic Sciences, University of Abomey-Calavi, Cotonou, Benin

F. K. Akinnifesi
Food and Agriculture Organization of the United Nations (FAO), Rome, Italy

F. G. Vodouhe
Laboratory of Analysis and Research on Economic and Social Dynamics (LARDES), Faculty of Agronomy, University of Parakou, Parakou, Benin

P. W. Chirwa
Department of Plant and Soil Sciences, Faculty of Natural and Agricultural Sciences, University of Pretoria, Pretoria, South Africa

T. O. Amusa
Department of Forest Resources Management, Faculty of Agriculture, University of Ilorin, Ilorin, Nigeria

P. W. Chirwa et al. (eds.), *Trees in a Sub-Saharan Multi-functional Landscape*,
https://doi.org/10.1007/978-3-031-69812-5_9

very few indigenous fruits and fruit tree products have been developed as commercial products in the African multi-functional landscapes. Prominently, there are five fruit tree species whose products have reached the international market and are already contributing to local community and national incomes, namely: *Adansonia digitata, Allanblackia spp, Sclerocarya birea, Vitellaria paradoxa, Ziphiphus mauritiana* and *Irvingia gabonensis*. The commercialization process of indigenous fruits and fruit products is hindered by various factors including the limited shelf life of these fruits and their seasonality and perishability, low domestication levels, low awareness and demand, inadequate research and development, policy and regulatory challenges, land use and agricultural practices, limited access to finance and investment. There is also limited development of the IFT value chain in Africa—few enterprises with no formal markets. Many IFT products are, therefore, unable to make it to the international market because of poor quality standards, low value addition, poor infrastructure and investment in market research and development.

Keywords Indigenous fruit trees · Commercialization potential · Domestication · Value addition · Nutrition · Rural livelihoods

9.1 Introduction

Africa's annual food import bill has more than tripled in the last few decades, and was estimated to have reached US$35 billion in 2015 (Woldemichael et al. 2017). Much of this imported food could be produced locally, creating much needed jobs and incomes for youth on the continent and smallholder farmers. The potential for commercializing Africa's indigenous fruit trees (IFTs) has been overlooked compared to other continents in the global south. The global fresh fruits market size was estimated at USD 551 billion in 2021 and is expected to reach USD 569 billion in 2022 according to Grandview Research report,[1] but only few countries in Africa participate in revenue from fruit export.

For over a millennium, farmers have relied on indigenous fruit trees (IFTs) for their food security, nutrition, social, cultural, economic and environmental benefits. These indigenous fruit trees possess a remarkable potential to transform the livelihoods of rural community dwellers across Africa where most of the population depends on agriculture for their subsistence (Akinnifesi et al. 2008a, b; Leakey et al. 2022). However, agricultural expansion and the quest to feed a growing global population, have created a disincentive to developing, cultivating and commercializing these indigenous species (Leakey et al. 2022). As a result, these nutritious and culturally valued, and commercially viable species were generally cleared, underdeveloped and neglected to make way for a few commodities and staple food crops. Despite this, many indigenous fruit trees have continued to grow in domesticated or semi-domesticated conditions on farms, homesteads and in the wild. They are

[1] https://www.grandviewresearch.com/industry-analysis/fresh-fruits-market-report#

known to bear edible fruits, nuts, seeds, and leaves that can either be eaten raw, cooked or processed into other products.

In Africa, the cultivation and commercialization of IFTs offer a promising avenue to improve nutrition, market diversification and export potential, cultural preservation and food sovereignty, value addition and agribusiness development, biodiversity conservation, climate resilience and at the same time enhance livelihood security and promote sustainable land use practices. As the continent grapples with the challenges of climate change (Eshetu et al. 2021), land degradation (Ziadat et al. 2022), and food insecurity (Bjornlund et al. 2022), harnessing the potential of commercial products from IFTs becomes therefore increasingly urgent. Indeed, the diverse agroecological zones across the continent provide a conducive environment for the cultivation of a wide variety of IFTs.

Over three decades of research and development across the continent, involving studies on ethnobotany, domestication, economic and livelihood assessment provide evidence that commercialization of IFTs and value-added products, offers opportunity for generating income, and improving food security and nutrition among community dwellers in Africa (Akinnifesi et al. 2008a, b; Leakey et al. 2022; Omotayo and Aremu 2020; Okullo et al. 2022).

This chapter reviews the commercial and nutritional potential of Africa's indigenous fruit trees and their products towards enhancing nutrition and rural livelihoods on the continent. It examines the nutritional and economic benefits offered by fruit tree cultivation, shedding light on the role of agribusiness and value chain development in transforming their subsistence cultivation into profitable enterprises. Drawing on empirical evidence and case studies, the chapter demonstrates how the strategic integration of fruit trees into farming systems can lead to income diversification, enhanced food security, and improved resilience to the impacts of climate change.

Additionally, it considers the importance of community-based approaches, partnerships with non-governmental organizations, and market linkages to ensure that the benefits of fruit tree commercialization reach those who need them most.

9.2 Indigenous Fruit Trees (IFTs) and Their Products in Africa

Africa has many fruit tree species, with sub-Saharan Africa alone reported to possess about 1200 edible fruit tree species (Awodoyin et al. 2015). This diversity is a great asset and a source of wealth for the subsistence of local populations in all tropical regions. Indigenous fruit trees (IFTs) are in abundance as wild in the forests, woodlands, savannah and agro-silvo-pastoralism systems of Africa and play very significant role in food, nutrition, income and livelihood security of millions of people, especially the poor (Sileshi et al. 2023). The fruits are rich in essential nutrients and antioxidants, enabling consumers to maintain a balanced diet (Lapeña et al.

2014). Indeed, they contribute to the daily diet of the people as they are good sources of vitamins and minerals. The fruits are consumed mostly as snacks, and in times of food shortage become the principal source of food, thereby fulfilling a much more substantial role in the daily diet of rural people (Akinnifesi et al. 2006). Snacks are important for children who need to eat more frequently than adults, and wild fruits and nuts are important sources of the micronutrients that are often deficient in cereal-based diets that most African families depend on (Akinnifesi et al. 2006). Community dwellers in African countries often depend on indigenous fruits to increase their cash income for purchasing agricultural inputs, buy food to supplement subsistence farm production or pay school fees (Akinnifesi et al. 2008b; McMullin et al. 2019. They are traded at local and urban markets and do yield substantial household incomes.

Through different projects, initiatives, research and approaches, domestication of indigenous fruit trees in different regions of Africa in the last three decades have focused on top priority species that have potential for market, nutrition, and food security (Leakey et al. 2017, Leakey et al. 2022). These include documentation work in West, Southern and Eastern Africa and the Sahel. The general documentation strategies and approaches have been described elsewhere (Leakey and Akinnifesi 2017, Akinnifesi et al. 2008a, b). However, studies on the potential for commercializing IFTs of African origin are limited, even though a few of them have market penetration beyond the locality or national borders.

9.3 Commercial Potential of Africa IFTs and Their Products

Indigenous fruit tree species play a crucial role in poverty alleviation in Africa (Nkosi et al. 2020). The trends in food markets have changed significantly in the last decades, with a growing interest toward indigenous, natural, wild-harvested, organically produced or sourced products, and fair trade. The demand for natural and organic products is increasing the market potential of African indigenous fruits and their tree products, which are still mostly grown in the wild and semi-wild conditions, without the use of fertilizers or chemicals. They are environmentally friendly and almost zero cost species. Indeed, indigenous fruit tree products rarely use fertilizers or chemicals because they mostly come from the wild; and can be considered as organic products with prospect of gaining market share and entrance into the multi-billion dollar growing global niche markets of natural product, especially when coupled with fair trade principles.

The market for most African indigenous fruit trees and products is still largely domestic. This is further characterized by underdeveloped products and short value chain for many of them. Nevertheless, few indigenous fruit trees have been able to penetrate global and continental export markets, notable among which are shea butter (*Vitellaria paradoxa*), amarula liquor and super oils from marula (e.g. maruline), baobab (*Adansonia digitata*) oil and Allanblackia (*Allanblackia* spp.) oil, that are traded and well sought internationally. Table 9.1 shows market potential of some indigenous trees producing marketable fruits, nuts, and seeds in Africa.

The section below further describes a few indigenous fruit tree products that are traded regionally and/or internationally with substantial potential for expanded markets.

*9.3.1 Shea Butter Tree (*Vitellaria paradoxa*)*

The global demand for shea butter, *Vitellaria paradoxa* (Shea butter or karite tree), is growing, and it is expected to exceed USD 3.5 billion by 2028 (Nanyang Technological University (NTU) 2023), and a quarter of the trade of shea for vegetable fat and oil is in Europe. Africa produces around 1.76 million tonnes of raw shea nuts a year, and more than half is used in domestic applications (Abdul-Mumeen et al. 2018). The growing zone of Shea butter tree is known as the "Shea belt," which comprises about 21 countries, including: Benin, Burkina Faso, Cameroon, Central African Republic, Chad, Ethiopia, Eritrea, Ghana, Guinea-Bissau, Côte d'Ivoire, Mali, Niger, Nigeria, Senegal, Sierra Leone, South Sudan, Sudan, Togo, Uganda, the Democratic Republic of the Congo, Kenya and Guinea. Of these, the largest shea butter-producing countries are Ghana, Nigeria and northern Uganda.

The shea butter or *karite* tree is the second most important oil crop in Africa after the oil palm tree. Shea butter is used extensively in the food, pharmaceutical, cosmetic industries and often as cocoa butter substitute by chocolate manufacturers and for margarine and baking purposes. According to Abdul-Mumeen et al. (2018), Nigeria and Mali are the largest producers of shea kernels in Africa. Nigeria accounted for more than 39 percent, while Mali accounted for almost 31 percent of global shea nuts production. Most of the shea butter produced in Nigeria is for domestic consumption. There is also undocumented trade of shea nuts across the country's borders with Benin, Togo and Ghana.

Bup et al. (2014) estimated that about 500 million shea trees grow in 21 countries in Sub-Saharan Africa with the potential of producing shea nuts worth about USD 150 million USD annually. The high production zone has a production potential of 70,000–300,000 tons annually. The intermediate potential zone produces about 10,000–70,000 tons per annum, while the low potential zone produces less than 10,000 metric tons per year (Bup et al. 2014). According to the Global Shea Alliance, Africa produces 600,000–900,000 metric tonnes of shea nuts annually. In 2020, it is reported that Nigeria produced 450,000 metric tonnes of shea nuts, with an estimated value of over USD 300 million, of which it exported USD 50 million (Emmanuel 2022). In 2021, it was estimated that 42,500 metric tonnes of shea nuts were exported, which is about 10% of the 425,000 metric tonnes produced.

In general, from year to year, most of the shea nuts in Nigeria are exported largely to neighbouring African countries, with only about 20% of the products consumed domestically. Shea nut was listed by the Nigeria Export Promotion Council (NEPC)

Table 9.1 Market potential of indigenous trees producing marketable fruits, nuts and seeds in Africa

Tree species	Domestic use	Commercial potential/ industrial use/ type of market	Total market volume (per annum)	Level of market/ industrial development	Top 2–5 countries with largest production/ export potential	References
Vitellaria paradoxa (Shea butter tree)	Edible cooking oil;	Local market for edible seeds, and seed oil. Lucrative global export market in cosmetics, pharmaceutics and food industries. For instance, its seed oil is used for making margarine, body lotion, ointments, chocolate formulations, etc.	Potential demand to exceed USD 3.5 billion by 2028. Africa produces around 1.76 million tonnes of raw shea nuts a year, and more than half is used in domestic applications	Well-developed market; local, regional and international market. Emerging infrastructure and industry in Africa	Nigeria, Burkina Faso, Mali, and Ghana.	Abdul-Mumeen et al. (2018)
Allanblackia spp.	Edible cooking oil and soap making;	Lucrative global export market for its oil for making butter, skin care, ointment, soap; potential alternative to diesel oil	The potential market for oil is more than 100,000 tons annually.	The scale of production and marker is relatively small, and is still under development. Unilever is the major buyer.	Ghana, Nigeria and Tanzania	Jamnadass et al. (2010) and Kyereh et al. (2021)
Adansonia digitata (Baobab)	Juice,	Cosmetics, skin care,	The global market for Baobab powder has been estimated at USD8.2 billion in 2022, and is projected to reach USD12.1 billion by 2030. Another study puts it at Baobab market value of baobab has been estimated at USD 2.1 billion in 2022.	Baobab market is huge, and it is exported to Europe, United States, the Americas and Asia	Harvested throughout Africa	Research and markets Market research future https://www.researchandmarkets.com/report/baobab-powder? More (2024)

Sclerocarya birrea (Marula)	Oil, powder, pulp, alcoholic beverages, beer, wine, juice, and jam	The most important export trade is Amarula cream liqueur, with other products, e.g. Amarula African gin, Marula oil, Marula vegan, and many other products.	Underexplored and unquantified. Amarula cream liqueur brand alone averages one million 9 litre cases per year from 2009 to 2019.	Scale is limited, but potential is huge.	Mostly South Africa and pockets of cases in other southern African countries.	Ham et al. (2008)
Ziphiphus mauritiana (jujube, masau)	Fruit, juice,	Pharmaceutical oil			Southern Africa	Pandey et al. (2023)
Irvingia gabonensis (African bush mango)	Fruit, kernel	The most important export trade is the kernel	The value of the exports of the kernels of bush mango from Cameroon to neighbouring countries including Nigeria, Gabon, Equatorial Guinea, and Central African Republic is estimated at US$260,000 per annum	Regional	Cameroon	Tchoundjeu et al. (2005)

as the 13th most valuable commodity in terms of its potential foreign exchange earnings for the country (Emmanuel 2022). Bup et al. (2014) indicated that shea butter from *V. paradoxa* represents the fourth most important export crop in Burkina Faso, only after gold, cotton, and livestock. It has the potential to contribute about USD six million annually to the national economy of Burkina Faso. In 2020, Ghana, Benin and Burkina Faso accounted for 90% of the over 47,000 tonnes of shea kernels imported by India from West Africa, an increase from about 34,000 tonnes imported in 2018 (Nanyang Technological University 2023). Shea nuts and its products are listed among the top ten Non-traditional Exports of Ghana.

The processing and sales of shea nuts represent significant income earning opportunities for rural women who are the main stakeholders in the production chain (Bup et al. 2014). Most of the shea produced in Africa are consumed locally as food condiments. The extracted oil from its kernels is marketed internationally, continentally and domestically, for its much-sought value for cosmetics such as lotion, pharmaceutics and in chocolate formulations. Most of the processed shea butter (85%) goes to the food industry, while the rest is used in the personal care industry for diverse cosmetics applications and formulas, such as hair treatment, anti-ageing and anti-wrinkle creams, face and body-moisturizing creams and lotions, skin treatment after sunburn; stretch mark prevention products during pregnancy; hair treatment for dry scalp; shaving and aftershave creams to reduce skin irritation; creams to ease muscle fatigue and pain; treatment products for insect bites and stings; baby care products against diaper rash.

Shea trees from the West and Eastern Africa produce slightly different shea butter in terms of consistency, texture and nutrient content, and are suitable for different uses in the cosmetics industry. In particular, compared to the East African shea, the West African shea butter is known to be rich in vitamin A; higher melting point, lower concentration of oleic acid; harder in consistency; and higher concentration. On the other hand, the East African shea butter is considered a more luxurious product and scarcer because of a limited supply. It is more yellow in colour, lower melting point, higher concentration of oleic acid, and soft and creamy texture (Lovett 2015).

In 2020, an analysis of the most promising export market potential in Asia and Africa, with consideration to exports from Benin, Burkina Faso, Mali, and Togo, was undertaken and prepared by Partnership for Natural Ingredients (PNI) for Global Shea Alliance (GSA) (Funt et al. 2020). The study included 10 Asian and 10 sub-Saharan African economies selected based on their 2019 gross domestic product. The supply factors considered included relative tariffs on shea kernels and butter, and on finished products; the preferential trade status between beneficiary countries and target export countries, as well as the overall enabling environment of target export countries. On the demand side, the factors considered included export of finished products, growth in exports of finished products, import of shea kernels and butter, import of finished products and import of products similar to shea butter by the target country. The findings showed that the most promising markets for the export of shea products are China, Japan, South Korea in Asia and South Africa, Angola, and Kenya in Africa. Although Nigeria, Ghana and Cote d'Ivoire showed more compelling export markets in Africa, but were excluded from final selection due to their own domestic shea markets (Funt et al. 2020).

As a result of this very lucrative prospect of shea butter in Africa, some large scale domestication and cultivation of shea trees are underway. For instance, as part of the Action for Shea Parklands "ASP" program, the Global Shea Alliance in partnership with the United States Agency for International Development (USAID) launched a massive project. The project, designed to develop a 50-seedling shea nursery, aims to promote and protect parklands, while advancing the industry's commitment to plant ten million trees over the next 10 years across West Africa (CBI 2020). As of 2020, about 8000 shea trees have been planted and about 100,000 seedlings have been raised as part of the Sustainable Shea Initiative, to sustainably expand the shea industry in Ghana, Benin, Côte d'Ivoire, Togo, Mali, Nigeria, and Burkina Faso (USAID 2020).

9.3.2 Allanblackia (Allanblackia Spp.)

Another tree of commercial interest is *Allanblackia* spp. (Clusiaceae family). It is dioecious and a genus of nine tree species indigenous to the humid forests of West, Central and East Africa regions. These nine species include: *A. floribunda, A. gabonensis, A kimbilensis, A. kisonghi, A. marienii, A. parvijlora, A. Stanerana, A. Stuhlmannii, and A. ulugurensis*. Allanblackia has several benefits, but the production of edible oil from its seeds is the economically most important use for household consumption, and with the high potential for improving nutrition and income to rural populations. The dry seeds contain about 67–73% of solid white fat. The seed oil is used for cooking, and the production of margarine. Its use in the manufacturing of ointments and soap, are being developed as a new agri-business in Ghana, Nigeria, Cameroon, and Tanzania (Kyereh et al. 2021). Allanblackia oil is well sought in the global food industry for its low trans-fats.

For generations, local farmers have retained Allanblackia trees on farmlands during forest clearing and have extracted oil from the seeds with similar properties to palm oil for use as edible cooking oil and for soap making. The Unilever PLC estimated the market potential of Allanblackia oil to be more than 100,000 tonnes annually provided that the right quality standards are met (Jamnadass et al. 2010). It has been projected that around 100,000 farmers could earn an average of 200 Euros per annum from planting the Allanblackia by 2020. Through their investments, Unilever PLC has created a stable market for the oil, and projected to expand into a new African industry with a market value exceeding $100 million (Shrestha and Akangaamkum 2008). *Allanblackia* seed oils have favorable physicochemical characteristics with up to 95% combined stearic and oleic acid, which have raised its profile in the food industry, including vegetable-based dairy products, ice cream, and spreads (Shrestha and Akangaamkum 2008).

Jamnadass et al. (2010) have documented harvesting of Allanblackia from the wild in the forests and semi-wild on farmlands in Tanzania involving 6500 collectors (44% women), with 54 collection centres in the Eastern Arc Mountains purchasing approximately 450 tonnes of seed in 2008. Tanzania, Ghana and Nigeria produce an average of 450, 110 and 60 tons of seeds, respectively, equivalent to 150, 40 and 20 tons of oil per annum (Kattah 2010). Much of the research on Allanblackia in the last decade has focused on how to increase the cultivation in order to meet the growing market demand.

9.3.3 *Marula (*Sclerocarya birrea*)*

Marula belongs to the Anacardiaceae family grown in the miombo woodlands of southern Africa and Sudano-Sahelian zones of West Africa. It is one of the top indigenous fruit trees in southern Africa (Akinnifesi et al. Aḳinnifesi et al. 2008a, 2008b). The seed kernel is an important source of edible oil. Marula fruit is used to extract two types of commercial liquid oils—one is unintegrated and extracted from the nut's shell and the other from the seed of the fruit. The marula oil market is a niche market, and is segmented into organic and conventional, based on nature, type, source, application, end use and sales channel (Data Bridge Market Research 2022). The Data Bridge Market Research (2022) has estimated the global marula oil market at USD 58.39 million by 2029 at an annual growth rate of 4.32% in the forecast period of 2022–2029.

Legodi et al. (2022) reported that in Limpopo province, the Pedi (Northern Sotho) local communities were able to recognize three varieties of marula trees and their fruits based on the aroma and flavour, namely, *morula o mobose*, *morula wa go baba* and *morula wa go nkga*. The marula-derived products are increasingly entering the market, either through the local rural communities or by private sector companies with the intent to improve the livelihood of the communities (Fig. 9.1). Meanwhile, large postharvest losses still occur in the marula value chain. For instance, only 30% of the collected fruits are taken by Distell SA to make Amarula Cream Liqueur, which has an alcohol content of 17% (*v/v*), while a significant percentage of the fruits are classified as substandard fruits and therefore not utilized and discarded as waste (Legodi et al. 2022).

A study by Shackleton and Shackleton (2005) in Bushbuckridge, South Africa showed the sales of marula fruits by rural women as an important aspect of their income generation (Table 9.3). The study further indicated that marula products were an integral part of the livelihoods and culture of the rural community dwellers in the area, many of whom were living below the poverty line and dependent on natural resources to meet a wide range of basic needs. Majority of households (90%) were reported to have collected marula fruit, primarily to brew beer but also for making jam and to sell. The average quantities of fruit sold per household over the season varied between 150 kg (Hokwe) and 1114 kg (Allandale), with an average of 764 ± I 56 kg (Table 9.2). About one third (31%) of households were selling fresh fruit to one or both of the commercial enterprises, but proportions varied from 3% in Hokwe to 63% in Edinburgh (Shackleton and Shackleton 2005). In the study,

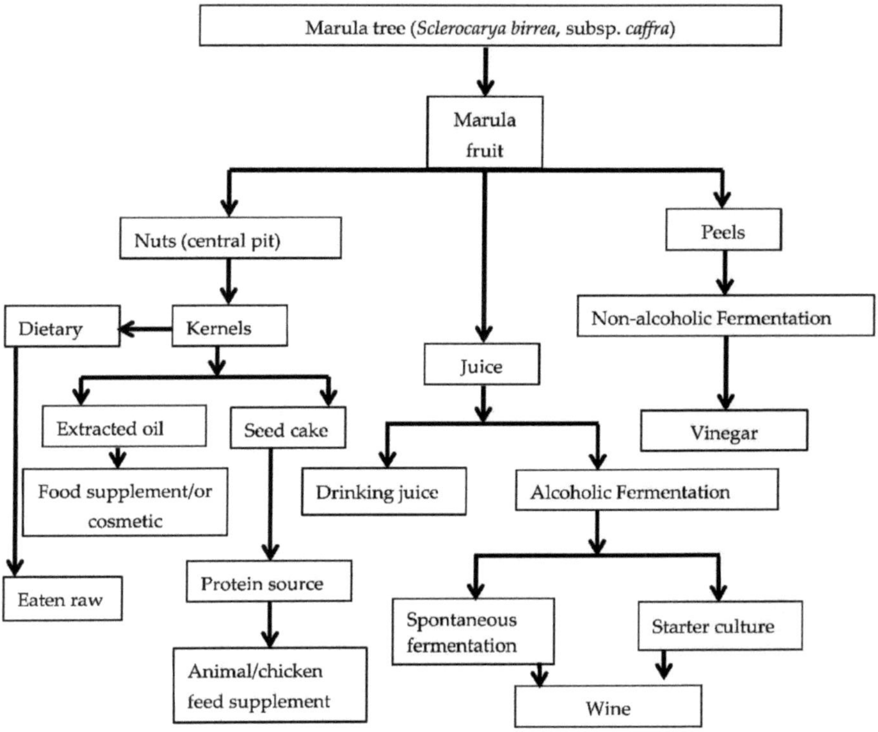

Fig. 9.1 Bioprocessing of marula tree to produce value-added products. (Source: Legodi et al. 2022)

Table 9.2 Sales of Marula fruits by rural women in Bushbuckridge, South Africa during 2002

	Allandale (n = 26)	Edinburgh (n = 26)	Hokwe (n = 26)	Rolle (n = 26)	All (n = 142)
% households collecting	89	89	92	89	90
Amount (kg) sold by selling households per season (mean ± SE)	1114 ± 623.1	722 ± 162	150 (1 hh)	510.0 ± 113	764 ± 156
Number of years selling (mean ± SE)	2 ± 0.2	4 ± 1.0	1.0	3 ± 1.0	2 ± 0.4
Price (R) per 80 kg bag (mean ± SE)	22 ± 1	16 ± 1	–	21 ± 1	18 ± 1
% selling to:	0				
Distell	0		1 hh	37.5	64.1
MDC	100.0	100.0	0	83.3	38.5
Net income (R) from fruit sales per selling household per season	335	163	38	146	194

Net income = gross income minus all costs except labour; hh = household; R = South African currency (Rand). Source: Shackleton and Shackleton (2005)

Table 9.3 Nutritional and mineral composition of fruits from selected indigenous fruit tree species

Species	Common name (English)	Part	Calcium	Magnesium (mg)	Potassium	Sodium	Copper	Phosphorus	Iron	References
			(mg/100 g)							
Adansonia digitata	Baobab	n.a.	275	232	1730	14.3	0.82	51	6.2	Stadlmayr et al. (2013)
		Pulp	1156	2090	28,364	188	–	450	58	*Saka et al. (2008)
Balanites aegyptiaca	Desert date		164	44	1110	–	0.43	51	13.8	Stadlmayr et al. (2013)
Dacryodes edulis	African pear	n.a.	50	47	443	0.8	0.52	60	1.7	Stadlmayr et al. (2013)
Sclerocarya birrea	Marula	n.a.	35	19	325	–	0.16	31	3.4	Stadlmayr et al. (2013)
		Pulp	16	27	309	4	–	3	1.7	
Syzygium guineense	Black plum	n.a.	67	34	–	3	–	–	7.9	Stadlmayr et al. (2013)
		Pulp	149	430	19,683	253		1081	136	*Saka et al. (2008)
Tamariundus indica	Tamarind	n.a.	192	66	933	30.2	0.42	87	3.1	Stadlmayr et al. (2013)
Uapaca kirkiana	Wild loquat	n.a.	17	39	375	10	–	15	11.0	Stadlmayr et al. (2013)
Vitex doniana	Black plum	n.a.	21	17	637	8	–	77	1.3	Stadlmayr et al. (2013)
		Pulp	132	1811	18,208	245	–	823	283	*Saka et al. (2008)

Ziziphus mauritiana	Jujube	n.a.	23	5	256	6	0.01	19	0.8	Stadlmayr et al. (2013)
		Pulp	135	507	17,318	226	–	2162	–	*Saka et al. (2008)
Parinari curatellifolia		Pulp	129	830	10,380	252	–	339	103	*Saka et al. (2008)
Strychnos cocculoides	Monkey orange	Pulp	60	1633	28,670	459	–	2106	60	*Saka et al. (2008)

Proximate analysis and Vitamins A & C

	Part	Dry matter (%)	Ash (%)	Protein (%)	Fat (%)	Fibre (%)	Carbohydrate (%)	Vitamin C (Vitamin A)	Energy value (kJ/100 g)	References
Adansonia digitata	Pulp	86.8	5.0	3.1	4.3	8.3	79.4	179.1 (21)	1480	*Saka et al. (2008)
	n.a	–	4.9	2.5	0.5	6.2	74.9	273	1380	Stadlmayr et al. (2013)
Dacryodes edules	n.a.	–	1.4	3.5	21.0	1.6	9.5	24.5 (–)	1010	Stadlmayr et al. (2013)
Balanites aegyptiaca	n.a	–	4.0	4.2	0.6	5.3	68.6	–	1300	Stadlmayr et al. (2013)
Irvingia gabonensis	n.a.	–	0.8	1.1	1.1	0.4	17.8	55.9 (–)	364	Stadlmayr et al. (2013)
Sclerocarya birrea	Pulp	15.0	1.0	–	–	–	–	- (35)	–	*Saka et al. (2008)
	n.a.	–	–	0.7	0.5	–	–	167	–	Stadlmayr et al. (2013)
Syzygium guineense	Pulp	22.1	4.1	5.4	31.2	17.6	42.1	11.9 (226)	1923	*Saka et al. (2008)
Strychnos cocculoides	Pulp	27.8	3.7	11.5	6.0	17.9	61.0	22.9 (22)	1390	*Saka et al. (2008)

(continued)

Table 9.3 (continued)

Proximate analysis and Vitamins A & C										
	Part	**Dry matter (%)**	**Ash (%)**	**Protein (%)**	**Fat (%)**	**Fibre (%)**	**Carbohydrate (%)**	**Vitamin C (Vitamin A)**	**Energy value (kJ/100 g)**	**References**
Tamarindus indica	n.a.	–	2.3	3.6	0.8	6.0	60.4	15.5 (–)	1160	Stadlmayr et al. (2013)
Uapaca kirkiana	Pulp	58.1	4.5	17.0	22.9	8.1	47.5	16.8 (67)	1897	*Saka et al. (2008)
	n/a.	–	0.8	0.3	0.4	2.1	28.7	16.8 (–)	523	Stadlmayr et al. (2013)
Vitex doniana	Pulp	26.5	3.4	5.7	2.6	10.2	78.1	19.6 (175)	1445	*Saka et al. (2008)
	n.a.	–	1.5	0.4	0.3	1.4	26.3	5.2 (–)	474	Stadlmayr et al. (2013)
Flacourtia indica	Pulp	19.2	5.7	4.2	3.6	5.7	80.7	10.2 (303)	1290	*Saka et al. (2008)
Ziziphus mauritiana	Pulp	14.8	10.1	4.1	9.5	3.4	73.0	13.6 (35)	1588	*Saka et al. (2008)
	n.a.	–	0.8	0.4	0.7	1.2	8.3	2.8–13.6	184	Stadlmayr et al. (2013)

*Malawi

about 70% of households indicated that they made marula beer (vukanyi), while only 19 households (15%) sold marula beer. Nuts were generally extracted from fruit that the household had already collected for beer or to eat fresh. Most households stored both nuts and kernels, especially nuts. The mean 'shelf-life' was indicated as about 1 year. Nearly half of households made marula jam from hard green fruits.

According to a report by the Organisation for Economic Co-operation and Development (OECD) Aid for Trade case study, Namibian rural women in the north central part of the country established the Eudafano Women's Cooperative (EWC) to market marula products for local and export markets (Arora 2020). By 2008 EWC reached membership of over 5000 women in 22 groups producing marula oil from wild trees, and by 2010, EWC had become the second largest producer of marula products in southern Africa (Arora 2020). With access to new markets, by 2000, rural marula producers were receiving over US $60,000 annually and by 2010 were receiving US $2.35 per kilogram of marula. According to Arora (2020) the interest in international markets for marula skyrocketed since the Namibian women organised to set up EWC in partnership with PhytoTrade, which did not only benefit the companies selling marula products, but has also had a major impact on local producers in Namibia. The demand shot up to 20 tonnes of oil worth over US $20 million in 2008 (Arora 2020). According to the report, Marula retail prices sold four times more than non-marula products, and marula went from a supplementary source of income to the primary source of income for many women in rural Namibian communities.

The major player on marula remains Distell with various products such as Amarula Cream Liquer, Amarula Vegan, Amarula African Gin, etc. (https://amarula.com/en-za/our-products/). Taunted as the most successful export of South Africa, the Amarula global sale of the liquor brand Amarula by Distell Group ranged from 1 to 1.4 million of 9-litre cases, averaging over one million of 9-liter cases from 2009 to 2019 (Conway 2021), and is said to have been exported to 103 countries. Less information is available on the market of oils and other marula-derived products, which are also entering the niche markets.

*9.3.4 Baobab (*Adansonia digitata*)*

The baobab (*Adansonia digitata,* Malvaceae), is a multi-purpose tree of the genus *Adansonia*, and is native to the sub-Saharan African countries. In South Africa it is restricted to the Limpopo and Mpumalanga provinces. The word "baobab "is derived from the Arabic *bu hobab*, meaning "fruit with many seeds". Nearly every part of baobab is useful: its fibrous bark is used for making ropes, weaving and clothing; its young leaves are edible and rich in proteins and Vitamins A, B1, B2, and C, and other minerals contents (Table 9.3), making it a rich source as food supplements; its fruit pulp (powder), considered by many as superfood, is concentrated in Vitamins especially Vitamin C and iron, and is used for making juice, beverages and alcoholic drinks, its roots are used for medicinal purposes (Jackson 2015); and the hollow trunk of large baobab tree provides shelter. The seeds are used for oil production and the capsule coat for soap making. Since ancient times, baobab's fruit pulp, seeds, leaves, flowers, roots and bark have been used locally in African countries for human consumption.

Baobab market was valued at USD 2.1 billion in 2022 and is projected to reach USD 6.2 billion in 2032 (More 2024). Since its approval as a novel food ingredient by European Union (EU) food regulators in 2008, the native African baobab tree has gained increasing market and media exposure (Jackson 2015). It is generally sold in fruit pods, oil, powder and pulp. The Baobab Powder market in the U.S. alone is estimated at US$2.2 Billion in 2022, while China, is forecast to reach a projected market size of US$2.7 Billion by 2030 (Global Baobab Powder Markets Report, 2024[2]). The leaves, seeds and fruits are rich in vitamins and are used for food supplements, thus creating huge potential for value addition, processing and export market and boosting local economies.

9.3.5 African Bush Mango (Irvingia Spp.)

The fruit of the African bush mango is eaten fresh, with a sweet and bitter taste that is cherished by local people. The most important product of the species is processed seed cake, which is widely used in soups and stews by Africans at home and in the diaspora, especially from West Africa. The kernels of both species are used as a condiment in the preparation of soups, increasing their viscosity and drawability (sliminess) (Ofundem et al. 2017). Research has shown that bush mango kernels can reduce obesity, control appetite, and reduce fat and cholesterol (Ngondi et al. 2005; Oben 2010; Ofundem et al. 2017). There are five species: *I. gabonensis, I. wombolu, I. excels, I. grandiflora, I. robur and I. smithii.*

Bush mango kernels is most commonly traded in the Southwest region of Cameroon (Nkwatoh et al. 2010; Ingram et al. 2012), contributing to diversifying local communities' livelihood opportunities (FAO 2009; Awono et al. 2013). The price of bush mango increases along the market chains, contributing to enhancing the livelihoods of the people involved (Ingram et al. 2009; Nkwatoh et al. 2010; Agbelade and Onyekwelu 2013; Ofundem et al. 2017). Tchoundjeu et al. (2005) estimated the value of the exports of the kernels of bush mango from Cameroon to neighbouring countries including Nigeria, Gabon, Equatorial Guinea, and Central African Republic, to be at US$260,000 per annum.

[2] Global Baobab Powder Markets Report (2024). Baobab Powder: Global Strategic Business Report. Global Industry Analysts, Inc. 313 p.

9.4 Nutritional Components of Indigenous Fruit Trees

Indigenous fruit trees are rich in nutrients and often act as sources of minor and major vitamins for people living in rural areas. Cemansky (2015) indicated that IFTs are more nutritious than many exotic fruits. They have shown that IFTs are good sources of energy, carbohydrates, protein, fats, vitamins and minerals. For instance, baobab pulp is very rich in Vitamin C (up to 500 mg^{-1} 100 g dry basis) with 80% of this vitamin bioavailable to the human body (Chadare et al. 2009). In addition, baobab pulp has an antioxidant capacity 30 times higher than that of Kiwi, which is among the reference antioxidant fruits, and 110 times more than orange.

Table 9.3 shows the levels of these benefits derived from different IFTs. There are within species differences in ranges of some of the vitamins and nutrients, which can be attributed to lack of uniformity in methodology and incomplete reporting (Sileshi et al. 2023). Stadlmayr et al. (2013) compiled data with 332 entries on nutrient composition from selected indigenous fruit trees in Africa (Table 9.3). The data used for the desk review included macronutrients (38%), followed by minerals (27%), fatty acids (14%), phytochemicals (8%), vitamins (5%), amino acids (5%) and 'other' (3%). *Adansonia digitata* had the highest number of publications (26), but more food entries were compiled for *D. edulis* (111) and *A. digitata* (78) and least for *V. doniana* (eight), U. kirkiana (three) and *S. guineense* (three).

9.5 Domestication of Indigenous Fruit Trees

Indigenous fruit trees are a valuable resource as they contribute to household needs in terms of income and food, and have an economic value that can contribute to exports and a country's Gross Domestic Product. IFTs are still harvested from the wild despite their food security potential (Muok 2019). With increased deforestation and forest degradation, most wild fruit trees are vulnerable to overexploitation and genetic impoverishment. Therefore, there is a need to prioritize and introduce them as resources for small-scale farmers to ensure their continued contribution to food and income security (Jusu and Cuni-Sanchez 2017).

The domestication of IFTs starts by identifying priority species that have economic potential (Nieminen et al. 2017). The prioritization of IFTs differs based on the geographic location (Jusu and Cuni-Sanchez 2017). Ethnobotanical studies and scorecard assessment methods are widely used for prioritizing IFTs with domestication and commercialization potential at a local level (Franzel et al. 2008; Jusu and Cuni-Sanchez 2017; Nieminen et al. 2017).

Leakey et al. (2022) documented the research and development works in the first three decades of indigenous fruit trees domestication globally and highlighted communal, socio-economic, legal and political factors that affect domestication of indigenous fruit trees. Akinnifesi et al. (Akinnifesi et al. 2008a, 2008b) have also documented efforts that have been put towards domesticating IFTs. Progress have been made in domesticating priority indigenous fruit tree species in Africa that have the potential for cultivation in the four regions of Africa namely: African Pear or Safouu (*Dacryodes edulis*) in the Humid Lowlands of West and Central Africa; Jujube (*Ziziphus mauritiana*) in Sahelian Drylands of West and North Africa; Macadamia (*Macadamia integrifolia*) in Highlands and Drylands of East Africa; and Marula (*Sclerocarya birrea*) in the Savannah and Miombo Woodlands of Southern Africa (Leakey et al. 2022). These focused on the state of management of the natural resources of the IFTs, their governance, management, ethnobotany, and commercial potential (in a broad sense of non-timber forest products).

9.6 Value Addition and Processing

Adding value to a product means enhancing its quality, functionality, or appeal by processing, branding, packaging, or differentiating it from competitors' products (Asamoah et al. 2023; Rundh 2016; Klimchuk and Krasovec 2013). This would boost the resource poor farmers' expected revenue and encourage them to embark on deliberate cultivation of IFTs (Awodoyin et al. 2015). For most of the indigenous fruit trees, standardized agro-techniques for production, harvesting, storing the value-added products and improved processing are not common or still at infancy stage. Value addition involves the development and commercialization of enhanced processing techniques and value-added products from fresh produce for market diversification of IFT. Furthermore, there is still a low market demand for many underutilized fruit trees. Lack of proper awareness and low market development is still a problem to be solved in Africa (Awodoyin et al. 2015). Moreover, in many countries, indigenous fruits are mainly marketed in local markets which are assumed to be low-income markets (Ham et al. 2008). It is difficult to state the adequacy of the domestic market supply of indigenous fruits in Africa owing to several factors, including geographical location, seasonal availability, market infrastructure, and consumer preferences. However, demands in these markets are generally low and limited benefit can be derived from them. Table 9.4 shows the results of indigenous fruit enterprise feasibility studies conducted in Malawi, Tanzania, and Zimbabwe by Jordaan et al. (2008).

9.7 Supply Chain for Different Products

The supply chain of many IFTs is short, consisting of collectors/transporters, wholesalers, retailers, and consumers (Karaan et al. 2005; Asamoah et al. 2023). The supply chain normally starts with farmers or local community members who collect fruits for home consumption and/or trading (Ham et al. 2008). They sell fruits at the roadside to other community members or to traders who transport the fruits to urban

Table 9.4 Financial analysis and feasibility assessment of indigenous fruit trees (IFTs) entreprise in Malawi, Tanzania and Zimbabwe

Indigenous fruit products	Gross production value (US$)	Total cost (US$)	Net income (US$)	Net profit (after tax) US$	%
	Income statement				
Juice concentrate, Malawi	107,400	61,700	45,700	29,700	27.7
Juice concentrate in Tanzania	137,359	108,187	29,172	18,962	13.8
Baobar in Zimbabwe	61,090	33,080	28,010	18,206	29.8
Jam in Zimbabwe	20,209	11,336	8873	5767	28.5
	Cash flow analysis				
	Cash flow	**Net present value (NPV)**		**Breakeven price (US$)**	
Juice concentrate, Malawi	Net positive for all months	Positive over selected period		8.5/20 kg can	
Juice concentrate in Tanzania	Net positive for all months	Positive over selected period		14.64/20 kg can	
Baobar in Zimbabwe	Net positive for all months	Positive over selected period		0.34/50 g bar	
Jam in Zimbabwe	Net positive for all months	Positive over selected period		1.22/41 g jar	

Adapted with modification from Jordaan et al. (2008)
[a]Profit as % of gross value of production

markets. These traders either function as wholesalers and sell to retailers or sell the fruits themselves as retailers. Farmers sometimes sell their fruits directly to the public at markets, effectively eliminating all middlemen (Kaaria 1998). Figure 9.2 shows an example of Non-Timber Forest Products (NTFPs) supply chain from Ghana as reported by Asamoah et al. (2023).

Meanwhile, IFTs are often characterized by mass fruiting seasons that typically last about 3 months. This seasonality can lead to excess supply over demand and hence a drastic fall in prices during the harvesting period (Ham et al. 2008). This is aggravated by a lack of market information that results in inefficient distribution of fruits. Poor linkage of producers with consumers and the market, and the lack of market information on IFTs have been identified as some of the major bottlenecks to the commercial development of indigenous fruit enterprises. The implication of long-distance transportation is that mechanical damage to the fruits is increased, thereby reducing their market value.

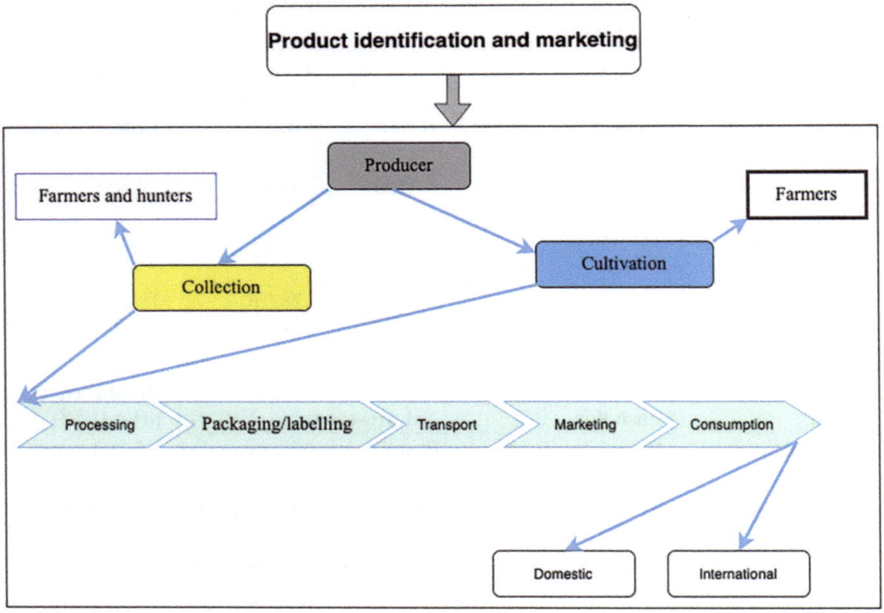

Fig. 9.2 Supply chain of NTFPs: an example from Ghana. (Source: Asamoah et al. 2023)

9.8 Marketing of Commercial Products (Informal, Formal Trade)

In Africa, IFTs commercialization activities are still less developed. Although domestication of IFTs—either as fully domesticated or semi domesticates (e.g shea in Mali, Burkina Faso and Nigeria; Marula in South Africa; Masuku in Malawi, Zimbabwe and Zambia, jujube in Sudan, Egypt and Ethiopia; Tamarind in Sudan, Nigeria and Tanzania; *Dacryodes edules, Irvingia gabonensis, Chysophyllum albidum* in Nigeria and Cameroun, etc.) has made significant progress, research on the commercialization potential of IFTs fruit products is limited. Available information is on the use, nutrition and consumption, economic analysis of wild collection, value addition and *ad hoc* market surveys on a few species. These studies have focused on priority setting, nursery, and propagation techniques, scaling up cultivation, adoption and domestication, as well as postharvest processing to expand market opportunities (Akinnifesi et al. Akinnifesi et al. 2008a, b, Leakey and Akinnifesi 2017; Leakey et al. 2017; Leakey et al. 2022). There is dearth of major publications on the commercial potential of IFTs in Africa beyond isolated case studies, and data on international trade are limited to a few species.

Traditionally, indigenous fruits were collected freely and consumed by the rural population. With time, informal trade became increasingly common when local people started to subsidize their household incomes. Rural dwellers in Africa rely on wild fruits to supplement their diet, nutrition and to generate cash income. A variety of indigenous fruits were increasingly sold in villages, growth points and district markets as well as city markets. However, despite an increase in the number of people trading in indigenous fruits, the marketing system has remained informal, with many economic inefficiencies (Ramadhani and Schmidt 2008). For example, traditional taboos at many sites limit the production and collection of the fruits for the market, thus hindering efforts towards commercialization. These traditional taboos were derived from a belief that fruits are a gift of God and are therefore meant to benefit the whole community. The taboos also protect the trees from unsustainable harvesting techniques. Therefore, taboos and the production/collection of IFTs and marketing are linked in a complex relationship. In addition, unsuitable transport is an important obstacle faced by producers, collectors, and retailers (Ramadhani and Schmidt 2008). Using buses designed to carry passengers rather than fruits increases the risk of damaging the fruits and incurs high losses, especially for fragile non-pericarpous species. Such conditions require specialized traders who can deal efficiently with the transport of fruits. Undoubtedly, the need to use intermediate dealers significantly increases marketing costs and lowers price efficiency.

Home consumption of IFT fruits is a very popular activity among all rural dwellers in Africa, to complement the consumption of staples. In Zimbabwe, a study by Mithöfer and Waibel (2003) on labor productivity of wild fruit collection from IFTs analysed the gross margins and returns to labour in collection, use and sale of products of *Uapaca kirkiana, Strychnos* sp. (*S. cocculoides* and *S. spinosa*), and *Parinari curatellifolia* were compared with other farming activities. The result showed that gross margins for collection of IFT products were lower than for livestock and crop production. However, returns to labour from collection and use of IFT products were considerably greater than from other activities including gardening and livestock rearing. Children are the main consumers of the fruits. Marketing of IFTs fruits is carried out by women who used the receipts to purchase household goods. While *U. kirkiana* fruits were more important in generating cash income than others, fruits of *P. curatellifolia* were important for home consumption during periods of food shortages. The authors concluded that the study indicates that collection of IFT products is an efficient labour allocation strategy in Zimbabwe.

In a recent study in Kaduna state of Nigeria, Omotayo et al. (2021) determined the socio-economic characteristics of shea butter (*V. paradoxa*) marketers, and profitability using stochastic profit frontier model. The authors found that marketers of Shea butter were mostly women (85%), with 12 years average marketing experiences. Marketing of shea butter was profitable. Marketing experience and education level of marketers significantly influence profit efficiency of Shea butter marketers. This means that marketers of shea butter have 52% chances of increasing profit efficiency. Profitability was constrained by lack of credit facilities, bad road infrastructures, inadequate extension officers, lack of storage facilities, and poor transport facilities. This agreed with the findings of Ayanrinde et al. (2023), indicating that lack of capital, high cost of transportation and lack of storage facilities were identified as a major constraint in Shea butter marketing.

The challenges faced by most Africa's indigenous fruit trees with commercialization potential include relatively low market demand, market and value chain development compared to exotics. The supply chain is generally short. Local traders mainly commercialized fresh fruits because fruit processing is still in its infancy. There are many problems along the supply and processing chains (Ham et al. 2008). There is lack of standardized agro-industrial techniques for quality fruits production (market oriented ideotypes, e.g. for size, taste and uniformity), harvesting and post-harvests, such as storage, branding, grading and other value-addition of fruit and product (Leakey et al. 2022, Akinnifesi et al. 2008a, b). The indigenous fruit marketing system is characterized by a lack of sophisticated product differentiation activities. At the producer and wholesaler levels, pre-sale processes such as washing, grading, sorting and packing do not exist. Traders believe that the fruits coming from trees are clean and that there is no need for further processing (Ramadhani 2002).

In most of Africa, indigenous fruits are not displayed as prominently as other produce in rural markets. In some instances, the other products are displayed on tables, while the indigenous fruits are sometimes kept in containers on the ground. Indigenous fruits are sold mostly to women and children as a snack that is often eaten fresh in the market (see Fig. 9.3). Indigenous fruits can therefore, be considered as products that consumers buy on impulse, in very small quantities, and that

Fig. 9.3 A local fruit market at Makenene market, Cameroon, showing safou (*Dacryodes edulis*), among other fruits and nuts (including coconut, orange, pineapple, and tomato, etc) (© Roger Leakey)

are consumed very soon after being purchased. People come to the market to buy other goods and see that indigenous fruits are present, and then purchase them. Retailers of *U. kirkiana* in Zimbabwe conduct simple processing activities such as sorting rotten fruit, dividing fruits into selling units, and sometimes washing. Other retailers, however, felt that the practice of washing and packaging in bags reduces the shelf life of the fruits (Ramadhani 2002). In Zimbabwe, traders consider that washing of *U. kirkiana* fruits could lead to fruit discoloration (Kadzere et al. 2002). In Malawi only some market traders grade their fruits based on fruit size and degree of ripeness, while in Zimbabwe collectors and traders of *Z. mauritiana* have been able to obtain better prices for their fruits by cleaning and packaging the fruit in plastic bags (Ham et al. 2008).

Akinnifesi et al. (2008a, b) have documented attempts to commercialize IFTs. Limited work has been published on species products commercialization at national, regional, and international level.

Despite its poor state of commercial development, indigenous fruit trading and marketing are important economic activities in most African countries (Kaaria 1998; Ramadhani 2002). IFTs are sold in both urban and rural markets, showing that consumers like the fruits and are willing to pay for them (Ham et al. 2008). In many African countries, many people still value indigenous fruits such as the purple safou (*Dacryodes edulis*) sold on the market. Ultimately, in these countries with vibrant informal markets, farmers need training to help them ensure a future for native fruit trees, which are rapidly disappearing from the wild.

9.9 Marketing Strategies

Limited marketing strategies for indigenous fruits have been developed due in part the lack of government support and information on their commercialization and uses. Instead, political actions were more developed on fruit tree biodiversity conservation. Scattered studies show that farmers' profit on fruits marketed depends on whether they market the fruits themselves, or sell to an intermediary. Although for collectors selling fruits to intermediary has the advantages for them to get their money in bulk, cut down the labour and transport costs and reduce perishability during transport, their presence prevents price setting according to supply and demand. Moreover, selling as a group allows fruit producers to reduce their risk, and puts them in a better position to negotiate prices rather than being price takers. Marketing with intermediaries sometimes enables collectors to have fixed prices and sometimes get advance payments or loans.

Elsewhere, indigenous fruit marketing is subject to high transaction costs in the markets. This is not unexpected in typical developing-country markets (Ham et al. 2008). Transaction costs include product losses, delays, costs of monitoring market transactions, negotiation or bargaining costs, transport and logistics inefficiencies, and poor market information. Transaction costs are better contended with through improved coordination and information flows in supply chains. This is a key challenge for economic development efforts in emerging markets. Supply and consumption is entirely directed at local markets, given the traditional status of the products and consumption habits. The local markets are assumed to be low-income markets, by and large, with limited ability to pay for greater sophistication and value addition that adds significantly to the marketing margins. Hence, there is a tendency to favour generic marketing of produce. The implication is, arguably, that limited benefit can be derived from much value addition in local markets; at best, the improvements would be incremental. Alternatively, the improvements could radically change the product on offer through processing (e.g. juices, jams, preserves, etc.) whilst retaining the product appeal. Another alternative may be to consider acceptability in higher-income markets and adapt the product accordingly (Karaan et al. 2005; Ham et al. 2008).

9.10 Opportunities for Fair Trade/Certification

There is an opportunity for marketing indigenous fruit tree products at international level, which is motivated by a growing trend that emphasizes variety and natural/organic products (Akinnifesi et al. 2008a, b). However, in order to penetrate the international market, producers and exporters need to follow required standards that clearly show the consumers the source of the products and management systems where they were obtained from. In most African countries, there are institutions that are mandated to share information on external markets. For example, in Zambia, The Zambia Development Agency, working with the Ministry of Commerce, provides such information and link exporters to these markets. For indigenous fruit trees and their products, market research may be required, to establish the niche of these fruit products in the USA, UK and other regions where such products are sought. In addition, assurance of importers, different companies in the value chain and consumers about the sustainability of the production systems need to be established, and this is done through the process of certification.

Certification is becoming very important in the natural products sector. Companies want to be able to trace the whole chain of natural products to be completely certain that products that claim to be natural, organic or fairly traded are indeed truly what they claim to be. However, very few low-income countries have the high degree of infrastructural and institutional development, strict quality control and sophisticated supply-chain management practices necessary to enter the international market with a new product (Agustino et al. 2011). Because of these concerns, there is a better prospect for small producers to access their products to niche markets and 'fair trade' opportunities than to mainstream markets.

The fair-trade system has been established to give small-scale rural producers in Third-World countries the opportunity to be supported to the point where they can sustainably and equitably engage First-World markets. Fair trade promotes workers' rights, protects the environment and sustains the ability of local producers to meet community needs (Ham et al. 2008). It promotes development opportunities for disadvantaged producers, especially women and indigenous people, and seeks to protect human rights and the environment (Agustino et al. 2011). It aims to secure a better deal for producers in the developing world through setting stable product prices that cover both their production costs and a premium that their organization can reinvest either in the business or social and environmental schemes in the wider community. It also raises awareness among consumers so that they can exercise their purchasing power positively and campaigns for changes in the rules and practice of conventional international trade.

The primary objective of fair trade is to develop access to niche markets in developed countries for producers in developing countries, who tend to be marginalized by conventional trading structures (Ham et al. 2008). This enables them to reach appropriate quantity and quality standards, as well as ensuring a fair price for their products. Fair trade eco-label guarantees consumers that certain criteria are met and also guarantees fair trading relations. It also aims at the introduction of a single international fair-trade label. However, unlike other certification systems, producers do not pay for their certification by Fairtrade Labelling Organizations (FLO) (Agustino et al. 2011).

Fair trade has the potential to support sustainable livelihoods and conservation efforts related to Indigenous Fruit Trees in Africa. However, it is important to continuously evaluate and strengthen these initiatives to ensure they effectively meet the needs of producers and communities while promoting environmental and social sustainability.

Fair trade also guarantees fair production conditions, including for small farmers' co-operatives- a democratic, participatory structure; for plantations/factories- the workers should have decent wages i.e. at least the legal minimum, good housing, where appropriate, minimum health and safety standards, the right to join trade unions, no child or forced labour and minimum environmental requirements. The FLO monitoring programme ensures that all the trading partners continue to comply with fair trade criteria, and that the individual producers really do benefit. The costs pertaining to the certification process therefore, prohibit smallholders from entering the 'green' markets.

9.11 Challenges to Commercialization

The precise definitions of commercialization depend on the scope, nature and goal. While some authors narrowly view it as increasing the proportion of marketed output; others broadly view it as a transition from subsistence towards market-oriented production (Zhou et al. 2013). Generally, commercialization is a process of shifting from a predominantly subsistence production for consumption towards market economy and profit-oriented production value chain. It is a process of bringing new products or services to the market. The key components of commercialization are its drivers, determinants, processes, strategies, indicators and effects - which can be synthesized into a conceptual model useful in planning, implementation and review of commercialization programs (Zhou et al. 2013).

The commercial potentials of indigenous fruit trees can be immensely promoted if cultivars can be adapted to commercially focused investment. In addition, the breeding, improvement and propagation of indigenous fruit trees need significant resources for the development of entirely new cultivars that can equally conform to the modern form of transportation and storage requirements (Arias et al. 2012). However, not much has been achieved in terms of their domestication and commercialization. Hence, domestication is a major step in the commercialization process of IFTs. Encouraging the domestication and conservation of IFT species in the agroecosystem can provide highly sustainable production systems that conserve soil fertility, micronutrients and biodiversity, as well as guarantee food security, climate change adaptation and mitigation (Awodoyin et al. 2015). Many indigenous fruits are still sourced from the wild and natural environments (Awodoyin et al. 2015; Nkosi et al. 2020; Pfukwa et al. 2020) which have generally limited their potential for higher yield and growth (Awodoyin et al. 2015) and made them to remain underutilised.

Increased commercial value of non-cultivated IFTs can lead to resource depletion (Belcher and Kusters 2004, Akinnifesi et al. 2008a, b). Overharvesting can also be triggered by the commercialization of species that have previously been used exclusively for local subsistence (Kala 2009). The key challenges to fruit tree sustainability include indiscriminate and illegal logging, low accessibility, and low acceptability, as well as inadequate research on their cultivation (Omotayo and Aremu 2020). As the rate of deforestation increases, wild fruit trees become prone to overexploitation and extinction. Therefore, the livelihood of rural people who largely depend on this natural resource is seriously threatened. This invariably will also reduce food availability to wild animals negatively impacting the food chain, and possibly triggering competition with humans for food produced on farmlands.

There is also an observed lack of access to market information, low level of acceptability and access, strong deficiencies in processing technologies and skills, and inherently weak value chains to propel the indigenous fruits to wider commercial markets (Omotayo and Aremu 2020). The efficiency of the fruit processing industry is further thwarted by lack of sustained supply of primary raw materials or intermediate products, and medium scale entrepreneurs complain about their dependence on imports. The industry depends on importation of exotic fruit concentrate only.

Fresh fruits are highly perishable and incur direct or indirect nutrient and quality losses along the market chain from production to consumption (Akinnifesi et al. 2008a, b). Indigenous fruits, mostly harvested from the wild, are generally sold or consumed fresh, and large proportions (up to 55%) of the collected fruits are wasted (Ramadhani 2002). Postharvest decay can be a serious problem in many fruits undergoing storage. Also, another major factor limiting the development of indigenous fruits is that they are characterized by short but intense fruiting seasons (Nasare et al. 2022; Ham et al. 2008).

9.12 Chapter Highlights

- The African continent is well endowed with a rich diversity of IFTs with fruits and products that contribute to improved nutrition and rural livelihoods. Many of these have additional potential to generate cash incomes, and developed into viable commercial products and enterprises. Unfortunately, very few of them have penetrated regional and international markets;
- Despite considerable progress in the last three decades on IFTs domestication research and development—mainly the economics of wild harvesting, ethnobotany, propagation and cultivation, IFTs fruits markets and commercialization in Africa are still poorly developed. The process of developing IFTs and their products for the international market will require addressing issues of quality and quantity.
- Markets within Africa are mostly through direct home consumption, opportunistic local markets—often ad hoc, characterized by irregular supply, short supply chain and low investment in value addition; thus making prices remain low;
- The high nutritive value of IFTs and the need for organic products at international scale provides an opportunity for IFTs. The opportunity for fruits and fruit tree products lies with the need for organic products and the consciousness of most of the international consumers for health products;
- Even with current low quantities of IFTs, through the Fairtrade, the IFTs have potential to fetch fair prices at international level. However, indigenous fruit trees still require that they get domesticated if their sustainable supplies were to be assured;
- The fresh fruits markets of IFTs are underdeveloped, and will depend on investment market development, sustainable supply, fruit quality and market chain. Until IFTs start to be produced in large quantities, with a broad range of products, their international markets will remain limited to a few that have already penetrated the market.
- On the other hand, there is a growing potential for the IFTs products in the food and cosmetic industries, with attractive international markets. More investment is need to widen the market opportunities for IFTs.

References

Abdul-Mumeen I, Beauty D, Adam A (2018) Shea butter extraction technologies: current status and future perspective. Afr J Biochem Res 13(2):9–2. https://doi.org/10.5897/AJBR2018.1007

Agbelade AD, Onyekwelu JC (2013) Poverty alleviation through optimizing the marketing of Garcinia kola and Irvingia gabonensis in Ondo State, Nigeria. ISRN For. 2013:5 pages, article ID 376247. https://doi.org/10.1155/2013/376247

Agustino S, Mataya B, Senelwa K et al. (2011) Non-wood forest products and services for socio-economic development. A Compendium for Technical and Professional Forestry Education. The African Forest Forum, Nairobi, 219 pp

Akinnifesi FK, Kwesiga F, Mhango J et al (2006) Towards the development of miombo fruit trees as commercial tree crops in southern Africa. For Trees Livelihoods 16:103–121

Akinnifesi FK, Sileshi G, Ajayi OC et al (2008a) Domestication and conservation of indigenous Miombo fruit trees for improving rural livelihoods in southern Africa. Biodiversity 9(1–2):72–74. https://doi.org/10.1080/14888386.2008.9712888

Akinnifesi FK, Sileshi G, Ajayi OC et al. (2008b) Accelerated domestication and commercialization of indigenous fruit and nut trees to enhance better livelihoods in the tropics: lessons and way forward. In: Akinnifesi FK, Leakey RRB, Ajayi OC et al. (eds) Indigenous fruit trees in the tropics: domestication, utilization and commercialization. World Agroforestry Centre: Nairobi. CAB International Publishing, Wallingford, pp 392–428

Arias RS, Borrone JW, Tondo CL et al (2012) Genomics of tropical fruit tree crops. Publications from USDA-ARS / UNL Faculty 893. https://digitalcommons.unl.edu/usdaarsfacpub/893

Arora M (2020) Eudafano women's cooperative and Marula tree: from local use to international recognition and benefit sharing. Intl J Integr Law Rev 1:62–74

Asamoah O, Danquah JA, Bamwesigye D et al (2023) Perceptions of commercialisation and value-addition of non-timber forest products in forest adjacent communities in Ghana. Discov Sustain 4:30. https://doi.org/10.1007/s43621-023-00146-6

Awodoyin R, Olubode O, Ogbu J et al (2015) Indigenous fruit trees of tropical Africa: status, opportunity for development and biodiversity management. Agric Sci 6:31–41. https://doi.org/10.4236/as.2015.61004

Awono A, Ingram V, Schure J et al. (2013) Guide for small and medium enterprises in the sustainable non-timber forest product trade in Central Africa. CIFOR, Bogor, 26 p. doi:https://doi.org/10.17528/cifor/004053

Ayanrinde FA, Oyewole SO, Anyanrinde OA et al (2023) Analysis of Shea Butter (Vitellaria paradoxa) Marketing in Ibarapa Central Local Government Area of Oyo state, Nigeria. J Appl Sci Environ Manag 27 No. 9. https://doi.org/10.4314/jasem.v27i9.13

Belcher B, Kusters K (2004) Non timber forest products commercialization: development and conservation lessons. In Kusters K, Belcher B (eds) Forest products, livelihoods and conservation. Case studies of non-timber Forest product systems. Volume 1—Asia. CIFOR, Indonesia, p 1

Bjornlund V, Bjornlund H, van Rooyen A (2022) Why food insecurity persists in sub-Saharan Africa: a review of existing evidence. Food Secur 14(4):845–864. https://doi.org/10.1007/s12571-022-01256-1

Bup DN, Mohagir AM, Kapseu C et al (2014) Production zones and systems, markets, benefits and constraints of shea (Vitellaria paradoxa Gaertn) butter processing. Oilseeds Fat Crops Lipids 21(2):D206. https://doi.org/10.1051/ocl/2013045

CBI (2020) The European market potential for shea butter. Posted 29.03.2022. https://www.cbi.eu/market-information/natural-ingredients-cosmetics/shea-butter/market-potential. Accessed 15.11.2023

Cemansky R (2015) Africa's indigenous fruit trees: a blessing in decline. Environmental Health Perspectives 123: A291–A296. 12 CID: https://doi.org/10.1289/ehp.123-A291

Chadare FJ, Linnemann AR, Hounhouigan JD, Nout MJR, Van Boekel MAJS (2009) Baobab food products: a review on their composition and nutritional value. Crit Rev Food Sci Nutr 49(3):254–274

Conway J (2021) Amarula liqueur's global sales volume 2009–2019. https://www.statista.com/statistics/308846/amarula-liqueur-global-sales-volume/

Data Bridge Market Research (2022) Global marula oil market – industry trends and forecast to 2029. https://www.databridgemarketresearch.com/reports/global-marula-oil-market

Emmanuel KA (2022) Nigeria: forget oil and gas, look at the potential of shea. The Africa report, posted August 8, 2022. https://www.theafricareport.com/228721/nigeria-forget-oil-and-gas-look-at-the-potential-of-shea/

Eshetu G, Johansson T, Garedew W et al (2021) Determinants of smallholder farmers' adaptation options to climate change in a coffee-based farming system of Southwest Ethiopia. Clim Dev 13(4):318–325. https://doi.org/10.1080/17565529.2020.1772706

FAO (2009) Immediate results for systemic support to NTFP based enterprise development in Cameroon. Available from: http://www.fao.org/forestry/enterprises/45716/en/

Franzel S, Akinnifesi FK, Ham C (2008) Setting priorities among indigenous fruit tree species in Africa: Examples from southern, eastern and western regions. In: Akinnifesi FK et al (eds) Indigenous fruit trees in the tropics: domestication, utilization and commercialization. World Agroforestry Centre/CAB International Publishing, Nairobi/Wallingford, p 1

Funt J, van der Van C, Iqbal R et al. (2020) Identification and prioritization of African and Asian export markets for West African Shea SMEs. https://globalshea.com/gsamain/storage/img/marqueeupdater/2020; Accessed Sept 28 2020

Ham C, Akinnifesi FK, Franzel F et al. (2008) Opportunities for commercialisation and enterprise development of indigenous fruits in southern Africa. In: Akinnifesi FK, Leakey RRB, Ajayi OC et al. (eds) Indigenous fruit trees in the tropics: domestication, utilization and commercialization. World agroforestry Centre: Nairobi. CAB International Publishing, Wallingford, p 254

Ingram V, Awono A, Schure J (2009) Hidden costs and values of NTFP market chains in Central Africa. Mobilisation and capacity building for small and medium enterprises in NWFP production and commercialisation in Central Africa project GCP/RAF/408//EC. CIFOR, ICRAF, SNV, COMIFAC, FAO, EU, Yaounde, 15 p

Ingram V, Ndoye O, Iponga DM et al (2012) Non-timber forest products: contribution to national economy and strategies for sustainable management. In: De Wasseige C, De Marcken P, Bayol N et al (eds) The forests of The Congo Basin: state of the forest 2010. Publications Office of the European Union, Luxembourg, p 137

Jackson S (2015) Baobab: the tree of life - an Ethnopharmacological review. Herbal Gram 108:42–53. https://www.herbalgram.org/resources/herbalgram/issues/108/table-of-contents/hg108-feat-baobab/

Jamnadass R, Dawson IK, Anegbeh P et al (2010) Allanblackia, a new tree crop in Africa for the global food industry: market development, smallholder cultivation and biodiversity management. For Trees Livelihoods 19(3):251–268. https://doi.org/10.1080/14728028.2010.9752670

Jordaan D du PS, Akinnifesi FK, Ham C et al. (2008) The feasibility of small-scale indigenous fruit processing enterprises in Southern Africa. In: Akinnifesi FK, Leakey RRB, Ajayi OC et al (eds) Indigenous fruit trees in the tropics: domestication, utilization and commercialization. World agroforestry Centre: Nairobi. CAB International Publishing, Wallingford, UK, p. 273

Jusu A, Cuni-Sanchez A (2017) Priority indigenous fruit trees in the African rainforest zone: insights from Sierra Leone. Genet Resour Crop Evol 64:745–760

Kaaria SW (1998) The economic potential of wild fruit trees in Malawi. MSc thesis, The University of Minnesota, Minneapolis

Kadzere I, Hove L, Gatsi T et al (2002) Current status of postharvest handling and traditional processing of indigenous fruits in Zimbabwe. In: Rao MR, Kwesiga FR (eds) Proceedings of the regional agroforestry conference on 'agroforestry impacts on livelihoods in southern Africa: putting Research into practice'. World Agroforestry Centre (ICRAF), Nairobi

Kala CP (2009) Aboriginal uses and management of ethnobotanical species in deciduous forests of Chhattisgarh state in India. J Ethnobiol Ethnomed 5:20

Karaan M, Ham C, Akinnifesi FK et al (2005) Baseline marketing surveys and supply chain studies for indigenous fruit Markets in Tanzania, Zimbabwe and Zambia. Research report of the world agroforestry Centre, Malawi and CPWild Research Alliance. see http://www.cpwild.co.za/DocsCPW.htm

Kattah C (2010) Developing a sustainable Allanblackia business in Africa. First Allanblackia Conference, Thinkers Village Hotel, Monrovia, Liberia 2–3 February 2010

Klimchuk MR, Krasovec SA (2013) Packaging design: successful product branding from concept to shelf. Wiley, UK

Kyereh D, Maňourová A, Hendre PS et al (2021) Diversity chemical composition, and domestication potential of Allanblackia parviflora. A Chev West Africa For 12:1758. https://doi.org/10.3390/f12121758

Lapeña I, Turdieva M, Noriega IL et al (2014) Conservation of fruit tree diversity in Central Asia: policy options and challenges. Biovers Internat, Rome, p 251

Leakey RRB, Akinnifesi FK (2017) Towards a domestication strategy for indigenous fruit trees in the tropics. In Leakey RRB (ed) Multifunctional agriculture achieving sustainable development in Africa. Elsevier Inc, pp 28–49. https://doi.org/10.1016/B978-0-12-805356-0.00016-7

Leakey RRB, Weber JC, Page T, Cornelius JP, Akinnifesi FK, Roshetko JM, Tchoundjeu Z, Jamnadass R (2017) Tree domestication in agroforestry: progress in the second decade (2003_2012). In: Leakey RRB (ed) Multifunctional agriculture: achieving sustainable development in Africa. Academic Press, pp 139–153

Leakey RR, Tientcheu-Avana ML, Awazi NP, Assogbadjo AE, Mabhaudhi T, Hendre PS, …, Manda L (2022) The future of food: domestication and commercialization of indigenous food crops in Africa over the third decade (2012–2021). Sustain For 14(4):2355. https://doi.org/10.3390/su14042355

Legodi LM, Lekganyane MA, Moganedi KLM (2022) Morula tree: from fruit to wine through spontaneous fermentation and the potential of deriving other value-added products. PRO 10:1706. https://doi.org/10.3390/pr10091706

Lovett PN (2015) Shea butter: properties and processing for use in food. In: Talbot G (ed) Specialty oils and fats in food and nutrition: properties, processing and applications Woodhead Publishing, pp 125–158

McMullin S, Njogu K, Wekesa B et al (2019) Developing fruit tree portfolios that link agriculture more effectively with nutrition and health: a new approach for providing year-round micronutrients to smallholder farmers. Food Secur 11:1355–1372. https://doi.org/10.1007/s12571-019-00970-7

Mithöfer D, Waibel H (2003) Income and labour productivity of collection and use of indigenous fruit tree products in Zimbabwe. Agrofor Syst 59:295–305

More V (2024) Baobab market research report information–market forecast till 2032. ID: MRFR/F-B & N/3482-HCR 110 pages. https://www.marketresearchfuture.com/reports/baobab-market-4913

Muok BO (2019) Potential and utilization of indigenous fruit trees for food and nutrition security in East Africa. Glob Adv J Agric Sci 8(2):40–49

Nanyang Technological University (NTU) (2023) Export of West Africa shea butter to India set to increase. https://www.ntu.edu.sg/cas/news-events/news/details/export-of-west-africa-shea-butter-to-india-set-to-increase. Posted 27 Feb 2023

Nasare LI, Stout J, Lovett P et al (2022) Determinants of shea (Vitellaria paradoxa C.F. Gaertn.) fruit yield: a review of research approaches and current knowledge. Scient Afr 17:e01371. https://doi.org/10.1016/j.sciaf.2022.e01371

Ngondi JL, Oben JE, Minka SR (2005) The effect of Irvingia gabonensis seeds on body weight and blood lipids of obese subjects in Cameroon. Lipids Health Dis 4:12. https://doi.org/10.1186/1476-511X-4-12

Nieminen R, Sørensen M, Theilade I (2017) Identification of indigenous fruits with export potential from Mukono district, Uganda: an assessment of two methods. Agrofor Syst 91:967–979

Nkosi NN, Mostert THC, Dzikiti S et al (2020) Prioritization of indigenous fruit tree species with domestication and commercialization potential in KwaZulu-Natal, South Africa. Genet Resour Crop Evol 67:1567–1575. https://doi.org/10.1007/s10722-020-00932-5

Nkwatoh AF, Labode P, Iyassa SM et al (2010) Evaluation of Irvingia (I. gabonensis and I. wombulu. Bail) trade; a major non-timber forest product in the Ejagham Forest Reserve of South West Cameroon. Ethiopian J Environ Stud Manage 3:70–77

Oben JE (2010) Methods and compositions using Irvingia gabonensis to reduce weight, control obesity, effect adipogenesis, lower PPAR-gamma, leptin, G3PDH and triglycerides levels and increase adiponectin levels. Patents US20100098792 A1

Ofundem T, Ndip NR, Abdon A et al (2017) Bush mango (Irvingia spp.): forest and on-farm resource availability and market chains in the southwest region of Cameroon, forests, Trees Livelih, 26:3, 170–182, DOI: https://doi.org/10.1080/14728028.2017.1283250

Okullo JB, Omujal F, Enuru T et al (2022) Farmers' use of indigenous fruit trees to cope with climate variability in the Lake Victoria Basin districts of Uganda. https://nru.uncst.go.ug/handle/123456789/6383

Omotayo AO, Aremu AO (2020) Underutilized African indigenous fruit trees and food–nutrition security: opportunities, challenges, and prospects. Food Energy Secur 9(3):e220. https://doi.org/10.1002/fes3.220

Omotayo AO, Ukamaka EE, Bosede OE et al. (2021) Profit efficiency of marketing Shea butter (Vitellaria paradoxa) in Kaduna state: applications of stochastic profit frontier model. Austr J Sci Technol 5(1, March)

Pandey J, Gaire S, Sharma K et al. (2023) Exploring the potentials of Ziziphus mauritiana Lam. seed kernel oil as pharmaceutical oil base: physicochemical characterization and ketoconazole soap formulation. https://doi.org/10.1016/j.heliyon.2023.e21034

Pfukwa TM, Chikwanha OC, Katiyatiya CLF et al (2020) Southern African indigenous fruits and their byproducts: prospects as food antioxidants. J Funct Foods 75:104220

Ramadhani T (2002) Marketing of indigenous fruits in Zimbabwe. Socio-economic studies on rural development, vol 129. University of Hanover, Germany

Ramadhani T, Schmidt E (2008). Marketing of indigenous fruits in Southern Africa. In: Akinnifesi FK, Leakey RRB, Ajayi OC et al (eds.). Indigenous fruit trees in the tropics: domestication, utilization and commercialization. CAB International Publishing/World Agroforestry Centre, Wallingford/Nairobi, p 224

Rundh B (2016) The role of packaging within marketing and value creation. Br Food J 2016(118):2491

Saka JKD, Kadzere I, Ndabikunze BK, et al (2008) Product development: nutritional value, processing and utilization of indigenous fruits from the Miombo ecosystem. In: Akinnifesi FK, Leakey RRB, Ajayi OC et al (eds) Indigenous fruit trees in the tropics: domestication, utilization and commercialization. World Agroforestry Centre/CAB International Publishing, Wallingford/Nairobi, pp 288–309

Shackleton S, Shackleton C (2005) The contribution of Marula (Sclerocarya birrea) fruit and fruit products to rural livelihoods in the Bushbuckridge District, South Africa: balancing domestic needs and commercialisation. For Trees Livelihoods 15(1):3–24. https://doi.org/10.1080/14728028.2005.9752504

Shrestha RB, Akangaamkum AD (2008) Novella partnership, a partnership for poverty reduction through sustainable enterprise development based on Allanblackia. SNV Ghana, Accra

Sileshi GW, Dagar JC, Akinnifesi FK et al (2023) Potentials of indigenous fruit trees in enhancing nutrition, income and biodiversity conservation in African agroforestry. In: Dagar et al (eds) Agroforestry for sustainable intensification of agriculture in Asia and Africa. Springer Nature Singapore, Singapore, pp 321–361. https://doi.org/10.1007/978-981-19-4602-8_11

Stadlmayr B, Charrondi`ere UR, Eisenwagen S et al (2013) Nutrient composition of selected indigenous fruits from sub-Saharan Africa. J Sci Food Agric 93 (11):2627–2636. https://doi.org/10.1002/jsfa.6196

Tchoundjeu Z, Atangana AR, Degrande A (2005) Indigenous methods in preserving bush mango kernels in Cameroon. Am J Appl Sci 2:1337–1342

USAID (2020). Promote, plant, protect: U.S. supports Shea parkland restoration initiative and women's economic empowerment in Ghana. Press release, Friday, August 21, 2020. https://2017-2020.usaid.gov/ghana/news-information/press-releases/promote-plant-protect-us-supports-shea-parkland-restoration

Woldemichael A, Salami A, Mukasa A et al (2017) Transforming Africa's agriculture through agro-industrialization. Africa Econ Brief 8(7):1–12

Zhou S, Minde IJ, Mtigwe B (2013) Smallholder agricultural commercialization for income growth and poverty alleviation in southern Africa: a review. Afr J Agric Res 8(22):2599–2608. https://doi.org/10.5897/AJAR11.1040

Ziadat FM, Zdruli P, Christiansen S et al (2022) An overview of land degradation and sustainable land management in the near East and North Africa. Sustain Agric Res 11(1):11–24. https://doi.org/10.5539/sar.v11n1p11

Chapter 10
Germplasm Systems of Multipurpose Indigenous Trees in Africa

Simon A. Mng'omba, Festus K. Akinnifesi, Crispen Marunda, Marie L. Avana-Tientcheu, Gudeta W. Sileshi, and Stephen Syampungani

Abstract There is an increased momentum in global efforts to expand tree planting through afforestation, reforestation, landscape restoration and agroforestry initiatives in response to climate change, biodiversity loss, land degradation, food and nutrition insecurity challenges. The success of these initiatives strongly depends on the availability and accessibility of planting materials of suitable tree species. Stakeholders' efforts to contribute to planting initiatives are constrained by multidimensional factors limiting access to quality germplasm. Tree germplasm refers to all types of planting materials including seeds, seedlings, wildlings and clonal materials. This chapter reviews tree germplasm management and policy issues that are likely to affect planting initiatives in Africa. Tree germplasm quality and supply depend on documented importance which drives germplasm demand at local, national and regional levels. Supply of good quality germplasm is underpinned

S. A. Mng'omba (✉)
Lilongwe University of Agriculture and Natural Resources (LUANAR), Lilongwe, Malawi
e-mail: smngomba@luanar.ac.mw

F. K. Akinnifesi
Food and Agriculture Organization of the United Nations (FAO), Rome, Italy

C. Marunda
Sustainable Timber Tasmania, Level 1, Hobart, TAS, Australia

M. L. Avana-Tientcheu
Département de Foresterie, Faculté d'Agronomie et des Sciences Agricoles, Université de Dschang, Dschang, Cameroun

African Forest Forum, United Nations, Nairobi, Kenya

G. W. Sileshi
Department of Plant Biology and Biodiversity Management, Addis Ababa University, Addis Ababa, Ethiopia

S. Syampungani
Department of Plant and Soil Sciences, Faculty of Natural and Agricultural Sciences, University of Pretoria, Pretoria, South Africa

Chair of Environment and Development, Oliver R. Tambo Africa Research Chairs Initiative (ORTARChI), Copperbelt University, Kitwe, Zambia

P. W. Chirwa et al. (eds.), *Trees in a Sub-Saharan Multi-functional Landscape*,
https://doi.org/10.1007/978-3-031-69812-5_10

by scientific knowledge, decision support tools appropriately designed for end users, institutional capacity and policy frameworks. We review the role of both human and infrastructure capacity development for actors in the germplasm supply chain. We highlight the importance of germplasm conservation and the need for developing, upgrading and fostering quality assurance measures through certification to reduce proliferation of low-quality germplasm. The Community Agroforestry Tree Seed Bank model, an innovative and time-tested approach is proposed for improving and ensuring a decentralised and sustainable tree germplasm supply, empowering smallholder farmers as key players in sourcing and managing quality germplasm in a manner that is in tandem with national germplasm supply systems. We recommend strengthening regulations for quality assurance and providing incentives to stakeholders adhering to germplasm quality standards. We recommend increasing investment for an efficient supply of good quality tree germplasm at community, national and regional levels for a range of benefits that include improving food security, rural livelihoods, and addressing climate change and environmental degradation challenges.

Keywords Certification · Community agroforestry tree seed bank · Germplasm · Propagules · Restoration · Seed systems · Tree nurseries

10.1 Introduction

There is always a need to plant and manage diverse tree species to meet the ever-increasing demand for forest products and services. The increasing advocacy of nature-based solutions to global warming and for abating deforestation has raised the profile of tree planting in international policy debates and high-level agreements (Seddon et al. 2021; Stevens and Bond 2023). Afforestation and reforestation were initially adopted globally and integrated into climate policies through the Kyoto Protocol's Clean Development Mechanism. Recently, the Trillion Trees initiative has captured the imagination of world leaders (Seddon et al. 2021; Stevens and Bond 2023). Several global and continental commitments have been made by governments to restore fragmented and degraded forests, seascapes and landscapes as part of the efforts to contain climate change, reduce land degradation, halt desertification and loss of biodiversity (Akinnifesi 2018; Giacomini et al. 2023; Sileshi et al. 2023).

Ambitious global restoration targets have been set such as the New York Declaration on forests to globally restore 350 million hectares of forests by 2030. Together with the United Nations Decade on Ecosystem Restoration (UNDER 2021–2030), such commitments and targets have triggered major investments and political momentum to increase tree cover and restore landscapes through massive tree planting and a well-managed natural regeneration. The African leaders at the Global Landscape Forum at COP21 in Paris in 2015 pledged to restore 100 million Ha of degraded landscapes by 2030. The African Forest Landscape Restoration initiative (AFR100) is being complemented by other Africa-led initiatives such as the African Landscapes Action Plan (ALAP), Africa Resilient Landscapes Initiative (ARLI), the Great Green Wall (GGW) initiative and the Global Evergreening Alliance.

To date, thirty-four African countries have already committed to restoring at least 129.5 million ha of degraded and deforested landscapes by 2030 (www.afr100.org, accessed on 24th March 2024). These aimed to increase the 'sustainability', resilience and recovery potential of degraded forests and agricultural landscapes (Akinnifesi 2018), and carbon storage, job creation and improve livelihoods (Bond et al. 2019). This includes sustainable productivity, biodiversity conservation, ecosystem services and the agility of the ecosystems to respond to shocks and natural disasters.

Some notable Africa's land restoration commitments in million hectares include: Ethiopia (15), Sudan (14.6), Cameroon (12), Mali (10), DR Congo (8), Tanzania (5.2), Kenya (5.1), Cote D'Ivoire (5), Nigeria (4), Malawi (4) and South Africa (3.6) (Akinnifesi 2018; Bond et al. 2019; Lewis et al. 2019; www.afr100.org).

The GGW alone has a target of restoring 100 million hectares of degraded lands and—a barrier of 'natural wonders' stretching across the entire width of Africa, covering 7775 km from Senegal on the Atlantic coast to Eritrea on the Red Sea coast, and a width of 15 km (Sileshi et al. 2023, www.thegreatgreenwall.org). The concept evolves from planting trees to water harvesting techniques, protecting greenery (including natural tree regeneration) and improving indigenous land use practices such as assisted natural regeneration. Such large initiatives demand a huge supply of good quality tree germplasm from a diverse range of species to achieve the ambitious targets.

The success of such large-scale landscape restoration initiatives, including through domestication of indigenous tree species, that have high potential to adapt to the African environment, is dependent on a steady supply of good quality planting materials (germplasm) for successful tree establishments or through effective and well-managed natural regeneration (Sileshi et al. 2023; Mansourian and Berrahmouni 2021; Kuyah et al. 2023). The genetic and physiological qualities (traits) of tree germplasm play an important role in afforestation, reforestation, agroforestry and other land restoration efforts (PATSPO 2018; Prasad 2023). Hence, enhancing access to high quality germplasm, free from diseases, damage, pests and other deformities, could ensure a high initial tree survival and optimal growth (PATSPO 2018; Prasad 2023).

Over the years, there have been several projects and programmes at local, national, and continental levels related to producing, handling, testing, promoting and distributing tree germplasm for plantation initiatives with multiple and varying objectives, involving various categories of stakeholders including smallholder farmers. Some of the reforestation projects in Africa were designed for earning carbon credits (Africa Forest Carbon Catalyst 2022).

Developing a robust tree germplasm supply system can support large-scale landscape restoration initiatives. Existing literature has emphasized that inadequate, low-quality planting material and ineffective germplasm delivery system have been major constraints to massive adoption of agroforestry and other tree plantation initiatives in Africa, especially for multipurpose indigenous trees (Akinnifesi et al. 2010; Akinnifesi 2018; Avana-Tientcheu et al. 2019; Nyoka et al. 2011a). Indigenous tree species considered as candidates for domestication are woody perennials found in natural ecosystems that provide a wide range of products for human livelihoods for food, fibre, oils, timber, medicine, energy, as well as the provision of environmental services such as soil conservation and amelioration, water catchment and regulation, windbreaks and climate change mitigation (Sileshi et al. 2007; Avana-Tientcheu et al. 2019). An evaluation of the seed supply system for indigenous tree species in Burkina Faso, Cameroon, Ghana and Kenya (four countries committed to increase tree cover) indicated significant gaps (Giacomini et al. 2023). This was attributed to inadequate political and institutional environment, low involvement of local communities in the seed systems, weak policies, capacity gaps in germplasm and tree improvement as well as low investment in tree seed systems. Afforestation, reforestation and successful scaling up of landscapes restoration and agroforestry hinge on the development of a sustainable tree germplasm supply system (Akinnifesi et al. 2008). Similarly, reviews of tree breeding and tree germplasm supply, and demand in Africa by the African Forest Forum showed a decreasing trend in research, development investments and widening gaps in the supply of good quality germplasm (Marunda et al. 2017).

This chapter highlights the contribution of research in identifying and improving indigenous MTPs, and to a lesser extent, adapted and naturalised exotic MPTs that can be planted for a variety of end-uses. It also explores options for tree germplasm management focusing on the demand, supply, distribution, and the conservation of indigenous multipurpose tree species as part of the Africa' approaches for sustainable management of forest resources as well as a strategy for inclusive restoration of degraded lands. The chapter also looks at national and regional policies to ensure intra-Africa supply and exchanges/transfer of good quality tree germplasm for successful tree establishment. Moreover, investment in research, management, improvement and enabling regulations along the tree germplasm value chain are important for successful establishment of MPTs in restoration and agroforestry initiatives.

10.2 Multipurpose Tree Germplasm

Tree germplasm also known as propagules (see Box 10.1) is a broad terminology which increases the complexity of tree germplasm management and offers an increased scope for exploring options available for tree growing. Over the last three decades many indigenous food and non-food tree species in many parts of Africa, described as multipurpose trees have been tested for various vegetative propagation techniques in the framework of tree domestication programme (Ky-Dembele et al. 2016; Leakey et al. 2022). Some of these species such as *Adansonia digitata* (Baobab), *Sclerocarya birrea* (Marula tree), *Vitellaria paradoxa* (Shea tree), *Allanblackia* species (Tallow tree), *Prunus africana* (African Plum tree), *Pterocarpus erinaceus* (African kino tree), *Pterocarpus santalinoides* (Mututi), *Tamarindus indica* (Tamarind) and *Ziziphus mauritiana* (Jujube) are classified as important pan-African priority species because they are collected, utilized and traded in at least two African regions. Others are valued at national and regional levels, for example *Uapaca kirkiana, Strychnos cocculoides* and *Flacourtia indica* in southern Africa (Akinnifesi et al. 2008, Akinnifesi 2009); *Cola* spp., *Dacryodes edulis, Irvingia gabonensis, Ricinodendron heudolotii,* in humid lowland of Central Africa (Kanmegne et al. 2017, Tchoundjeu et al. 2008); and *Anogeissus leiocarpus, Detarium microcarpum, Khaya senegalensis, Parkia biglobosa* in West Africa and Sahel (Kalinganire et al. 2008, Tekhlehaimanot 2008). Growth and development cycles for many of these indigenous tree species are characterised by long juvenile phases if derived from seeds or any juvenile planting material. Recent research works have demonstrated that this juvenile phase can be substantially reduced by vegetative propagation techniques (Tsobeng et al. 2017).

Box 10.1: Defining Tree Germplasm

Tree germplasm refers to all types of planting materials which include seeds, seedlings, wildlings and clonal (vegetatively propagated) materials such as cuttings, grafts and marcots (Kuyah et al. 2023; Akinnifesi et al. 2008). Among the different types of propagules, seed is the most common, easy to handle and cheap to produce for many tree species (Akinnifesi et al. 2008; Mng'omba et al. 2008).

Vegetative propagation (asexual propagation) has a very long use in forestry, agroforestry and horticulture as a reproduction method where vegetative parts of a plant (root, scion, stem, flowers or leaf) are used to produce new plants through grafting, budding, cutting, air layering (marcotting) and micropropagation (tissue culture). These methods are used to produce clonal planting materials (true to type) hence capturing superior traits such as desired fruit size and taste, and tree growth (tree dwarfing). The use of mature materials in vegetative propagation enables precocious fruiting in fruit tree species (Mng'omba et al. 2016, Kalinganire et al. 2008). However, the application of vegetative propagation techniques requires some technical knowledge and skills to successfully produce high-quality propagules for tree planting. Micropropagation (*in vitro* propagation) requires, in addition to specialised skills, an aseptic laboratory environment to produce plantlets from plant cells, tissues and organs (explants) (Hartman et al. 2017).

An assessment of the growth, flowering and fruiting of rooted cuttings, grafted trees and seedlings of *Allanblackia floribunda* in Cameroon showed that grafted trees started flowering and fruiting three (3) years after planting, while trees derived from seedlings started flowering and fruiting after 4 and 6 years, respectively, while rooted cuttings did not flower more than eight years after planting (Tsobeng et al. 2017; Akinnifesi et al. 2009). In the Sahel, grafted fruit trees started fruiting three (3) months, 2 and 3 years after planting respectively for *Ziziphus mauritiana*,

Vitellaria paradoxa and *Adansonia digitata*, while the seedlings flowered and fruited after 3, 10 and 20 years respectively (Kalinganire et al. 2008). Flowers and/or fruits from such early fruiting grafted trees while in the nursery must be removed to allow complete wound healing at the graft union as well as adequate vegetative growth to support a heavy fruit load later in orchards (Mng'omba et al. 2016).

Tree seeds are also important for vegetative propagation such as grafting as the seedlings (raised from seed) are usually used as rootstocks. Seedlings develop normal and functional roots unlike cuttings and marcots (Mng'omba et al. 2008). Tree propagation technique used to produce propagules has implications on growth, development, multiplication, genetic integrity, ease of distribution and dissemination pathways, costs and field performance (Akinnifesi et al. 2009, Tsobeng et al. 2017). The use of seed enables natural genetic variation due to random pollination, allowing selection and improvement of superior trees within a large population to be achieved (Dawson and Were 1997).

In addition, germplasm multiplication, may involve the rejuvenation of germplasm accessions, which may involve sowing seeds, or planting vegetative materials, for the purpose of enhancing the production and harvesting of the seeds or planting materials with the same genetic characteristics as the original population.

In Africa, there have been several studies on MPT germplasm production and management (Akinnifesi et al. 2010; Garrity et al. 2010; Tchoundjeu et al. 2010). This led to two main types of MPTs for germplasm development based on project-led tree species identification and prioritisation by scientists for specific and preferred end-use. The first type, for instance, included tree species used for fodder, fertilisation, nuts, fruits and fuelwood, and which are usually fast growing. The second type of MPT are trees on farmland often retained, nurtured, and managed by farmers for diverse uses such as timber, bark, roots, flowers, fruits, nuts, seed oils, edible leaves, fodder and others. This categorisation helps in understanding and articulating the research, management and policies in regard to tree germplasm development.

10.3 Research on Tree Germplasm in Africa

There have been several research studies (see Table 10.1) and development work aimed at scaling up conservation, management and domestication of indigenous multipurpose tree species in Africa. Such studies included determining seed traits, propagation methods, seedling growth and survival, stock plant management, and seedling response to water and genetic variation (Table 10.1). Much of the research on tree germplasm focused on testing and characterization, but rarely targeted seed quality traits, except for the collection and germination. Other studies were on established multipurpose tree trials derived from different propagules and such studies included provenance-progeny trials, determining genetic or phenotypic variations for best utility of the propagules in terms of tree survival, growth, and fruit productivity as well as responses to edaphic-climatic factors, management, pest

Table 10.1 Assessment of seed of multipurpose trees in Africa using provenance-progeny trials

Species and major uses	Type of provenance/ progeny trial & Sites	No. of Provenances tested	Site & Country	Parameters measured	References
Uapaca kirkiana (fruit)	Field trial. Germplasm from Malawi, Tanzania, Mozambique, Zambia and Zimbabwe	16	Makoka, Malawi	Field establishment, growth, survival	Akinnifesi et al. (2004)
Vachelia abyssinica (craft, firewood, charcoal, fertility, edible gum)	Nursery assessment	4	Debub Omo, Ethiopia	Seed traits and seedlings growth	Tsegaye et al. (2021)
Sclerocarya birrea (fruit)	Field trial: Germplasm collection from Malawi, Tanzania, Mozambique, Zambia and Zimbabwe.	22	Mangochi, Malawi	Growth, survival, fruiting (precocity)	Nyoka et al. (2015)
Adansonia digitata	Field trial	6	Banguineda, Cinzana & Samanko, Mali	Growth, leaf production	Kalinganire et al. (2008) and Akinnifesi et al. (2008)
	Fiel trials	25	Sadore, Niger	Survival, growth	Kalinganire et al. (2023)
	Field trial and nursery assessment	30	Samanko, Mali	Leaf morphology	Korbo et al. (2012)
Vitellaria paradoxa (shea butter, medicines)	Field trial	5	Gonse, Burkina Faso	Growth	Kalinganire et al. (2008), Akinnifesi et al. (2008), and Bayala et al. (2009)
Tamarindus indica, Adansonia spp., *Ziziphus mauritiana, Vitellaria paradoxa*	Field Trials Burkina Faso, Mali, Senegal	13	Gonse, Dinderesso, Burkina Faso	Growth	Kalinganire et al. (2008)
Faidherbia albida (soil fertility, fodder, shade)	Nursery assessment	6	Debub Omo, Ethiopia	Seed traits and seedlings growth	Tsegaye et al. (2021)

(continued)

Table 10.1 (continued)

Species and major uses	Type of provenance/ progeny trial & Sites	No. of Provenances tested	Site & Country	Parameters measured	References
F. albida (soil fertility, fodder, shade)	Southern and Eastern Africa provenances	5	Kenya	Seedling response to water; genetic variation	Koech et al. (2016) and Dangasuki et al. (1997)
Balanites aegyptiaca (fruit, leaves, fodder, flowers, medicine)	Nursery assessment	4	Debub Omo, Ethiopia	Seed traits and seedlings growth	Tsegaye et al. (2021)
Cordia africana (timber, woodlot)		12	Ethiopia	Seed traits and seedlings growth	Derero et al. (2017)
Khaya senegalensis (timber)			Burkina Faso	Seed traits and seedlings growth	Ky-Dembele et al. (2014)
Senegalia senegal (Arabic gum, food additive, woodlot, fodder)	Field trial	4	Bambey, Dahra, Senegal; Cinzana, Segou, Mali	Growth and survival	Diatta et al. (2021)

infestation and rhizobia inoculation (Tchiechoua et al. Tchiechoua et al. 2020, Tsobeng et al. 2017).

Research studies for many indigenous multipurpose tree species have been ongoing. These include provenance trials, tree improvement on farmland, propagation systems, field management, economic and market analyses. The seeds from many indigenous tree species are still harvested from the wild as planting of such tree species is uncommon. Some challenges that stalled domestication and utilization of tree species (Leakey and Akinnifesi 2008) include (1) research and development have been done on a few tree species; (2) limited investments in scaling up research findings and hence their slow application and limited use to conclude furthering domestication process; (3) slow pace and low interest in funding long-term research (multi-year cycles) that could allow evaluation of indigenous tree species on farmland and better inform their cultivation and (5) to evaluate them on farmland and limited awareness of the contributions of indigenous MPTs among many age groups. Also, the end-users of MPTs are diverse and so are the end products and services. The target end users are usually rural farmers whose needs are continuously changing, who are risk-averse and may take long time to adopt new technologies (Tembani et al. 2014).

There have been some research studies on germplasm supply, quality and delivery of indigenous tree species in Africa (Lillesø et al. 2021; Dawson et al. 2013; Nyoka et al. 2011a). Akinnifesi et al. (2009) reported greater fruit yield, early tree growth and better dry matter allocation in *U. kirkiana* derived from marcots than those derived from grafted trees eight years after establishment. Mwase et al. (2010) reported wide genetic variation within and between *U. kirkiana* populations derived from seeds using morphological and amplified fragment length polymorphism (AFLP) markers. Tsobeng et al. (2017) observed in Cameroon greatest height for *Allanblackia floribunda* trees derived from seedlings as compared with those from vegetatively propagated materials although trees generated from grafts started flowering and fruiting 3 years earlier than the others. Tree growth and performance are influenced by the origin and progeny of the seeds they derive from, and this is with respect to tree growth, form, crown size, seed production, tolerance to stresses, and nutrient content and other factors. Seeds collected from the same individual tree exhibit progeny with distinctive characteristics.

Lillesø et al. (2021) reported a broken chain in the delivery of tree germplasm to the growers by NGOs and other project implementers. They reported that commercial seed dealers and private tree nurseries were left out in the network despite their enormous potential to ensure sustainable supply of good quality germplasm. Consequently, there was an emphasis on germplasm quantity rather than quality (Lillesø et al. (2021).

The testing and improvement of tree germplasm led to the development of agroforestry and other afforestation and reforestation initiatives in Africa. For agroforestry, early testing and assessment of seeds from numerous multipurpose woody species used for improved fallows and alley cropping development began in the 1980's. These also include cultivars and accessions, especially for nitrogen fixing leguminous trees used for soil fertility improvement as well as many indigenous food-bearing (fruit, seed and nut) trees (see Table 10.2).

10.3.1 History of Priority Tree Species Identification for Different End-Uses

The main groups of MPTs widely planted in Africa are listed in Table 10.2. These are usually found in the forests, scattered trees on-farms and riverbanks, protected areas, homesteads and other places. They serve different important roles in landscapes where they grow and contribute to the tapestry of trees in landscapes and ultimately to restoration efforts. Sileshi et al. (2007) listed the roles of MPTs as follows: provision of food, medicines, fibres, timber, energy and fodder; erosion control, wind breaks, micro-climate modification, mitigating desertification, carbon sequestration, control of crop pest, biodiversity conservation and cultural services.

Table 10.2 Important indigenous fruit tree species for sub-Saharan Africa

Indigenous fruit tree species	Region / country	References
Artocarpus communis, Chrysophyllum albidum, Dacryodes edulis, Garcinia cola, Irvingia gabonensis, Ricinodendron heudelotti, Treculia africana	**West and Central Africa** Cameroon, Nigeria, Equatorial Guinea, Gabon, DR Congo, Ghana	Franzel et al. (2008) and Tchoundjeu et al. (2006, 2008)
Balanites aegyptiaca, Bussea occidentalis, Cola lateritia, Heritiera utilis, Parinari excelsa, Pentaclethra macrophylla	Sierra Leone	Jusu and Cuni-sanchez (2016)
Adansonia digitata, Balanites aegyptiaca, Lannea microcarpum, Parkia biglobosa, Prosopis africana, Saba senegalensis, Sclerocarya birrea, Tamarindus indica, Vitellaria paradoxa, Ximenia americana, Ziziphus mauritiana	**West Africa (Sahel region)** Guinea, Senegal, Mali, Burkina Faso, Niger	Kalinganire et al. (2023), Raebild et al. (2011), and Franzel et al. (2008)
Adansonia digitata, B. aegyptiaca, Berchemia discolor, Borassus aethiopium, Carissa edulis, Cordeaux edulis, Myrianthus holstii, Parinari curatellifolia, Sclerocarya birrea, T. indica, Strychnos cocculoides, Vangueria madagascariensis, Vitellaria paradoxa, Z. mauritiana	**Eastern Africa** Burundi, Ethiopia, Kenya, Sudan, Tanzania, Uganda, Ethiopia Rwanda	Teklehaimanot (2008) and Jama et al. (2007)
A. digitata, P. curatellifolia, S. birrea, S. cocculoides, Uapaca kirkiana, Z. mauritiana	**Southern Africa** Malawi, Mozambique, Zimbabwe, Zambia & Tanzania	Akinnifesi et al. (2008)

Research on MPTs has been to some extent considerably done in West and Central Africa (Cameroon, Nigeria and Ghana) and East and Southern Africa (Kenya, Malawi, Zambia and Zimbabwe), where priority IFTs were identified (Table 10.2). Identification of priority indigenous multipurpose tree species was done with the aim of domesticating them on farmland. The domestication process described as participatory was a farmer-driven and market-led process aimed at building on farmers' knowledge and market information for selecting priority tree species, their integration in human-managed landscape for improved quality and quantity production to meet households needs and market demand (Tchoundjeu et al. 2010; Akinnifesi et al. 2006; Simons and Leakey 2004). This involved, for each identified priority species, selection of superior trees and improvement based on end-users' preferred traits for enhanced utilization and marketing of selected products (Akinnifesi et al. 2006). Selection of superior trees used a participatory strategy (Fig. 10.1) with the help of farmers, traders, and other end- users to select superior genotypes based on preferences and market demand for priority species' products (Leakey and Akinnifesi 2008).

The collection and characterization of superior genotypes was followed by development of suitable propagation techniques and nursery management practices to produce improved germplasm for integration into farming systems. The updated list of priority species for each African region has been reviewed in Leakey et al.

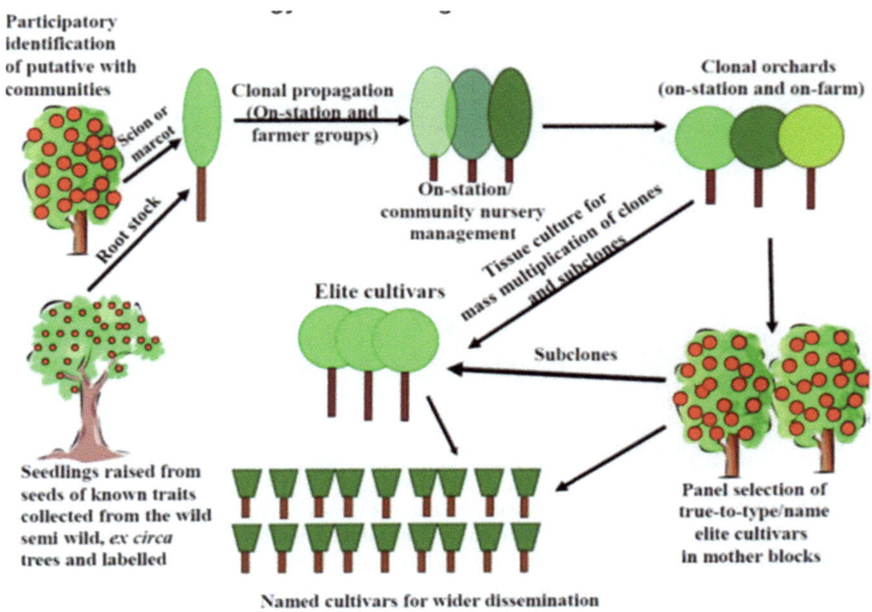

Fig. 10.1 Strategy for domesticating indigenous fruit trees in Southern Africa using participatory clonal propagation approach. (Akinnifesi et al. 2006)

(2022) with highlights of progress achieved for their germplasm development and deployment to support their domestication pathways.

This community led participatory domestication strategy based on clonal propagation has been documented for southern Africa (Fig. 10.2) as well as in West and Central Africa.

In both regions, vegetative propagation techniques such as grafting and marcotting have addressed the long juvenile phase challenge exhibited by many trees derived from seedlings. Asaah et al. (2014) reported a shortened juvenile phase (4 years) to fruiting of Allanblackia floribunda in Cameroon when grafted but takes 12 years to fruiting when derived from seedlings (juvenile material).

10.3.2 Case Studies of Timber sp., Acacia sp., Vachellia sp. and Senegalia sp.

The African savannas have a diversity of timber species with high commercial values. There are about 62 timber species in dry forests and woodlands in sub-Saharan Africa including *Afzelia* spp., *Combretum* spp., *Entandrophragma* spp., *Khaya spp.* and many others (Sitoe et al. 2010). Over the years, these species have been under pressure for harvesting due to limited diversity of timber species in these

Fig. 10.2 Selection, propagation and establishment of *Uapaca kirkiana* elite cultivars in Malawi (**a**) participatory rural appraisal with communities, (**b**) rural community members indicating plausible fruit sizes that could be identified in the wild, (**c**) a tree branch of identified putative cultivar in the wild for clonal propagation, (**d**) ICRAF agroforestry tree nursery at Makoka, Malawi, (**e**) *U. kirkiana* provenance trial at Makoka, Malawi and (**f**) *U. kirkiana* clonal fruit orchard. (Akinnifesi et al. 2006, 2008)

environments (Syampungani 2008) hence an increased risk of extinction of the timber species with commercial values in the African dry forests and woodlands (Swart and Caro 2002). In view of this, there is a need to develop appropriate silvicultural systems to support the development and sustainability of such species.

African acacias are important MPTs in many African countries as they provide gum arabic (exudate from tree stems or branches), fodder for both livestock and wildlife, fencing, charcoal and firewood. Gum arabic is one of the oldest articles of trade from North, East and West Africa and contributes significantly to the economies of the African countries with natural and artificial plantings of the species. The main gum producing species also planted are *Senegalia senegal* (formerly *Acacia* species) and *Vachellia (*syn. *Acacia) seyal. Vachellia* sp. is well-known for providing initial ground cover after disturbance hence allowing the climax species to recover. Extensive plantations of *Senegalia senegal* are found in Cameroon, Ethiopia, Mali, Kenya, Nigeria and Sudan.

Faidherbia albida is a leguminous woody tree widely distributed throughout Africa and the Middle East and promoted in many African countries for fodder and for improving soil condition (Barnes and Fagg 2003). *Vachelia sp.* or *Senegalia sp* and *F. albida* trials and planted stands have been established in several African countries (Tsegaye et al. 2021; Koech et al. 2016).

In Africa, the vast majority of (leguminous) trees that have been successfully used in agroforestry as fertilizer trees, for fodder and in rotational woodlots are mainly exotic or introduced species. A number of them were introduced many years ago and have already been naturalised, so will not be covered in this chapter (which are outside the scope of this chapter). However, a few of them are native to Africa, such as *Faidherbia albida* and *Tephrosia vogellii*. These fertilizer trees are generally established from seeds. For some of these species, seeds require pre-treatments before sowing, while others readily germinate once sown. Seed pre-treatments for the studied fertiliser tree species include: (i) soaking in water at room temperature (*G. sepium*) or at hot or boiling temperature (*F. albida*) for those with hard seed coats; as well as (ii) mechanical treatments (nicking, cracking and scratching seeds).

10.4 Actors in Tree Germplasm, Collection, Assemblage and Distribution

Tree seed supply systems are composed of actors (governments, organisations, individuals) involved in different roles and functions (including breeding, multiplication, testing, processing, storage and marketing) towards seed production and distribution to meet different tree planting objectives. Such actors interact and influence germplasm availability, access, distribution, and use along the tree germplasm system (Avana-Tientcheu et al. 2019). Graudal et al. (2021) defined a tree seed system as the way that actors organize the collection and distribution of tree seeds and seedlings (i.e., from seed collection to the growers' planting of seedlings). Such systems may be *ad hoc* seed collection or a coordinated collection of quality seeds, based on standard procedures. According to Graudal et al. (2021), a well-organized seed system should have good established and maintained seed sources, germplasm quality control measures, good seed storage and nursery propagation facilities, and efficient delivery pathways to end-users.

A recent analysis (Fig. 10.3) indicated that tree germplasm supply systems in the Tropics could be categorized as 'formal' and 'informal'. The formal tree germplasm system involves public and private organisations with designated roles in production, distribution and regulation, while the informal system is composed of private households, farmers and NGOs sharing among themselves planting materials from a broad range of tree species for provision of a variety of products and services. The formal system could be referred to as the centralized government model, and the informal system comprising both the NGO and the decentralized models according to the description made by Avana-Tientcheu et al. (2019) for tree seed deployment systems in Central and Western African countries.

The informal system, which generally involves local smallholder farmers, private enterprises or agri-seed dealers, NGOs, nursery operators and some governmental organisations has been instrumental in supporting the distribution of tree germplasm in many African countries (Graudal et al. 2021). According to Lillesø et al. (2011), large quantities of germplasm are from the informal seed sector. National governments play an important role in regulating and controlling seed supply systems. The National Tree Seed Centres (NTSCs) in many African countries are mandated to regulate tree seed systems and provide advisory services to potential tree growers. However, very strict regulations (control) can hinder the development of vibrant seed supply systems. Roshetko and Verbist (2000) recognized three pathways in which improved tree germplasm can be made available to farmers. These include (i) distribution of seed/germplasm from national and regional research centres, community-based NGOs and the private sector; (ii) dissemination of selected seeds from farmers and farmer groups; and (iii) direct diffusion through informal farmer-to-farmer exchange.

The review of tree germplasm management showed that communities are active players in tree germplasm collection and distribution. Local farmers have indigenous knowledge on many multipurpose tree species since most of the species (identified in the priority species exercises) are from their surrounding woodlands, forests or managed landscapes. This allows them to be key players in the collection and distribution systems (see Fig. 10.3). With the drive to increase tree planting for land restoration and climate change mitigation, there is need for policies and strategies to empower farmers to identify seed zones, facilitate their certification and protection by state actors, get capacitated and involved in monitoring flowering, fruiting and

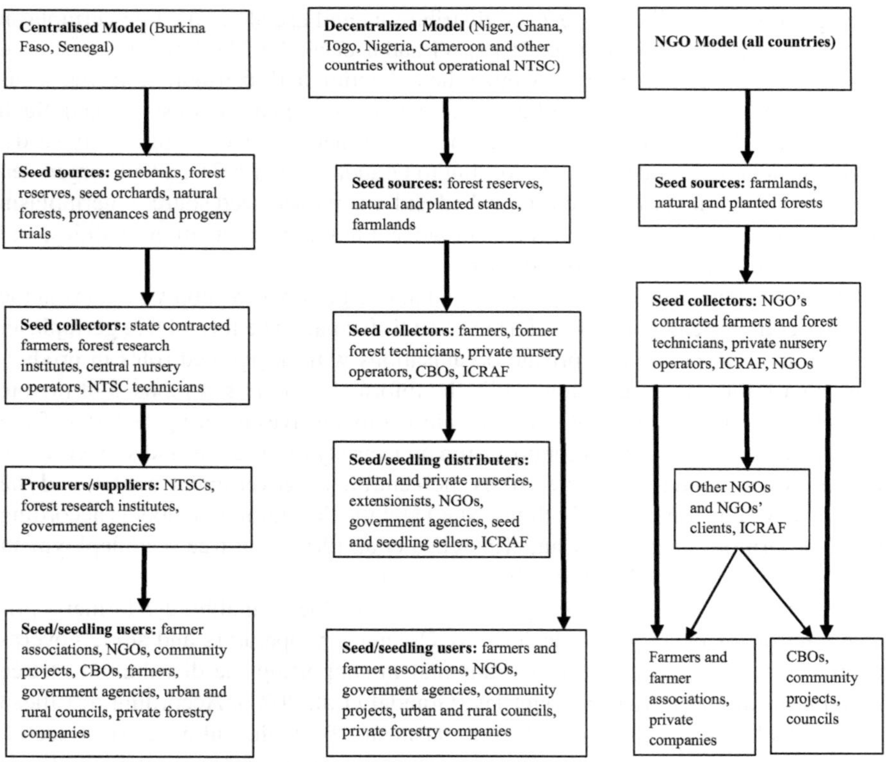

Fig. 10.3 Tree germplasm procurement and deployment pathways in West and Central African countries. (Avana-Tientcheu et al. 2019 adapted from Nyoka et al. 2015)

seed production for the identified superior stands (citizen science using mobile phone technology), guide the development of farmers' appropriate germplasm management technologies, and support suitable investment to formalize CATS Bank model.

Research and development on tree germplasm in Africa for selected priority species, has contributed to enhanced availability of tree germplasm through the following initiatives: (i) building stakeholder partnerships and model tree seed systems; (ii) strengthening breeding programmes for indigenous tree species; (iii) collecting and conserving genetic resources and making available diverse tree germplasm to support delivery and use; and (iv) supporting the development of enabling environment for the effective supply of tree seeds and seedlings (Leakey et al. 2022; Graudal et al. 2021). For example, some initiatives of the FTA programme contributed to the availability of tree germplasm to potential growers in Africa as well as to build capacity of people and local organizations in propagation techniques, farm management and other skills (Graudal et al. 2021). These initiatives include: (i) the Africa Orphan Crop Consortium aimed to address production gaps for 101 lesser-known food tree species through the development and delivery of improved tree germplasm to farmers; (ii) the 'fruit tree portfolio' established in 17 localities in East Africa, which involved working with local communities to determine a list of food tree species and support the production of their germplasm in local nurseries owned by farmers' groups, nursery businesses, and schools (Graudal et al. 2021); and (iii) the Rural Resource Centers (RRCs), which have become centres of training and demonstration on tree germplasm to communities (Odoi et al. 2019). Taking the example of the Rural Resource Centres (RRCs) and their satellite nurseries, they have contributed to the delivery of hundreds of thousands of vegetatively propagated planting material from a diverse range of trees for the diversification of cocoa, coffee and other production systems in order to supply quality food, sequester carbon and enhance resilience of people and landscapes (Avana-Tientcheu et al. 2019; Degrande et al. 2015).

In African countries, different types of nurseries are involved in the production and delivery of tree germplasm, from industrial and semi-industrial centralized nurseries managed under NTSCs to small-scale nurseries managed by training and research institutions, individual and community based and non-governmental organizations. The performance of many of the nursery operators involved in tree germplasm production strongly depends on their communication networks, their technical and financial capacity. An assessment done a few years ago indicated that many of them are not skilled in appropriate methods of seed collection, handling and processing for the production of improved planting materials (Degrande et al. 2015; Takoutsing et al. 2014). Consequently, planting materials produced and disseminated are always sourced from nearby natural and planted stands without proper documentation to trace their origins, monitor their genetic quality and guide their use to ensure climate-resilient plantation outcomes (Avana-Tientcheu et al. 2019).

Whilst NTSCs have been effective in distributing seeds for large scale commercial planting mainly for exotic species used in commercial forestry (through upstream linkages with tree breeding and seed orchard management) they have not been effective in distributing indigenous tree species. Many NTSCs lack reliable funding mechanisms to sustain and adapt their supply capacity to meet the increasing demand for diversified and quality tree germplasm in Africa (Avana-Tientcheu et al. 2019; Jalonen et al. 2017). This gap could be considered partly filled by international organisations through programmes and projects such as the FTA initiative as well as seed exchanges amongst farmers and other actors involved in the informal and decentralized germplasm delivery networks. Although significant progress has been made by African governments to strengthen national capacity for tree germplasm production and delivery in support to their ambitious commitments for forest landscape restoration, the current situation indicates that the contribution of the informal germplasm delivery approach to the overall tree germplasm supply system remains high in several African countries.

Tree germplasm supply to the farming communities is largely done through financial support from the development partners (projects) in many African countries. In many NGOs and government-led tree planting programmes in Africa, where tree seedlings are usually distributed free of charge to farmers, germplasm selection is usually guided by low price, hence incentivizing mass production of low quality and insufficiently diverse planting materials, resulting in constraining the achievement of restoration objectives (Jalonen et al. 2017). Countries such as Ethiopia have received new funding through the Norwegian International Climate and Forest Initiative (NICFI) to improve tree seed supply. For instance, the funded Provision of Adequate Tree Seed Portfolio in Ethiopia (PATSPO) was designed to improve the productivity and resilience of forest landscape restoration in the country. The emphasis in all these initiatives and investments was to increase the capacity of sectoral stakeholders including NTSCs and institutional and policy-enabling environments to respond adequately to the ambitious tree seed demand. In many African countries, reforms are underway to strengthen the technical and institutional capacity of the NTSCs for improved supply of quality, quantity and diversified tree germplasm to meet the increased national and regional demand. Such countries include Burkina Faso, Cameroon, Ethiopia, Kenya, Niger, Uganda, Ghana, Mali among others (Giacomini et al. 2023; Jalonen et al. 2017).

10.4.1 Community Agroforestry Tree Seed Bank (CATS Bank) Model

The scaling up of MPTs cultivation depends on the availability of affordable and good quality germplasm to smallholder farmers through strengthening decentralized informal seed supply systems. This entails training smallholder farmers to collect, conserve, manage and trade good quality germplasm in a way that ensures genetic integrity. Such training must ensure skills for efficient establishment, management of nurseries; tree establishment and management on-farms; seed harvest and post-harvest handling; and the germplasm marketing through germplasm and

Box 10.2: CATS Bank Model

To address the problem of accessing tree germplasm, Community Agroforestry Tree Seed Banks (CATS Bank) were introduced in Malawi and Mozambique (Akinnifesi et al. Akinnifesi et al. 2009). This model (Fig. 10.4) was designed as a farmer-led scaling up agribusiness platform for developing sustainable and profitable agroforestry tree seed supply system as an essential pillar of the large promotion of agroforestry in Malawi and Mozambique. The model was developed to promote and scale up decentralized germplasm supply systems in an efficient way. The CATS Bank approach aimed at (i) scaling up the platform built to generate multiple streams of impact from a range of integrated agroforestry practices; (ii) develop and test a framework for sustainable supply and diffusion of tree germplasm and document lessons learnt; and (iii) build capacity of smallholder farmers and development actors. It involves training smallholder farmers to collect, conserve, manage and trade good quality tree germplasm from their own farms.

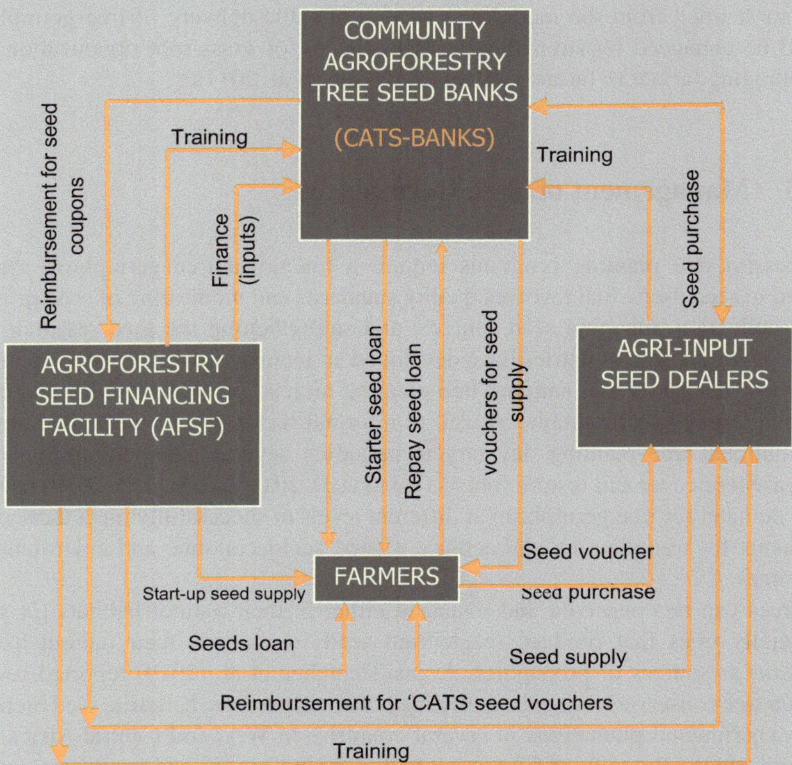

Fig. 10.4 Conceptual framework of the community agroforestry tree seed bank (CATS Bank) approach. (Akinnifesi 2009)

Table 10.3 Community Agroforestry Tree Seed Bank sales to Agro-dealers by country

Country	Number of CATS Banks sold seeds to agro-dealers	Number of farmers sold seeds to agro-dealers	Amount seed traded (Kg)	Amount of money realized (US$)
Malawi	27	160	4000	3400
Mozambique	26	124	3200	2700
Total	53	284	7200	6100

Source: CATS Bank Project Progress Report July to December 2011 (Njoloma, pers. Comm., ICRAF, unpublished data)

tree products standards. The CATS Bank approach was developed and piloted for smallholder farmers (see Box 10.2).

An assessment of the CATS Bank model showed that 67% of farmers involved generated income that contributed to household income (Table 10.3) and increased the number of months of food security (Njoloma J, unpublished data). The model had a potential to scale up seed supply, especially for agroforestry tree species. One of the lessons learned from the model was that sustainable delivery of tree germplasm could be enhanced by strengthening the capacity of grass-root organization and encouraging farmer to farmer exchanges (Nyoka et al. 2011a).

10.5 Management of Tree Germplasm

Successful tree planting programs require a functional tree germplasm system linked to seed users, that involves quality standards and monitoring of seed production, collection, mapping seed-sources, and strengthening the seed research systems. Many African countries have developed an inclusive and integrated approach to meet their set annual national tree planting targets, involving private and commercial forestry sector, public as well as informal sector. For instance, community and national tree planting day targets including several other commitments to increase tree cover and restore forests (Bond et al. 2019; Lewis et al. 2019) hence a high demand for tree germplasm at different levels to successfully meet these commitments for tree planting and achieve related socioeconomic and environmental outcomes.

Trees can be conserved and managed either in their natural habitats (*in situ*), especially trees that produce recalcitrant seeds or outside their natural habitat (*ex situ*) as sources of propagules. Avana-Tientcheu et al. (2019) reported several *ex-situ* tree conservation stands, including botanical gardens, herbaria, seed orchards and experimental plantations in several countries in West and Central Africa. For example, in the Republic of Congo, more than 80 indigenous and exotic tree species are conserved in a 20 Ha arboretum located in Brazzaville, while in Cameroon, a mosaic of natural and planted forest reserves distributed according to national agro-ecological zonation ensures the conservation of the genetic resources of the country rich diversity of MPTs promoted for a variety of tree planting purposes. These sources of tree germplasm (gene banks, mother blocks, breeding seedling orchard) are critical in seed production and must be well-established and maintained to ensure good quality seeds free from pests and diseases, attaining good genetic, physiological and physical standards (Lillesø et al. 2021). A good seed production can lead to an improved seed system.

Tree seeds are easily stored in small spaces such as cold rooms unlike other propagules such as marcots, scions and cuttings which require large spaces and immediate use (Mng'omba et al. 2007). Orthodox seed can be preserved in cold storage (4 °C) and in liquid nitrogen, known as cryopreservation (−196 °C) for several years (Hartmann et al. 2017). Other potential propagule storage methods include *in vitro* and embryogenic axes. Also, a partial desiccation technique has been successful for extending the longevity of recalcitrant seeds from some indigenous MPTs. According to Sacande et al. (2004), 60% of priority sub-Saharan tree species for Forest Genetic Resources (SAFORGEN) indicate good potential for medium to long term conservation, indicating their likelihood for desiccation tolerance. CIFOR-ICRAF maintains a tree seed gene bank in Nairobi, Kenya under the Genetic Resources Unit (GRU). In 2021, the GRU had a wide diversity of about 190 species comprising wild, semi-domesticated and domesticated species (www.genebanks.org/genebanks/icraf, accessed 26 February 2024). The GRU reported availability of 8246 fruit tree accessions in its gene bank in 2021, but no vegetatively propagated fruit tree accessions in cryopreservation or *in vitro* as a safety duplicate. Also, there were 6213 seed accessions for MPTs and 1161 seed accessions held in long-term storage and safely duplicated at two levels, although none for vegetatively propagated materials (www.genebanks.org/genebanks/icraf).

10.6 Policy Issues on Tree Germplasm

National, regional and global policies on tree germplasm are vital instruments for effective and sustainable delivery of tree germplasm. However, it has been observed that regional and global policies are likely to have a weak implementation at national level. According to Giacomini et al. (2023), legal support, political will and adequate investment in research and capacity building create an enabling environment to enhance the seed supply system. They reported the existence of some policies for native tree species for restoration in Cameroon, Ghana and Kenya despite weak execution and implementation (Giacomini et al. 2023). Although policies on tree germplasm may exist in many African countries, weak implementation could be a common challenge in most of them. Moreover, in some African countries, tree seed quality and access have received little attention at high-level policy and planning (Giacomini et al. 2023; Jalonen et al. 2017).

Good policies promote adoption of trees on farms, hence increasing the demand for germplasm. Ajayi and Place (2012) reported an increase in fruit production, marketing and processing in Kenya due to favourable policies (no government interference) and promoting tree planting on farms. Also, the positive tax incentives for trees grown on farms in the 2005–2006 budget in Kenya promoted tree planting. Another example is in Niger, where land ownership in the forest code indicated that beyond a certain tree cover threshold, farmlands were declared forest and conserved as a public domain. Hence, farmers were not taking care of trees in their farmland in fear of increasing tree cover and losing their land ownership when declared a forest (Ajayi and Place 2012). Again, they reported a requirement for approval and licence fees when felling and selling some parkland tree species on their own farmland hence farmers were compelled to remove all tree seedlings from their farms before they were noticed. When such polices were revised and relaxed, there was an increase of trees on farms (Ajayi and Place 2012). Policies should have the primary objectives of ensuring genetic quality and identity, having documented seed sources, investing in tree improvement and propagation research, developing a production and distribution systems with appropriate controls for quality, mobilizing the participation of local stakeholders in decentralized tree collection and marketing. Unlike commercial species (such as pines, eucalypts, acacias) whose genetic quality is known from decades of research and international collaboration, the genetic quality of indigenous tree species is not known. Nyoka et al. (2011b) recommended the need to certify tree germplasm based on whether the species has been genetically improved, selected from the wild (seed zones) or truth-in labelling source of seed, and this certification depends on information generated from research, and the level of awareness on the importance of genetic quality and freedom from pests and diseases.

Public institutions such as NTSCs have the responsibility to enforce standards and certification including phytosanitary standards for high quality germplasm supply. Certification of vegetatively propagated germplasm (especially grafted trees) is equally important as graft failure due to incompatibilities (growth irregularities) between rootstocks and their scions threatens field establishment of grafted tree. Graft failure might also occur some years in the field after an initial successful tree establishment (Pina and Errea 2005). There is a lack of standards and inspection procedures for grafted trees in many African countries, hence proliferation of grafted trees with weak graft unions and deformities. Due to the lack of regulations, dishonest practices exist among nursery operators where seedlings are sold as grafted trees because the latter fetch high prices. This is common with vendors selling tree germplasm in the streets (also known as mobile tree nursery), who take advantage of absence of regulation and skilled inspectors. If quality is to be assured for vegetatively propagated materials, control mechanisms should be put in place and actors are adhering to the quality control measures. One way to guarantee optimum field performance of tree germplasm in the absence of research is to use seed zones. A more efficient seed zoning system for tree regeneration, could be sourcing seeds from natural populations near the intended areas of planting. This ensures some degree of site species matching and guarantees some success in tree planting.

Most of the studies on MPTs are conducted through international finances and benchmarking of the quality is already in place. To upscale the use of MPTs, goals could be set at national levels to strengthen research on tree germplasm (seed source) development. As noted by Akinnifesi (2018), forest and land restoration failed in the past due to lack of knowledge on how to select the right species, plant and manage them. This is particularly so for trees which do not have a long history of research, limited information on seed sources and inappropriate propagation methods. The land restoration initiative offers a good opportunity for state players and communities to come together and develop technologies and systems that work for tree growers.

10.7 Conclusions and Recommendations

The key conclusions from our review of the literature are that meeting the growing tree germplasm demand is very challenging as NTSCs, national herbaria and other NGO institutions cannot fulfil the ever-increasing tree germplasm demand. The demand for tree germplasm is unpredictable for small nursery operators due to the lack of formal markets. The limited domestic resources and research have constrained the management and use of gene banks (provenance trials, seed multiplication farms and mother blocks) as viable sources of germplasm, especially propagules such as scions, marcots, cuttings as well as recalcitrant and desiccation intolerant seeds that are difficult to conserve. Similarly, there are limitations in use of other seed storage methods such as liquid nitrogen, *in vitro* culture and cold storage facilities to preserve seed viability for a longer period. Inadequate seed quality controls and regulations are leading to supply of low-quality tree germplasm. The limited technical capacity among local nursery operators and NTSCs, has remained a challenge to conservation of many tree species and germplasm supply and delivery system in Africa.

In view of these challenges, it is recommended that quality controls of tree germplasm, including standards, certification as well as regulated tree germplasm or seed market must be put in place. We propose the use of voluntary germplasm certification scheme that could ensure germplasm affordability and availability to potential growers while improving and sustaining the quality standards, especially where regulatory tree germplasm certification has not worked. Promotion of contract tree seed production approach can improve tree germplasm supply, but seed quality has been compromised at times. There is the need to invest in research to finalise many stalled or inconclusive studies on indigenous fruit tree provenances assembled in many experimental plots and farmlands. There is also the need to promote assisted natural regeneration if the AFR100 targets are to be met. We propose the use of indigenous multipurpose tree species in afforestation and reforestation initiatives for high integrity carbon credits through conservation of biodiversity.

We also propose the use of CATS Bank approach as an ideal model to increase tree germplasm access where it has not yet been implemented. Also, RRCs with mother blocks is an ideal approach to improve availability of tree germplasm. There is need to capacitate small tree nursery operators for increased germplasm supply as local nursery operators are unable to propagate many IFTs. Access to credits (currently, limited or no access), and value addition could incentivize indigenous multipurpose tree conservation on farmland.

10.8 Chapter Highlights

In this chapter, the lack of regulations (voluntary or regulatory tree germplasm certification) has been a challenge to support tree germplasm value chain and improve tree germplasm quality and availability. Favourable policies and by-laws can promote tree planting of indigenous MPTs on farms. Despite the presence of NTSC, NGOs and other related institutions, CATS Bank approach, RRCs and nursery operators in tree germplasm management and distribution the demand for tree germplasm is still higher than the supply. Further research in long-term tree germplasm in support of tree domestication and scaling up are further warranted, including provenance trials, clonal propagation and proper management of field gene banks and tree seed orchards. Also, limited vegetative propagation (grafts and marcots) skills among decentralized nursery operators has reduced the availability of true to type propagules for fruit trees.

References

Africa Forest Carbon Catalyst (2022) Catalysing carbon projects in Africa, living carbon series, covering the power of natural climate solutions for communities, biodiversity and the planet

Ajayi OC, Place F (2012) Policy support for large-scale adoption of agroforestry practices: experiences from Africa and Asia. In: Nair PKR, Garrity D (eds) Agroforestry—the future of global land use, vol 9. Springer, Heidelberg, pp 175–201

Akinnifesi FK (2009) Community Agroforestry Tree Seed (CATS) Banks. Unpublished project Report. World agroforestry Centre, ICRAF-Southern Africa Regional Programme, Lilongwe, Malawi

Akinnifesi FK (2018) Africa's forest landscape restoration gathers momentum. Nature and Faune 32(1):3–10

Akinnifesi FK, Kwesiga FR, Mhango J, Mkonda A, Chilanga T, Swai R (2004) Domesticating priority miombo indigenous fruit trees as a promising livelihood option for smallholder farmers in southern Africa. Acta Horticul 632:15–30

Akinnifesi FK, Kwesiga F, Mhango J, Chilanga T, Mkonda A, Kadu CAC, Kadzere I, Mithofer D, Saka JDK, Sileshi G, Ramadhani T, Dhliwayo P (2006) Towards the development of miombo fruit trees as commercial tree crops in southern Africa. For Trees Livelihoods 16:103–121

Akinnifesi FK, Sileshi G, Mkonda A, Ajayi OC, Mhango J, Chilanga T (2008) Germplasm supply, propagation and nursery management of Miombo fruit trees. In: Akinnifesi FK, Leakey RRB, Ajayi OC, Sileshi G, Tchoundjeu Z, Matakala P, Kwesiga FR (eds) Indigenous fruit trees in the tropics: domestication, utilization and commercialization. CAB International Publishing, Wallingford, pp 341–368

Akinnifesi FK, Mng'omba SA, Sileshi G, Chilanga TG, Mhango J, Ajayi OC, Chakeredza S, Nyoka BI, Gondwe FMT (2009) Propagule type affects growth and fruiting of *Uapaca kirkiana*, a priority indigenous fruit tree of southern Africa. HortScience 44(6):1662–1667

Akinnifesi FK, Ajayi OC, Sileshi G, Chirwa P, Chianu J (2010) Fertilizer trees for sustainable food security in the maize-based production systems of East and Southern Africa region: a review. Agron Sustain Dev 30:615–629

Asaah E, Degrande A, Tchoundjeu Z, Biloso A (2014) Agroforestry and tree domestication in Central Africa. In: de Wasseige C, Flynn J, Louppe D, Hiol Hiol F, Mayaux (eds) The forests of The Congo Basin – State of the forest, Weyrich, pp 185–195

Avana-Tientcheu ML, Marunda C, Fongnzossie EF, Kemeuze VA, Ngueguim JR, Mutta D (2019) Tree germplasm management systems and their potential for sustaining plantation forestry in West and Central Africa. African J Rural Develop 4(1):33–63

Barnes RD, Fagg C (2003) *Faidherbia albida*. Monogram and annotated bibliography. Oxford Forestry Institute, Department of Plant Sciences, University of Oxford ISBN: 0857041564

Bayala J, Ouedraogo SJ, Ong C (2009) Early growth performance and water use of planted West African provenances of *Vitellaria paradoxa* C.F. Gaertn (karité) in Gonsé. Burkina Faso Agroforest Syst 75:117–127. https://doi.org/10.1007/s10457-008-9167-9

Bond WJ, Stevens N, Midgley GF, Lehmann CER (2019) The trouble with trees: afforestation plans for Africa. Trends Ecol Evol 34:963–965

CIFOR-ICRAF Genetic Resources Unit (2021). www.genebanks.org/genebanks/icraf. Accessed 26 Feb 2024

Dangasuki OG, Seurel P, Gudu S (1997) Genetic variation in seed and seedling traits in 12 African provenances of *Faidherbia albida* (Del.) A. Chev. at Lodwar. Kenya Agroforest Syst 37:133–141

Dawson I, Were J (1997) Collecting germplasm for trees: some guidelines. Agrofor Today 9(2):6–9

Dawson IK, Guariguata MR, Loo J, Weber JC, Lengkeek A, Bush D, Cornelius J, Guarino L, Kindt R, Orwa C, Russell J, Jamnadass R (2013) What is the relevance of smallholders' agroforestry systems for conserving tropical tree species and genetic diversity in circa situm, *in situ* and *ex situ* settings? A review. Biodiver Conserv. https://doi.org/10.1007/s10531-012-0429-5

Degrande A, Tchoundjeu Z, Kwidja A, Fouepe FG (2015) Rural resource centres: a community approach to extension. Note 10. GFRAS good practice notes for extension and advisory services. GFRAS: Lindau. https://www.g-fras.org/en/good-practice-notes/rural-resource-centres.html. Accessed 26 Feb 2024

Derero A, Tsefaye G, Woldemariam Z (2017) Variation in seed traits and seedling vigour of *Cordia africana* Lam. Provenances in Ethiopia. J For Res 28:925–933. https://doi.org/10.1007/s11676-016-0317-z

Diatta O, Sarr MS, Hansen JK, Diallo AM, Nielsen LR (2021) Survival and growth of *Acacia senegal* (L.) Wild. (*Senegalia senegal* (L.) Britton) provenances depend on the rainfall at the site of origin. Ann For Sci 78(82):4–16. https://doi.org/10.1007/s13595-021-01098-5

Franzel S, Akinnifesi FK, Ham C (2008) Setting priorities among indigenous fruit tree species in Africa: examples from southern, eastern and western Africa Regions. In: Akinnifesi FK, Leakey RRB, Ajayi OC, Sileshi G, Tchoundjeu Z, Matakala P, Kwesiga F (eds) Indigenous fruit trees in the tropics: domestication, utilization and commercialization. CABI, Oxfordshire, pp 1–27

Garrity DP, Akinnifesi FK, Ajayi OC, Sileshi GW, Mowo JG, Kalinganire A, Larwanou M, Bayala J (2010) Evergreen agriculture: a robust approach to sustainable food security in Africa. Food Secur 2:197–214

Giacomini FL, Prempeh JA, Jalonen R, Vinceti B, Ekue M, Bosshard E, Burslem DFRP, Kettle CJ (2023) Seeding African forest and landscape restoration: evaluating native tree seed systems in four African countries. Diversity 15(9):981. https://doi.org/10.3390/d15090981

Graudal L, Lillesø J-PB, Dawson IK, Abiyu A, Roshetko JM, Nyoka I, Tsobeng A, Kindt R, Pedercini F, Moestrup S, Jalonen R, Thomas E, McMullin S, Carsan S, Hendre P, Kettle C, Li Y, Jamnadass R (2021) Tree seed and seedling systems for resilience and productivity. FTA highlights of a decade 2011–2021 series. Highlight No. 2. The CGIAR research program on Forests, Trees and Agroforestry (FTA), Bogor

Hartmann HT, Kester DE, Davies FT, Geneve RL, Wilson SB (2017) Hartmann and Kester's plant propagation principles and practices, 9th edn. Pearson Education, Inc, New York. ISBN-13:978-0-13-448089-3

Jalonen R, Valette M, Boshier D, Duminil J, Thomas E (2017) Forest and landscape restoration severely constrained by a lack of attention to the quantity and quality of tree seed: insights from a global survey. Conserv Lett 00:e12424. https://doi.org/10.1111/conl.12424

Jama BA, Mohamed AM, Mulatya J, Njui AN (2007) Comparing the "Big Five": a framework for the sustainable management of indigenous fruit trees in the drylands of East and Central Africa. Ecol Indic 8:170–179. https://doi.org/10.1016/j.ecolin.2006.11.009

Jusu A, Cuni-Sanchez A (2016) Indigenous fruit trees in the African rainforest zone- insights from Sierra Leone. Genetic Resources and Crop. Evaluation 1–16

Kalinganire A, Weber JC, Uwamariya A, Kone B (2008) Strengthening rural livelihoods through domestication of indigenous fruit trees in the parklands of the Sahel. In: Akinnifesi FK, Leakey RRB, Ajayi OC, Sileshi G, Tchoundjeu Z, Matakala P, Kwesiga FR (eds) Indigenous fruit trees in the tropics: domestication, utilization and commercialization. CAB International Publishing, Wallingford

Kalinganire A, Savadogo P, Ky-Dembele C, Bush D, Raebild A, Tougiani A, Muchugi A, Tata Traore FT, Kone B (2023) Evidence of early genetic variation in survival and growth traits of baobab (*Adansonia digitata* L.) – the way forward for domestication and breeding. Agrofor Syst 97:1221–1231. https://doi.org/10.1007/s10457-023-008337-z

Koech G, Ofori D, Muigai AWT, Muriuki J, Anjarwalla P, De Leeuw J, Mowo JG (2016) Variation in the response of eastern and southern Africa provenances of *Faidherbia albida* (Delile A. Chev) seedlings to water supply: A greenhouse experiment. Glob Ecol Conserv 8:31–40

Korbo A, Sanou H, Ræbild A, Jensen JS, Hansen JK, Kjaer ED (2012) Comparison of East and West African populations of baobab (*Adansonia digitata* L.). Agrofor Syst 85:505–518. https://doi.org/10.1007/s10457-011-9464-6

Kuyah S, Buleti S, Dimobe K, Nkurunziza L, Moussa S, Muthuri C, Oborn I (2023) Farmer managed natural regeneration in Africa: evidence for climate change mitigation and adaptation in drylands. In: Dagar JC, Gupta SR, Sileshi GW (eds) Agroforestry for sustainable intensification of agriculture in Asia and Africa. Springer Nature, pp 53–88

Ky-Dembele C, Tigabu M, Bayala J, Oden PC (2014) Inter- and intra-provenances variations in seed size and seedling characteristics of *Khaya senegalensis* A. Juss in Burkina Faso. Agroforest Syst 88:311–320. https://doi.org/10.1007/s10457-014-9684-7

Ky-Dembele C, Bayala J, Kalinganire A, Traoré FT, Koné B, Olivier A (2016) Vegetative propagation of twelve fodder tree species indigenous to the Sahel, West Africa. Southern For 1-8:185–192. https://doi.org/10.2989/20702620.2016.1157980

Leakey RRB, Akinnifesi FK (2008) Towards a domestication strategy for indigenous fruit trees: clonal propagation, selection and the conservation and use of genetic resources. In: Akinnifesi FK, Leakey RRB, Ajayi OC, Sileshi G, Tchoundjeu Z, Matakala P, Kwesiga FR (eds) Indigenous fruit trees in the tropics: domestication, utilization and commercialization. World Agroforestry Centre/CAB International Publishing, Nairobi/Wallingford, pp 28–49

Leakey RRB, Tientcheu Avana ML, Awazi NP, Assogbandjo AE, Mabhaudhi T, Henndre PS, Degrande A, Hlahla S, Manda L (2022) The future of food: domestication and commercialization of indigenous food crops in Africa over the third Decade (2012–2021). Sustain For 14:2355. https://doi.org/10.3390/su14042355

Lewis SL, Wheeler CE, Mitchard ETA, Koch A (2019) Regenerate natural forests to store carbon. Nature 568:25–28

Lillesø J-BL, Graudal L, Moestrup S, Kjær ED, Kindt R, Mbora A, Dawson I, Muriuki J, Ræbild A, Jamnadass R (2011) Innovation in input supply systems in smallholder agroforestry: seed sources, supply chains and support systems. Agrofor Syst 83:347–359

Lillesø J-PB, Dawson IK, Graudal L, Jamnadass R (2021) Quality seed for tree planting: supporting more effective agroforestry and forest landscape restoration by learning from crop integrated seed system development. ICRAF policy brief 54. World Agroforestry, Nairobi, Kenya

Mansourian S, Berrahmouni N (2021) Review of forest and landscape restoration in Africa. FAO and AUDA-NEPAD, Accra. https://doi.org/10.4060/cb6111e

Marunda CT, Avana-Tientcheu ML, Msanga HP (2017) Situational analysis of tree breeding and tree germplasm supply in Africa: underpining sustainable forest management. African forest forum working paper series. Nairobi

Mng'omba SA, du Toit ES, Akinnifesi FK (2007) Germination characteristics of tree seed: spotlight on southern African tree species. Tree For Sci Biotechnol 1:81–88

Mng'omba SA, Akinnifesi FK, Sileshi G, Ajayi OC, Chakeredza S, Mwase WF (2008) A decision support tool for propagation Miombo indigenous fruit trees of southern Africa. Afr J Biotechnol 25:4677–4686

Mng'omba SA, Sileshi G, Nyoka BI (2016) *Uapaca kirkiana* phenological phases and effect of fruit thinning on fruit traits in Malawi. Acta Hort 1119. https://doi.org/10.17660/ActaHort.2016.1119.12

Mwase WF, Akinnifesi FK, Stedje B, Kwapata MB, Bjørnstad A (2010) Genetic diversity within and among southern African provenances of *Uapaca kirkiana* Muell. Arg using morphological and AFLP markers. New For 40:383–399

Nyoka BI, Mng'omba SA, Akinnifesi FK, Ajayi OC, Sileshi G, Jamnadass R (2011a) Agroforestry Tree Seed Production and Supply Systems in Malawi. Small-scale Forestry 10:419–434

Nyoka BI, Ajayi OC, Akinnifesi FK, Chanyenga T, Mng'omba SA, Sileshi G, Jamnadass R, Madhibha T (2011b) Certification of agroforestry tree germplasm in Southern Africa: opportunities and challenges. Agrofor Syst 83:75–87. https://doi.org/10.1007/s10457-011-9393-4

Nyoka BI, Chanyenga T, Mng'omba S, Akinnifesi FK, Sagona W (2015) Variation in growth and fruit yield of populations of *Sclerocarya birrea* (A. Rich) Hochst. Agrofores Syst 89(3):397–407

Odoi JB, Buyinza J, Okia C (2019) Tree seed and seedling supply and distribution system in Uganda. Small-scale Forestry 18:309–321. https://doi.org/10.1007/s11842-019-09420-w

PATSPO (2018) A guide to seed quality. World Agroforestry Centre (ICRAF). ICRAF Ethiopia Office

Pina A, Errea P (2005) A review of new advances in mechanism of graft compatible-incompatible in Prunus spp. Sci Hortic 106:1–11

Prasad SR (2023) Testing seed for quality. In: Dadlani D, Yadava K (eds) Seed Science and Technology. Springer Singapore, pp 299–334. https://doi.org/10.1007/978-981-19-5888-5_1

Raebild AA, Jensen JS, Ouedraogo M, de Groote S, van Damme P, Diallo BO, Bayala J, Sanou H, Kalinganire A, Kjaer ED (2011) Advances in domestication of indigenous fruit trees in the West African Sahel. New For 41(3):297–315

Roshetko J, Verbist B (2000) Tree Domestication, lecture notes in support of training course on agroforestry for improved land-use and livelihood systems in Southeast Asia. ICRAF, Bogor

Sacandé M, Pritchard HW, Dullo EM (2004) Seed science and technology needs of SAFORGEN trees for conservation and sustainable use. Pl Gen Res Newslet 139:53–58

Seddon N, Smith A, Smith P, Key I, Chausson A, Girardin C, House J, Srivastava S, Turner B (2021) Getting the message right on nature-based solutions to climate change. Glob Chang Biol 27:1518–1546. https://doi.org/10.1111/gcb.15513

Sileshi GW, Akinnifesi FK, Ajayi OC, Chakeredza S, Kaonga M, Matakala PW (2007) Contributions of agroforestry to ecosystem services in the miombo eco-region of Eastern and Southern Africa. Afr J Environ Sci Technol 1(4):068–080

Sileshi GW, Dagar JC, Kuyah K, Datta A (2023) The Great Green Wall initiatives and opportunities for integration of dryland agroforestry to mitigate desertification, Agroforestry for Sustainable Intensification of Agriculture in Asia and Africa. Sustainable agriculture and food security. Springer Nature Singapore Pte Ltd. pp 175–206. https://doi.org/10.1007/978-981-19-4602-86

Simons AJ, Leakey RRB (2004) Tree domestication in tropical agroforestry. In: Nair PKR, Rao MR, Buck LE (eds) *New Vistas* in agroforestry: a compendium for 1st World Congress of agroforestry. Springer, Cham, pp 167–181

Sitoe A, Chidumayo E, Alberto M (2010) Timber and wood products. In: Chidumayo E, Gumbo D (eds) The dry forests and woodlands of Africa. Earthscan Publishing, Washington, DC, pp 131–153

Stevens N, Bond WJ (2023) A trillion trees: carbon capture or fuelling fires? Trends Ecol Evol S0169-5347(23):00246-X. https://doi.org/10.1016/j.tree.2023.09.015

Swart MW, Caro TM (2002) Effect of selective logging on tree and understory regeneration in miombo woodland in western Tanzania. Afr J Ecol 41:75–82

Syampungani S (2008) Vegetation change analysis and ecological recovery of the Copperbelt Miombo Woodlands. PhD thesis, Stellenbosch University, South Africa

Takoutsing B, Tchoundjeu Z, Degrande A, Asaah E, Tsobeng A (2014) Scaling-up sustainable land management practices through the concept of the rural resource centre: reconciling farmers' interests with research agendas. J Agricult Educ Exten 20:463–483. https://doi.org/10.1080/1389224X.2014.913984

Tchiechoua YH, Kinyua J, Ngumi VW, Odee DW (2020) Effect of indigenous and introduced arbuscular mycorrhizal fungi on growth and phytochemical content of vegetatively propagated *Prunus africana* (Hook. f.) Kalkman provenances. Plan Theory 9:37

Tchoundjeu Z, Asaah EK, Anegbeh P, Degrande A, Mbile P, Facheux C, Tsoberg A, Atangana AR, Ngo-Mpeck ML, Simons AJ (2006) Putting participatory domestication into practice in West and Central Africa. For Trees Livelihoods 16:53–70

Tchoundjeu Z, Atagana A, Asaah E, Tshobeng A, Facheux C, Foundjem D, Mbosso C, Degrande A, Sado T, Kanmegne J, Mbile P, Tabuna H, Anegbeh P, Useni M (2008) Domestication, utilisation and marketing of indigenous fruit trees in West and Central Africa. In: Akinnifesi FK, Leakey RRB, Ajayi OC, Sileshi G, Tchoundjeu Z, Matakala P, Kwesiga FR (eds) Indigenous fruit trees in the tropics: domestication, utilization and commercialization. CAB International Publishing, Wallingford, pp 171–185

Tchoundjeu Z, Degrande A, Leakey RRB, Nimino G, Kemajou E, Asaah E, Facheux C, Mbile P, Mbosso C, Sado T, Tsobeng A (2010) Impacts of participatory tree domestication on farmer livelihoods in West and Central Africa. For Trees Livelihoods 19:217–234. https://doi.org/10.1080/14728028.2010.9752668

Teklehaimanot Z (2008) The role of indigenous fruit trees in sustainable dryland agriculture in Eastern Africa. In: Akinnifesi FK, Leakey RRB, Ajayi OC, Sileshi G, Tchoundjeu Z, Matakala P, Kwesiga FR (eds) indigenous fruit trees in the Tropics: Domestication, utilization and commercialization. CAB International Publishing, Wallingford, pp 204–223

Tembani M, Madhibha T, Marunda CT, Gapare WJ (2014) Sustaining and improving forest genetic resources for Zimbabwe: lessons from 100 years. Int For Rev 16(6):615–632

Tsegaye M, Lemage B, Hido A (2021) Seedling performance of different provenances of selected indigenous tree species in Debub Ari District, Southern Ethiopia. Global J Earth Environ Sci 6:38–43

Tsobeng A, Asaah E, Tchoundjeu Z, van Damme P, Ofori D, Jamnadass R (2017) Growth, flowering and fruiting of stecklings, grafts and seedlings of *Allanblackia floribunda* Oliver (Clusiaceae). Agric Syst 91:259–270. https://doi.org/10.1007/s10457-016-9925-z

Chapter 11
Management of Biological Invasions in Multifunctional Landscapes of Agroecological Ecosystems

Abdullahi A. Yusuf

Abstract Alien invasive plant, and insect species are introduced through different ways into Africa and are established throughout the continent providing new host for native and non-native insect pests and pathogens in natural and plantation forests. About 128 species of herbivores and pathogens have been reported in both native and non-native forests on the continent. Thus, threatening biodiversity and exerting more pressure on food security, local economy and calls for intensive management interventions. Available management strategies include mixed tree forests, adaptation of silviculture practices, sentinel plantings, biocontrol, semiochemicals, genetic breeding, geographical information services and large data. Legislation against invasive species is enforced with reference to the international plant protection convention by phytosanitary organisations at global, continental, and national levels following the international plant protection convention. There is an identifiable gap and the need to establish regional agencies that will oversee and consolidate management at the regional level. The challenges of biosecurity policies, their lack and capacity of implementation between member states as well as the inclusion of public awareness and indigenous knowledge for effective control are to be considered in the management of biological invasions.

Keywords Biodiversity · Diversity · Environmental issues · Invasive species · Landscape · Pathways

---

A. A. Yusuf (✉)
Department of Zoology and Entomology, University of Pretoria, Pretoria, South Africa

Forestry and Agricultural Biotechnology Institute, University of Pretoria, Pretoria, South Africa
e-mail: aayusuf@zoology.up.ac.za

P. W. Chirwa et al. (eds.), *Trees in a Sub-Saharan Multi-functional Landscape*,
https://doi.org/10.1007/978-3-031-69812-5_11

11.1 Introduction

Trees are well known for their multiple usages by different people across generations and societies; especially in the regulation of conditions and the supply of wood products and other goods which are important in nutrition and food security. However, native or indigenous trees and forests are threatened by invasive species. Invasive or alien species have been defined in many ways, with most of the definition based on the degree of naturalisation and the rate of spread or the negative impact on habitat and biodiversity (Skočajić and Nešić 2020). In the context of this chapter, we will define alien invasive species (AIS) as species introduced voluntarily or involuntarily into a territory outside their natural range, whose introduction or spread threatens or negatively affects biodiversity and ecosystem services. According to IUCN (2012); biological invasions caused by AIS, are one of the threats to biodiversity by fragmenting and destroying natural habitats (1992 United Nations Convention) and to species extinction (Genovesi and Shine 2004). Further, they alter ecosystem services and induce the disappearance of native species through various mechanisms such as hybridisation, predation and competition (Simberloff et al. 2013).

Biodiversity on the other hand is the variability of living organisms and the ecological complexes of which they are part, including diversity within species (genetic diversity), between species and ecosystem. Biodiversity is vital to the humankind because it is essential for sustainable production of agricultural and other food products, including the provision of a basis for the evolution or deliberate genetic improvement of new useful plant varieties, biological support for production through the biotope, pollinators and predators, and—a wider range of ecological services provided by agro systems (landscape protection, soil protection, water cycle and air quality, air). It is the basis of agriculture and especially of crop and animal production, it includes crop and livestock species, varieties and breeds within them, and it also includes components that support agricultural production. Species-level components that support ecosystem services include; termites, earthworms, fungi and some bacteria that contribute to plant nutrient availability and cycling through the decomposition of organic matter (CBD 2008) and symbioses. These processes are the basis on which agriculture thrives, especially of crop and animal production. Within the context of agrobiodiversity; plant genetic resources comprising all cultivated or spontaneous plants of agronomic, economic or ecological interest, have become rare or endangered. Consequently, this among other factors has led to increased nutrition and food insecurity affecting the majority of the population in sub-Saharan Africa (SSA). Plant genetic resources are facing extinction threats by several factors (Fig. 11.1), including the erosion of biodiversity, the destruction of habitats (conversion and fragmentation), biological invasions, pollution (air, water and soil), overexploitation of species and climate change (disappearance, migration).

The negative environmental impact on both biodiversity, food, nutrition and socioeconomic security as a result of AIS have a profound impact, especially for

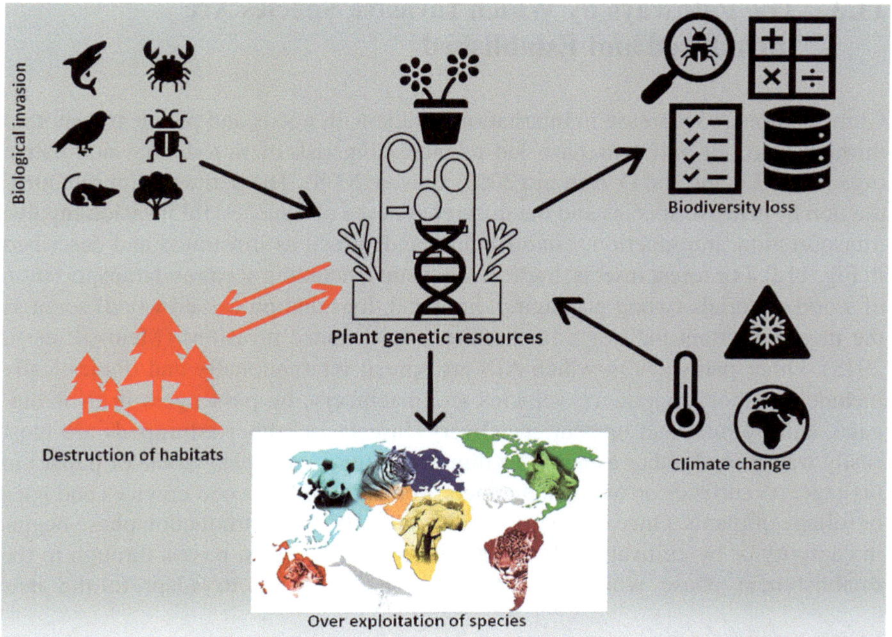

Fig. 11.1 Factors threatening plant genetic resources globally. Direction of the arrow indicates if the threat is inwards (adding to) or outwards (taking away). Bi-directional arrows indicate both effects

Africa where it is pegged at 1.2 billion USD dollars per annum (Pratt et al. 2017). However, this figure is a gross underestimation due to recent introductions and weakness of quarantine and importation prevention and management at the porous borders within the continent. Hence, AIS must be well managed and conserved for the preservation of a living heritage, maintenance of diversity and economic development.

The impacts AIS have on biodiversity, necessitates preventive and mitigating measures which are enshrined in different conventions, regulations, policies and action plans adopted at the national, regional and international levels such as the Sustainable Development Goals (SDGs) and the African Union's Agenda 2063. One of the targets of Goal 15 of the SDG is to combat the introduction of invasive species, reduce their impact, control and eradicate them. While Goal seven of the African Union's Agenda 2063 is aiming at an environmentally sustainable and climate resilient economic community for sustainable natural resources management and biodiversity conservation.

In the context of this chapter, we consider environmental and social impact as, those negative and or positive effects in the environment which in the long or short term alters/distorts the well-being of all components in it.

11.2 The Pathways by Which Invasive Species Are Introduced and Established

Globalisation and increase in international trade with goods and people transported through land, air and seas have led to increasing risk of introducing non-native organisms (Levine and D'Antonio 2003; Brasier 2008). The pathways for the introduction of invasive species and the different phases of a successful invasion involve transportation, introduction, establishment and spread as illustrated and described in Fig. 11.2. For forest insects, trade in live plants including seeds and transportation of wood materials (wood packaging, firewood, logs and processed wood) serve as the most important pathways facilitating long distance invasions (Meurisse et al. 2018). Other pathways in which AIS are spread internationally and domestically include transport containers, vehicles and machinery, by passengers in their luggage, and by mail and or courier delivery. Insects and other arthropods are most easily transported either as adult, or immatures (eggs, nymphs, larvae or pupae) in luggage, on surfaces or within containers, vehicles, within wood crevices and gaps or foliage of plants. Once transportation is successful, the introduction phase begins in captivity or by cultivation. If the species survives, it then passes through to the establishment phase which is dependent on its ability to adapt to the new

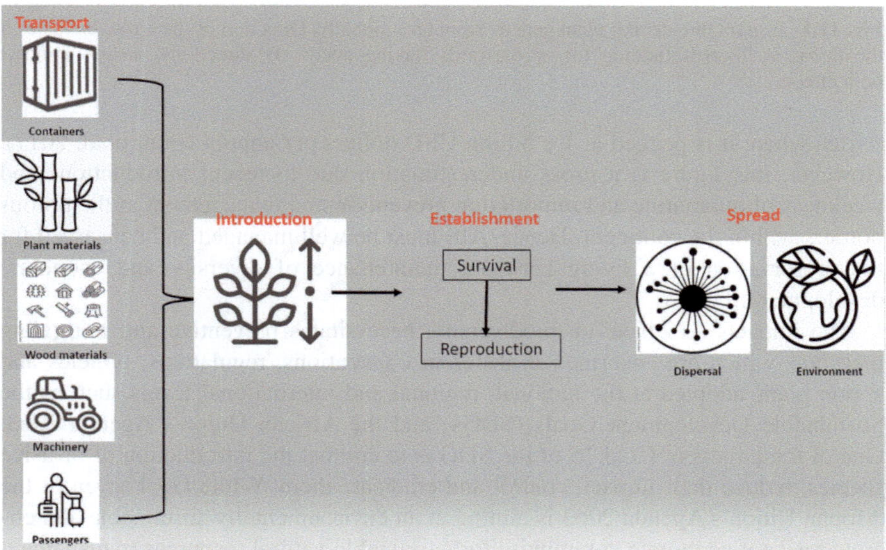

Fig. 11.2 Pathways of invasive species introduction and important phases of a successful biological invasion. After successful transportation, invasive species get introduced, and established if they survive and reproduce. Successful establishment leads to dispersion and spread into the environment if no management or containment method is applied. Note, the thicker the red box, the more important the transportation mode is to the importation and movement of invasive species. (Modified from Meurisse et al. (2018), Robertson et al. (2020), Skočajić and Nešić (2020))

environment and reproduce. If not contained, the species will disperse and spread to the environment (Robertson et al. 2020; Skočajić and Nešić 2020, Fig. 11.2).

Even though these pathways and their impacts in spreading invasive species have been indicated, a more complicated picture looms as a result of climate change. For instance, Gullino et al. (2022) mentioned that changes in climate from winter conditions to temperate is sufficient to assist the establishment of invasive plant pests where they would ordinarily not have been established. In general, pest risk will increase in agricultural and forest ecosystems under climatic change scenarios in the arctic, boreal, temperate and subtropical regions making Africa more at risk of these invasions.

11.3 An Overview and Geographical Distribution of Invasive Plant Species and Pests of Trees in Africa in Relation to Emerging Environmental Issues

The rapidly expanding demand for plant resources coupled with the decrease in natural forest areas are driving the expansion of plantation forests which has so far reached about 187 million ha globally (FAO 2021). In Africa, these plantation forests consist of both native and non-native species like eucalypts, pine, teak, black wattle, gamhar (*Gmelina arborea*), rubberwood, neem, African whitewood, African teak (iroko), baobab, thorn trees and tropical almond (Nair 2001; Graziosi et al. 2019) planted due to their rapid growth, favourable properties for timber, fuel, paper and cellulose, as well as protection from soil erosion and flooding. Although plantation forests are expanding, they are faced by serious losses incurred due to pests and diseases (Hurley et al. 2017) threatening their sustainability. This is largely due to the rapid increase in the introduction of invasive species that become pests of forest trees (Wingfield et al. 2008; Liebhold et al. 2012; Hurley et al. 2016), and the increase in new host associations between herbivorous or pathogenic species and non-native tree hosts (Paine et al. 2011; Wingfield et al. 2011).

About 128 species of native and non-native insect herbivores and pathogens have been reported to be associated with indigenous, exotic and natural forest plantations in Africa (Fig. 11.3, Tables 11.1, 11.2, 11.3 and 11.4). Of the 99 insect pest species belonging to the orders Coleoptera, Lepidoptera, Hemiptera and Blattodea, majority (82%) are native affecting broadleaves, conifers or both (Fig. 11.3, Table 11.2). Only 28 (18%) introduced insect species have been reported on plantation forests made up of exotic plants like eucalyptus, pines, neem and Cupressus (Table 11.1). Compared to insects pests, pathogens showed a different trend with majority (70%) been introduced and found in plantations than in indigenous (8%) or both (22%) forests on broadleaves (54%) than on conifers (23%) (Fig. 11.3). Most of the pathogens are from the fungal divisions Ascomycota, Basidiomycota, Oomycota and Proteobacteria (Tables 11.3 and 11.4).

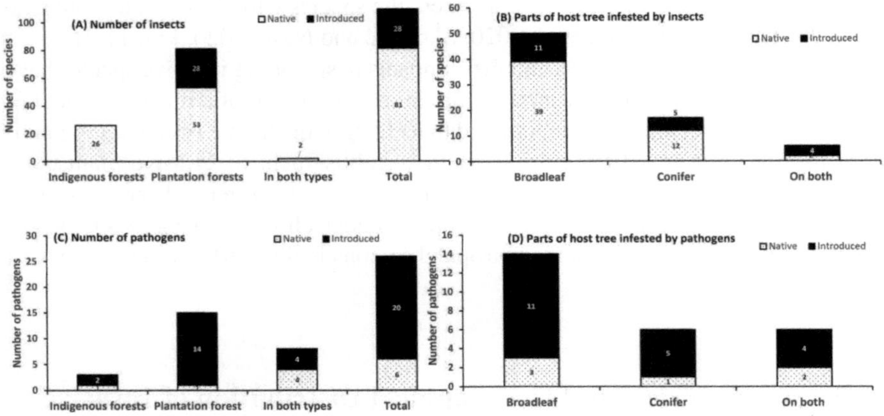

Fig. 11.3 Summary of the number of native and introduced insect pests and pathogens in different categories of forests and the affected plant parts in Africa. Insect pests and parts of trees they infest (**a** and **b**), pathogens, and parts of tree they infest (**c** and **d**). (Source: FAO 2009)

Insect pests and pathogens of trees in Africa have been reported from planted and natural forests across the different regions (eastern, western, central, northern and southern) of the African continent (Tables 11.1 and 11.2, Fig. 11.4a, b). Countries with confirmed and reported presence of at least one invasive insect pest include; Benin Republic, Democratic Republic of Congo, Uganda, Ghana and Zambia. While, Morocco, Sudan and Zimbabwe have three, four and five invasive insect pests, respectively. Countries with the highest number of reported invasive insect species are South Africa, Kenya and Malawi (Fig. 11.4a). In terms of regional spread, pests of eucalyptus like the long-horned beetle, oriental yellow scale are found mostly in West Africa with the eucalyptus gall wasp, longhorn borers and sirex wood wasp common in north and southern Africa (Table 11.1). The tamarind seed beetle is a pest that attacks *Vachellia spp.* and termites are widespread on the continent.

About 66 species of native insect pests have been recorded in different types of plantations in Africa (Table 11.2) from 16 countries (Fig. 11.4b). South Africa has the highest number (23) of pest species, followed by Ghana (22) and Malawi with 13 species. The productivity especially timber production of forest plantations made up of *Eucalyputus spp*., *Tectona grandis*, *Adansonia digitata*, Milicia regia, and *Triplochiton scleroxylon* in west Africa are affected by pests such as long-horned beetle, bark beetles, Psyllid, the African teak gall bug and the Iroko gall bug (Table 11.2). The quince wood borer *Coryphodema tristis* with a wide range of native hosts has recently become a major pest of eucalypt plantations in South Africa (Adam et al. 2013). The wattle mirid *Lygidolon laevigatum* (Hemiptera: Miridae) is a major pest of the exotic black wattle, *Acacia mearnsii* in forestry

Table 11.1 Non-native insect pests reported from plantation forests of *Eucalyptus spp., Pinus spp., Azadirachta indica, Cupressus spp.* and *Tectona grandis* in Africa listed in alphabetical order

Species	Common name	Host plant	Countries and regions reported
Analeptes trifasciata	Long-horned beetle	*Eucalyptus spp.*	West Africa
Aonidiella orientalis	Oriental yellow scale	*Azadirachta indica*	West Africa
Apate terebrans	Shot-hole borer	*Eucalyptus spp.*	Benin, Ghana, Zambia
Blastopyslla occidentalis	Eucalyptus psyllid	*Eucalyptus spp.*	South Africa
Caryedon serratus	Tamarind seed beetle	*Vachellia spp.*	Widespread
Cinara cronartii	Black pine aphid	*Eucalyptus spp.*	Kenya, South Africa, Zimbabwe
Cinara cupressi sensu lato	Cypress aphid	*Cupressus, Juniperus spp., Widdringtona spp.*	Central, East, North and Southern Africa
Ctenarytaina eucalypi	Blue gum pysllid	*Eucalyptus spp.*	Kenya, South Africa
Eulachnus rileyi	Pine needle aphid	*Pinus spp.*	Central, East, North and Southern Africa
Glycaspis brimblecombei	Redgum lerp pysllid	*Eucalyptus spp.*	East, North, and Southern Africa
Gonipterus sp.2	Eucalyptus snout beetle	*Eucalyptus spp.*	Kenya, Madagascar, Malawi, Mozambique, Rwanda, South Africa, Tanzania, Uganda, Zimbabwe
Heteropsylla cubana	Leucaena psyllid	*Luecaena, Mimosa spp.*	East, North, and Southern Africa
Hyblaea puera	Teak defoliator	*Tectona grandis*	East Africa, Southern Africa
Hylastes angustatus	Pine bark beetle	*Pinus spp.*	South Africa, Zimbabwe
Hylurgus ligniperda	Red-haired pine bark beetle	*Pinus spp.*	South Africa
Leptocybe invasa	Bluegum chalcid	*Eucalyptus spp.*	East, North, and Southern Africa
Ophelimus maskelli	Eucalyptus gall wasp	*Eucalyptus spp.*	North Africa, Southern Africa
Orthotomicus erosus	Mediterranean pine engraver Beetle	*Pinus spp.*	South Africa
Phoracantha recurva	Eucalyptus longhorn borer	*Eucalyptus spp.*	North Africa, Southern Africa
Phoracantha semipunctata	Eucalyptus longhorn borer	*Eucalyptus spp.*	North Africa, Southern Africa
Pineus boerneri	Pine wolly aphid	*Pinus spp.*	East Africa, Southern Africa
Pineus pini	Pine woolly aphid	*Pinus spp.*	East, North, and Southern Africa
Pissodes sp.	Deodar weevil	*Pinus spp.*	South Africa
Sirex noctilio	Sirex wood wasp	*Pinus spp.*	North Africa, Southern Africa

(continued)

Table 11.1 (continued)

Species	Common name	Host plant	Countries and regions reported
Spondyliaspis plicatuloides	Shell lerp psyllid	*Eucalyptus spp.*	Southern Africa
Thaumastocoris peregrinus	Bronze bug	*Eucalyptus spp.*	Kenya, Malawi, South Africa, Uganda, Zimbabwe
Trachymela tiincollis	Eucalyptus tortoise beetle	*Eucalyptus spp.*	South Africa
Various Blattodea	Termites	*Eucalyptus spp.*, *Pinus spp.*, *Cupressus spp.*	Widespread

Sources: Nair (2001), Roux et al. (2012), Bush et al. (2016), Hurley et al. (2017), Graziosi et al. (2019), FAO (2009) and https://www.cabi.org

plantations in South Africa, where it causes irregular branching and defoliation (Govender 2002). Pests like the black borer, seed beetle, termites are common in plantations all over the continent, while the winter moth *Operophtera brunata* is only found in North Africa within Quercus forests (Table 11.2).

The presence of microbial pathogens known to cause disease of trees have become an increasing threat to forestry in Africa. It is often difficult to detect and trace the origins of these pathogens due to their cryptic nature and the fact that they are mostly non-native (Fig. 11.4, Table 11.4). However, there is the presence of pathogens on indigenous trees reported from Ethiopia, Uganda, Kenya, Madagascar, Tanzania, and Democratic Republic of Congo with South Africa harbouring the highest with eight reports (Fig. 11.4c). One such common pathogen is the *Armillaria* root rot pathogen reported to have been introduced by early settlers with potted plants from Europe (Coetzee et al. 2001). This disease has now spread to Ethiopia, Kenya and Tanzania in East Africa and the Island of Madagascar (Table 11.3, Fig. 11.4c). Tree pathogens diseases on the continent have been reported more on plantations made up of non-indigenous trees than those of natural forests (Table 11.4). These have been reported from all over the continent with the following countries; Tanzania and Zambia reporting a pathogen each, Uganda two pathogens, Democratic Republic of Congo and Madagascar three each, Kenya, Ethiopia, six and seven respectively, while South Africa has 23 pathogens (Fig. 11.4d). Pathogens show the same trend as with insect pests with most of them discovered in non-indigenous forests of *Eucalyptus spp.* and *Pinus spp.* that later spread to other countries including the islands (Table 11.4) (Graziosi et al. 2019). The information provided here about pests and pathogens of forests in Africa is by no means exhaustive because of the limited number of studies conducted in most African countries. This is despite efforts by the FAO and the requirement by countries to provide quantitative data on the impact of insect pests and diseases in their forests (FAO 2009).

Agroecosystems in Africa are crucial in the resilience and adaptation of landscapes due to their contribution in mitigating climatic variables and carbon storage. However, forests in Africa are shrinking at a rate of 0.5% annually (FAO 2015) and are expected to be vulnerable to climate change (Wheeler and Von Braun 2013)

Table 11.2 Native insect pests of indigenous and exotic trees (*Acacia mearnsii, Eucalyptus spp., Tectona grandis, Milicia excelsa, Triplochiton scleroxylon etc*) listed in alphabetical order found in natural and plantation forests and their common names reported in Africa

Species	Common name	Host plant	Countries and regions reported
Achaea lienardi	Eastern cape moth	*Acacia mearnsii*	South Africa
Afromurzinia lutescens	Tiger moths	*Eucalyptus spp.*	Ghana, South Africa
Agrotis spp.	Cutworms	*A. mearnsii, Eucalyptus spp., Pinus spp.*	South Africa
Anacridium melanorhoden	Sahelian tree locust	*Acacia senegal, Balanites aegyptiaca*	Senegal
Anaemerus tomentosus		*Eucalyptus spp.*	Ghana, Zambia, Nigeria
Analeptes trifasciata	Long-horned beetle	*Eucalyptus spp., Tectona grandis, Adansonia digitata*	West Africa
Anape venata	African silk moth	*Triplochiton scleroxylon*	Ghana, Nigeria
Apate monachus	Black borer	*T. grandis, Azadirachta indica, Terminalia spp.*	Widespread
Apate terebrans	Shot-hole borer	*Eucalyptus spp., T. grandis, Terminalia spp., Khaya,*	Benin, Ghana, Zambia
Ascotis selenaria reciprocaria	Giant looper	*Eucalyptus spp., T. grandis*	Ghana, Nigeria
Biston (Buzura) spp.	Looper	*Eucalyptus spp.*	Southern Africa
Bruchidius uberatus	Seed beetle	*Vachellia spp.*	Widespread
Caryedon serratus	Tamarind Seed beetle	*Tamarindus spp., Acacia spp.*	Widespread
Chaetastus tuberculatus	Ambrosia beetle	*Eucalyptus spp.*	Ghana
Chaliopsis junodi	Wattle bagworm	*A. mearnsii*	South Africa
Cleora herbuloti	Pine looper	*Eucalyptus spp., Pinus spp.*	South Africa
Colasposoma spp.	Leaf beetle	*A. mearnsii, Eucalyptus spp., Pinus spp.*	South Africa
Coryphodema tristis	Quince borer	*Eucalyptus spp.*	South Africa
Diacrisia infestigatorum	Tiger moths	*T. grandis*	Ghana
Diclidophlebia eastopi	Psyllid	*Triplochiton scleroxylon*	West Africa
Doliopygus conradti	Ambrosia beetle	*Eucalyptus spp.*	Ghana
Doliopygus erichzoni	Ambrosia beetle	*T. grandis*	Ghana
Dulchia/Euproctis fasciata	Banded euproctis	*Eucalyptus spp.*	Ghana, Nigeria, DRC
Ellimenistes laesicollis	Beaded weevils	*Eucalyptus spp., Grevilea robusta*	South Africa
Euproctis spp.	Tussock moth	*A. mearnsii, Eucalyptus spp., Pinus spp.*	Ghana, South Africa

(continued)

Table 11.2 (continued)

Species	Common name	Host plant	Countries and regions reported
Euproctis terminalis	brown tail moth	*A. mearnsii, Eucalyptus spp., Pinus spp.*	South Africa
Eutricha capensis	Cape lappet moth	*A. mearnsii, Eucalyptus spp., Pinus spp.*	South Africa
Gonimbrasia tyrrhea	Zigzag Emperor	*A. mearnsii, Eucalyptus spp., Pinus spp.*	South Africa
Gonometa podocarpi	Wild silk moth	*A. mearnsii, Eucalyptus spp., Pinus spp.*	East Africa
Gynanisa maia	Speckled emperor moth	*A. mearnsii*	South Africa
Helopeltis anacardii	Mirid	*Eucalyptus spp.*	Malawi
Holocerina smilax	Variable prince (Emperor moth)	*Pinus spp.*	South Africa
Hypopholis sommerii	Wattle chafer	*A. mearnsii, Eucalyptus spp., Pinus spp.*	South Africa
Hypothenemus eruditus	Bark beetle	*T. grandis*	Ghana
Hypothenemus pusillus	Bark beetle	*T. grandis*	Ghana
Hypsipyla robusta	Mahogany shoot borer	*T. grandis*	East Africa, West Africa
Lamprosema lateritialis	Leaf tying moth	*Presicoosis elata*	Ghana, Malawi
Lygidolon laevigatum	Brown wattle mirid	*A. mearnsii*	South Africa
Manowia spp.	Eumasticidae grasshopper	*Pinus spp.*	Malawi
Monochelus calcaratus	Wattle chafer	*A. mearnsii*	South Africa
Narosa viridana	Slug caterpillar moth	*Eucalyptus spp.*	Zambia
Neocleosa spp.	Looper	*Eucalyptus spp.*	South Africa
Nudaurelia cytherea	Pine Emperor	*A. mearnsii, Eucalyptus spp., Pinus spp.*	South Africa
Oemida gahani	Long-horned beetle	*Cupressus*	West Africa, Kenya
Operophtera brunata	Winter month	*Quercus afares*	Algeria, Tunisia
Orgyia basali	Tussock moth	*Eucalyptus spp., T. grandis*	Ghana, Nigeria
Orygmophora mediofoveata	Shoot borer	*Nauclea diderrichii*	West Africa
Pachypasa capensis	Emperor moth	*A. mearnsii, Eucalyptus spp., Pinus spp.*	South Africa
Parasaissetia nigra	Black scale	*Eucalyptus spp.*	Ghana
Phymateus kazschi	Milkween locust	*Eucalyptus spp.*	Nigeria, Ghana
Phytolyma fusca	African teak gall bug	*Milicia excelsa*	West Africa
Phytolyma lata	Iroko gall bug	*Milicia regia*	West Africa

(continued)

Table 11.2 (continued)

Species	Common name	Host plant	Countries and regions reported
Planococcoides njalensis	Cocao mealybug	*T. grandis*	Ghana
Platypus cylindrus	Oak pinhole borer	*Quercus suber*	Algeria
Platypus lintzi	Ambrosia beetle	*Eucalyptus spp.*	Ghana
Pseudobunaea irius	Emperor moth	*A. mearnsii, Eucalyptus spp., Pinus spp.*	South Africa
Saissetia coffease	Helmet scale/coffee scale	*Eucalyptus spp.*	Ghana
Sphenoptera chalcichroa	Silver tree borer	*V. nilotica, V. arabica*	Sudan
Sphenoptera fulgens	Root-boring beetle	*Prosopis chilensis, others*	Sudan
Strepsicrathes rhothia	Tortricid moth	*Eucalyptus spp.*	Ghana
Thaumetopoea pityocampa	Processionary moth	*Pinus spp.*	N Africa
Thaumetopoea wilkinsoni	Processionary moth	*A. mearnsii, Eucalyptus spp., Pinus spp.*	South Africa
Various		*A. mearnsii, Eucalyptus spp., Pinus spp.*	Widespread
Various Blattodea	Termites	*A, mearnsii, Eucalyptus spp., Cupressus spp.*	Widespread
Xanthisthisa tarsispina	Pine looper	*Pinus spp.*	Malawi
Xyleborus perforans	Island pinhole borer	*Eucalyptus spp.*	Ghana

Sources: Nair (2001), Roux et al. (2012), Bush et al. (2016), Hurley et al. (2017), Graziosi et al. (2019), FAO (2009) and http://www.cabi.org

because of deforestation resulting from anthropogenic activities including logging for timber and domestic uses such as charcoal which accelerates desertification. To safeguard ecological and socioeconomic benefits from forests, especially for smallholders who are more vulnerable, afforestation then becomes necessary but without its inherent risks of introducing invasive pests and pathogens as shown earlier. Under current conditions, we saw the spread of pests and pathogens on the continent. A similar or even worse scenario is anticipated with changing precipitation patterns (floods already experienced all over the continent) and mean global temperature rises of between 2–8 °C (IPCC 2021). Changes in both temperature and precipitation are important environmental variables that favour the establishment and spread of insect pests and pathogens. These coupled with the lack of proper quarantine and weak cross boarder biosecurity, makes Africa vulnerable to invasive species as seen recently in the introduction and establishment of the fall army-worm *Spodotera frugiperda* and the South American tomato pinworm *Phthorimeae (Tuta) absoluta* introduced in 2018 and 2006 respectively (Peter et al. 2023; Adams et al. 2023).

Table 11.3 Native pathogens reported on plantation forests in Africa with their common names listed in alphabetical order

Species	Common name	Host species	Countries and regions reported
Ascomycota			
Lasiodiplodia theobromae		*Boswellia, Eucalyptus spp., others*	Ethiopia, South Africa, Uganda
Chrysoporthe spp.	Chrysoporthe canker	*Eucalyptus spp., Syzygium cordatum*	Kenya, Madagascar, Mozambique, South Africa
Colletotrichum spp.	Anthracnose	Eucalyptus spp., Proteaceae, others	Ethiopia, South Africa
Ceratocystis spp.	Wattle wit	*Eucalyptus spp., Acacia mearnsii*	Democratic Republic of Congo, Eastern and Southern Africa
Botryosphaeriaceae spp.	Botryosphaeria cankers	*Pinus, Acacia spp.*	South Africa
Basidiomycota			
Armillaria spp., A. mellea	Armillaria root rot	*Eucalyptus, Pinus spp., others*	Ethiopia, Kenya, Madagascar, South Africa, Tanzania
Erythricium salmonicolor	Pink disease	*Eucalyptus, Acacia, Podocarpus*	Ethiopia, South Africa
Oomycota			
Phytophthora cinnamomic	Phytophthora root rot	*A. mearnsii, Eucalyptus spp.*, Proteaceae	Kenya, South Africa

Sources: Nair (2001), Roux et al. (2012), Bush et al. (2016), Hurley et al. (2017), Graziosi et al. (2019) and http://www.cabi.org

11.4 Impacts of Invasive Plant Species and Pests of Trees

As highlighted earlier, the movement of AIS are facilitated by anthropogenic activities which forms an important part of global trade, and one of the most severe environmental problems facing humanity (Smith-Ramesh 2017). Besides serious environmental damage, the spread of invasive species can also cause economic losses in agricultural areas (Holmes et al. 2009; Drechsler et al. 2016). Therefore, continuous monitoring of invasive species, as well as mapping spread corridors and delineating areas already invaded, is of increasing importance for the purposes of nature conservation and agriculture (Follak et al. 2018).

Forecasts for areas that are potentially sensitive to invasion are also significantly important. Keane and Crawley (2002) reported that one of the several reasons cited for the success of invasive species outside their native range is what was called the enemy release hypothesis, which states that "natural enemies are important regulators of populations and plants can achieve higher densities when growing in locations without these enemies". Coupled with the increasing frequency of international

Table 11.4 Invasive pathogens reported on plantation forests in Africa with their common names listed in alphabetical order

Species	Common name	Host species	Countries and regions reported
Ascomycota			
Diplodia sapinea	Sphaeropsis blight	*Pinus*	East and Southern Africa
Aulographina eucalypti	Corky Leaf Spot	*Eucalyptus spp.*	Kenya, South Africa, Zambia
Fairmaniella leprosa	Fairmaniella leaf spot	*Eucalyptus spp.*	South Africa
Mycosphaerella spp.	Mycosphaerella leaf blotch	*Eucalyptus spp.*	East and Southern Africa
Kirramyces epicoccoides	Phaeophleospora leaf spot	*Eucalyptus spp.*	Kenya, South Africa
Teratosphaeria zuluense	Coniothyrium stem canker	*Eucalyptus spp.*	Ethiopia, Madagascar, Mozambique, South Africa, Uganda
Chrysoporthe spp.	Chrysoporthe canker	*Eucalyptus spp, Syzygium cordatum*	Kenya, Madagascar, Mozambique, South Africa
Colletotrichum spp.	Anthracnose	Eucalyptus spp., Proteaceae, others	Ethiopia, South Africa
Fusarium circinatum	Pitch canker	*Pinus spp.*	South Africa
Cylindrocladium spp.	Cylindrocladium leaf blight	*Eucalyptus spp.*	Kenya, Democratic Republic of Congo, South Africa
Botryosphaeriaceae spp.	Euc. Botryosphaeria canker	*Eucalyptus spp.*	Ethiopia, Kenya, South Africa, Uganda
Holocryphia eucalypti	Endothia canker	*Eucalyptus spp.*	South Africa, Uganda
Ceratocystis fimbriata	Ceratocystis wilt	*Eucalyptus spp.*	Democratic Republic of Congo, South Africa, Uganda, Côte d'Ivoire
Ceratocystis spp.	Wattle wit	*Eucalyptus spp., Acacia meansii*	Democratic Republic of Congo, East and Southern Africa
Botryosphaeriaceae spp.	Botryosphaeria cankers	*Pinus, Acacia spp.*	South Africa
Botryosphaeriaceae spp.	Grevillea stem canker	*Grevillea robusta*	Kenya, Uganda
Teratosphaeria gauchensis	Eucalyptus stem canker	*Eucalyptus spp.*	Kenya
Basidiomycota			
Armillaria spp., A. mellea	Armillaria root rot	*Eucalyptus, Pinus spp., others*	Ethiopia, Kenya, Madagascar, South Africa, Tanzania
Erythricium salmonicolor	Pink disease	*Eucalyptus, Acacia, Podocarpus*	Ethiopia, South Africa

(continued)

Table 11.4 (continued)

Species	Common name	Host species	Countries and regions reported
Quambalaria eucalypti	Quambalaria blight	*Eucalyptus spp.*	Ethiopia, South Africa
Amylostereum areolatum		*Pinus spp.*	South Africa
Phakopsora myrtacearum	Eucalyptus rust	*Eucalyptus spp.*	South Africa
Puccinia psidii	Eucalyptus rust	*Eucalyptus spp.*	South Africa
Oomycota			
Phytophthora cinnamomic	Phytophthora root rot	*A. mearnsii*, *Eucalyptus spp.*, Proteaceae	Kenya, South Africa
Pythium splendens	Pythium root rot	*Eucalyptus spp.*	South Africa
Proteobacteria			
Ralstonia solanacearum	Bacterial canker	*Eucalyptus spp.*	South Africa
Pantoea spp.	Bacterial canker	*Eucalyptus spp.*	South Africa

Sources: Nair (2001), Roux et al. (2012), Bush et al. (2016), Hurley et al. (2017), Graziosi et al. (2019) and http://www.cabi.org

trade and exchanges in recent years, a large number of exotic plants have been established on the continent (see Table 11.2).

11.4.1 Impact on Biodiversity Conservation

Invasive plant species and pests of trees pose substantial threats to biodiversity and ecosystem services delivery (Freer-Smith and Webber 2017). The occurrence and spread of invasive plant species and pests of trees is increasing, and so is Africa's susceptibility to biological invasion (Graziosi et al. 2019). Over the last three decades, studies have demonstrated the negative impacts of invasive species on native biodiversity from genetic to landscape level, and today, these invasions are considered to be one of the major drivers of biodiversity loss and species extinction (Didham et al. 2005; Blackburn et al. 2019).

One hundred and sixty-four species of invasive plants belonging to 110 genera and 47 species have been documented in East Africa (Witt et al. 2018). The dormant group are the woody trees and shrubs (74 species, 45%), herbaceous plants (46 species, 28%) and climbers (23 species, 14%). These species were introduced intentionally for ornamental, hedge and agricultural purposes. However, members of the family Apocynaceae, Euphobiaceae, Fabaceae have been reported in Kenya, Rwanda, Uganda, Tanzania and Ethiopia. They are known to form large and dense

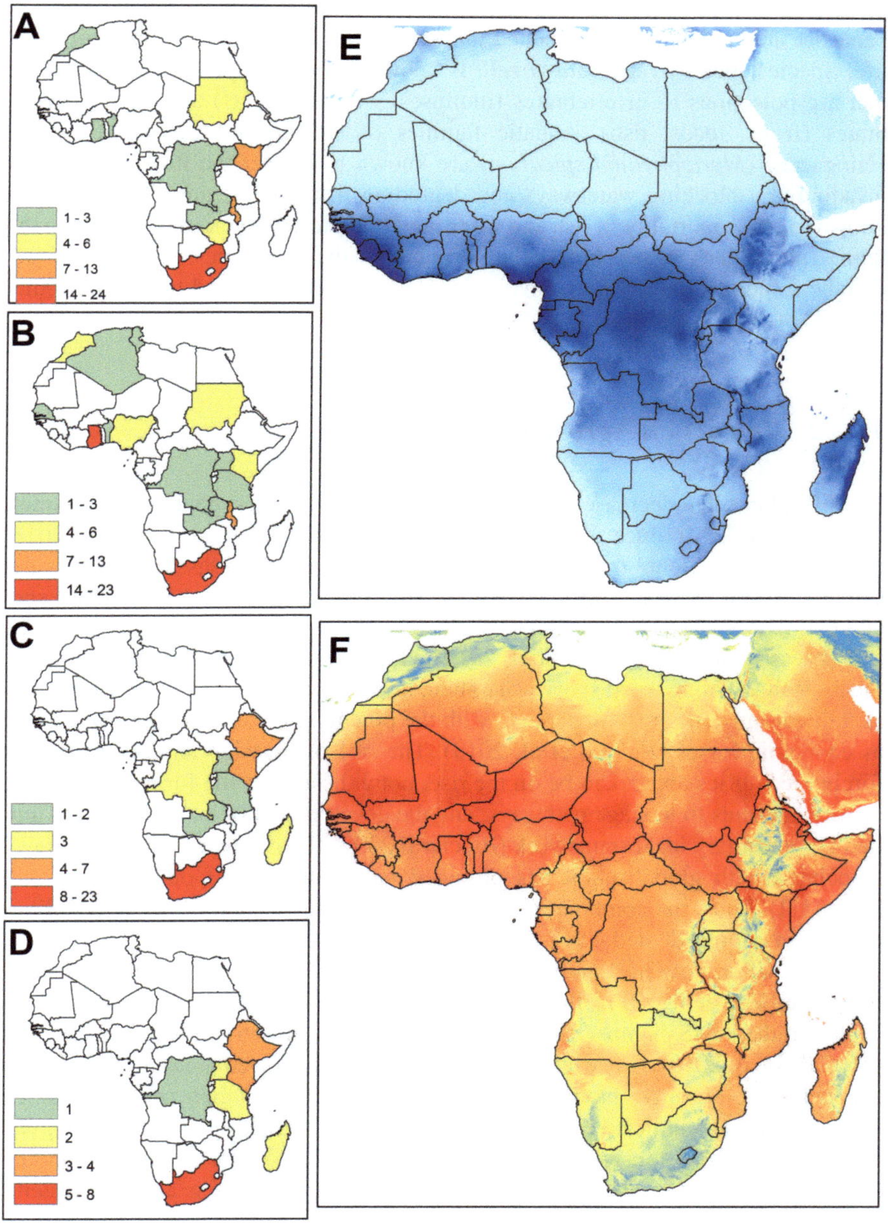

Fig. 11.4 (**a**) Geographical distribution and numbers of invasive insect pests, (**b**) native insect pests in natural, and (**c**) plantation forests, (**d**) pathogens of indigenous and non-indigenous trees, reported in Africa under current, (**e**) mean annual precipitation (mm) and (**f**) mean temperatures (°C) conditions as obtained from www.worldclim.org

thickets in various habitats and have the tendency to displace native species. Fabaceae are known to form impenetrable stands that inhibit the movement of wildlife. Species such as *Tephrosia vogelli* is known to be allelopathic and has leaves that are poisonous to invertebrates (molluscs, worms, insects) and aquatic vertebrates (frogs, toads, fish). Aquatic families of Araceae (*Pistia stratiote) and* Halogaceae (*Myriophyllum spicalum)* are known to impact the flow of water by invading and blocking waterways and degrading fish nesting sites. They are also known to cause mortality of fish and macroinvertebrates; with indirect impact on the fauna (particularly water birds) that depend on these aquatic species for food. Climbers such as the family Convolvulaceae have been reported to smother other flora (Witt et al. 2018).

11.4.2 Impact on Agroecosystem Viability

Invasive plants can decrease provision of ecosystem services, change fire regimes, cause crop losses, reduce the recreational value of natural areas (Vilà et al. 2011) and affect international trade. Ecological impacts from invasions have been widespread (Crystal-Ornelas and Lockwood 2020). These impacts vary from native species extinctions (Blackburn et al. 2019), decreased abundances (Bradley et al. 2019), and fitness reductions (Nunes et al. 2019). Globally and on the African continent, the most threatening weed species are alien, implying that biological invasions rank as a threat in environmental change, with the potential to cause ecological, as well as economic losses (Bellard et al. 2013). The rate of invasion and occupation of foreign territories appears to be on the rise, notwithstanding the formation of biosecurity agencies to manage pathways, and the development of an international unified phytosanitary legal framework (Ricciardi 2007). The increased trade and travel associated with globalisation has provided new pathways of dispersal for AIS (Fig. 11.2). Thus, the combination of new invasion threats and host vulnerabilities present an unprecedented AIS risk profile (Brasier 2008).

Since AIS are fundamentally opportunistic species with a broad environmental tolerance (Vilà et al. 2011), they cannot be sought as indicators of the ecosystem's health; rather, AIS hamper the natural functioning and structure of ecosystems (McCary et al. 2016). However, their establishment in invaded areas is a warning to the wellbeing of native biodiversity (Ehrenfeld 2010), and their presence can be indirectly used to indicate the safety of ecosystems. Predicting and preventing the introduction of AIS to a specific region is considered as one of the most important cost-effective means of monitoring and managing habitats and ecosystems (Broennimann and Guisan 2008).

11.4.3 Impact on Rural Livelihoods

Invasive exotic species can have large detrimental economic impacts on human enterprises such as fisheries, agriculture, grazing and forestry. Globally, the costs associated with such impacts of AIS have been estimated at USD 1.4 trillion per year; approximately 5% of global GDP (Pimentel et al. 2001). Invasive exotic species affect both natural and semi-natural ecosystem integrity and have global impacts mainly on ecosystems goods and services ultimately affecting the livelihoods of local households (Wise et al. 2012). A good example is the pastoral dry lands of Ethiopia where *Prosopis spp.* was intentionally introduced in the late 1970s to tackle the land problem of desertification (Sertse and Pasiecznik 2005). The plant has both positive and negative effects in areas where it has rapidly spread (Ilukor et al. 2016; Rogers et al. 2017; Tilahun et al. 2017). On the positive side, the plant is a source of income for pastoral households. The local people produce charcoal from the plant as a source of income (Mehari 2015; Ilukor et al. 2016). *Prosopis spp* has also ecological benefits in the region. Earlier studies in the area documented the plant's role in regulating microclimate, improving soil fertility, controlling soil erosion and desalinizing soils (Ayanu et al. 2015). However, *Prosopis spp* also have negative effects including its depressive effect on beneficial native species (van Wilgen et al. 2001). Furthermore, it easily covers paths, homesteads, water sources, crops and invades pasturelands causing its deterioration thereby severely affecting households' income from livestock production (Wakie et al. 2016; Zeray et al. 2017; Ayanu et al. 2015; Rogers et al. 2017).

The knowledge on the potential impacts of invasive species on rural livelihoods and well-being is still limited, particularly since it is the land and waters that are most affected by AIS (Ribeiro et al. 2019). However, it is assumed that the harmful impacts on ecosystem goods and services automatically translate into negative effects on human well-being. Yet, AIS is frequently integrated into local livelihoods, both as managed species, as well as exploitation of wild invasive populations (de Neergaard et al. 2005). Many scientists have indicated that limited work has been undertaken at the local level to explore the role both negative and positive of AIS in livelihoods and informal economic activities (de Neergaard et al. 2005; Siges et al. 2005). This gap is of particular concern to researchers working on household livelihoods systems since with increasing awareness of the negative impacts of AIS there is greater effort to control the plants (Shackleton et al. 2007). However, some of these invasive tree species are widely used by poor communities who have been shown to adapt to biodiversity change, including the establishment of non-native species, such that these species become integrated into their way of life (Pfeiffer and Voeks 2008).

There are a number of possible pathways whereby AIS becomes integrated into local livelihoods. Anthropogenic activities such as agriculture, trade, recreation, and development are key drivers and pathways behind the introduction of invasive alien species (Hulme 2009). Some plant species have been accidentally introduced as stowaways through trade and tourism activities, for example, *Parthenium hysterophorus* which has invaded several Asian and African countries (Adkins and Shabbir 2014; Makoni 2020). Some plant species are also purposely introduced for various reasons, such as provision of timber, shelter, hedging, fodder, ornamental use and bee fodder and as a subset of these become invasive. Globally, there are now over 751 invasive trees and shrubs introduced for a variety of reasons (Rejmánek and Richardson 2013) which also reflects the complexities of colonization and how the preferences of European settlers have shaped plant diversity (Bennett and Kruger 2015; Hirsch et al. 2020; Lenzner et al. 2022). Increasingly, many researchers have shown how invasive plants are causing negative impacts on ecosystem services, human well-being, livelihoods, health and societal stability and even causing conflict (Mazza and Tricarico 2018; Pratt et al. 2017; van Wilgen et al. 2017; Shackleton et al. 2017b, 2019; Bajwa et al. 2019; Rai and Singh 2020; Reynolds et al. 2020) and ecological integrity through disturbing biodiversity and ecosystem stability (Blackburn et al. 2014; Witt et al. 2018; Linders et al. 2019).

Furthermore, some invasive alien plant species can provide services and have functions that are not readily available from local species (Shackleton et al. 2017a). This combination of benefits and costs can foster conflicts of interest (Zengeya et al. 2017) on how invasive alien plants should be used and controlled, especially in developing countries where invasions are widely spreading. Despite an increasing understanding of the implications of invasive alien plants, the literature still generally focuses on well-established invasive plants, with relatively little information available on emerging invasive alien plants that are undergoing rapid spread. Generally, it is agreed that a lack of detailed assessments of the benefits and costs of invasive alien species can directly affect management options (Vimercati et al. 2022) and that delayed assessments of an emerging invasive species affect management practices.

11.5 Invasive Plant Species and Pests of Trees in Natural Woodlands Versus Agroecosystems: Inferences for Implementation of Multifunctional Landscapes

We have reviewed in section ii how invasive plant species and pest of trees are introduced outside their native range (Fig. 11.2). While sect. iii we highlighted the distribution and diversity of insect pests and diseases of forestry on the African continent and provided an overview of what the scenario would look like under changing climatic conditions (Tables 11.1, 11.2, 11.3, 11.4 and Fig. 11.4). In all instances, we saw how AIS alters ecological dynamics in forests not only by introducing pests and disease, but also offering new host and niches for native insect pests. Thus, requiring more management and control strategies, the consequence of which is an unprecedented decline in biodiversity through habitat destruction, monocultures, and the use of pesticides.

The concept of multifunctionality of landscapes is an important paradigm within sustainable development which brings together sustainability, agricultural and environmental policies, and international trade (Fig. 11.5b) and can save costs (Brandt

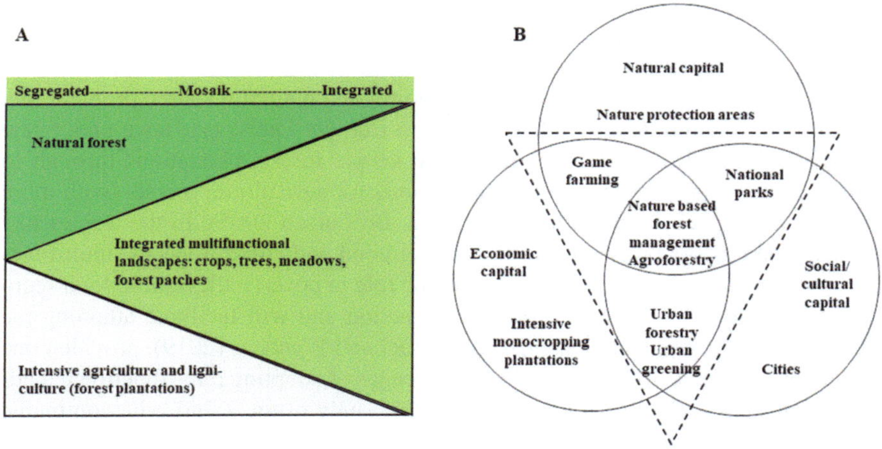

Fig. 11.5 (**a**) Landscape segregation and integration as a continuum and (**b**) different land use forms and systems placed in the sustainability-triangle. (Adapted from Bo Larsen Bo Larsen 2005)

and Vejre 2004) when targeted at large areas such as forests. Bo Larsen (2005) illustrated the multifunctional capability of forests and trees in a landscape with different land use forms (Fig 11.5b). From this, one could see forests and trees as the three pillars of sustainability with natural, economic and socio-cultural capitals overlapping. The multifunctionality of forests can be improved through integrating nature into silviculture, forest into human settlements, landscapes (agro-silviculture) and co-management by integrating people as custodians (Bo Larsen 2005). Forests, if well managed, have the ability to support biodiversity and recover from disturbances when maintained as a mosaic of connected heterogeneous habitats (Bengtsson et al. 2003).

The general aims of multifunctional landscapes are to help overcome negative consequences through the adoption of different land uses relevant to local ecological conditions against a segregated approach. It thus provides a balanced option to conservation or fibre production in managed plantations due to its consideration of habitat, human needs and other services (Gustafasson et al. 2012).

11.5.1 An Appraisal and Examples on Why the Concept of Multifunctionality Should Be Embraced in the Management of Invasive Species in African Forests

For a multifunctionality approach to be successful there has to be heterogeneity in terms of variability of patterns and patches in the landscape at multiple spatial scales, which links invariably to increased biodiversity within the agroecosystem

(Benton et al. 2003) all nested in sustainable agriculture. For an agricultural activity to be sustainable, it has to integrate biological and ecological processes, minimise the use of non-renewable inputs (that are harmful to the environment), use knowledge and skills of farmers and solve agricultural and natural resource problem in a collaborative manner (Pretty 2008). While in Europe, landscapes have undergone changes from their earlier usage to natural versus forests plantations through to mosaic, and recently integrated functional units for agriculture, forests (with more than one tree species) and cities (Fig. 11.5a, Bo Larsen 2005). In the tropics like Africa, where people rely on forests for livelihood and cultural values, multifunctional forests will play a vital socio-economic role in poverty alleviation, food security, gender equality, decent work, climate action and will facilitate attaining the Sustainable Development Goals (SDGs). Israel and Wynberg (2019), provided one of the few examples of the benefits and challenges of adopting multifunctional landscape concepts in South Africa with their case study of the Tshidzivhe communal land. In the Tshidzivhe territory, residents use the grassland for agriculture and there exists forest plantation which has been a source of resentment and conflict over the years. There also exists orchards, grazing, and harvesting food, medicine and wood. This example brings forth the need for effective consultations and integration of traditional uses and practices when establishing multifunctional forests, since user behaviour is critical to protecting and conserving forests (Araia and Chirwa 2019; Araia et al. 2020). In addition, greater consideration for gender issues is needed when establishing multifunctional landscapes beyond which is currently practised (Westholm and Ostwald 2020).

11.6 Resilience Approaches Against Invasive Plant Species and Pests of Trees in African Multifunctional Landscapes

Invasive species in multifunctional landscapes in Africa can be managed at each of the pathways (transportation, introduction, establishment, and dispersal) indicated in Fig. 11.2. The following eight key forms of management has been identified in Robertson et al. (2020): i. pathway management, ii. interception, iii. Limits to keeping, iv. secure keeping, v. eradication, vi. complete reproductive removal, vii. Containment and viii. Suppression. While the first four relates to improved biosecurity surveillance especially at the border which targets the arrival stage, the others relate to post border surveillance. The use of bated traps, sentinel trees, sniffer dogs, predatory wasps, acoustic detection, laser vibrometry, citizen science, genetic identification, and remote sensing have since replaced or complemented visual inspections (Poland and Rassati 2019). Thus, solving the problem of invasive species requires an interdisciplinary approach using expertise in biology, ecology, chemistry, economics and social sciences. The extent to which this can be achieved is often influenced by cost, capacity, field testing and research.

Pre-boarder pathway management involves measures like cleaning shipping materials and packaging, enforcing ballast waste management and monitoring of horticultural supply chains. Interception is achieved through surveillance of imports and border inspections with the aim of detecting species upon arrival. Eradication of an invading population is achieved through habitat unsuitability where a species fails to establish due to improper climate or lack of hosts which leads to extinction. This is governed by 'Allee' effects which refers to the correlation between population size and individual fitness of the population/species. When a strong Allee effect exists, eradication is achieved without killing all individuals using pesticides/herbicides, as they will go extinct without intervention (Liebhold and Tobin 2008).

Containment and suppression of invasive species involves several management strategies which should be targeted and precise, a process referred to as Precision Pest Management (PPM) by Slippers et al. (2020). These involve the integration of conventional forest management techniques (cultural, chemical), biocontrol, chemical analyses, sequencing, data collections using smart technologies (drones and Geographical information) and sharing data across fields for the development of Integrated Pest Management tools (IPM). There is also the need to enhance and apply strict biosecurity protocols and legislations at the national, regional and continental levels. Biosecurity policies and capacity to enforce theme are lacking in over 94% of countries on the African continent (Sileshi et al. 2019). A summary and description of the different PPM strategies for the management of invasive species in African forestry are presented in Table 11.5.

11.7 Insights into National/Regional Policies Relating to the Management of Invasive Plant Species and Pests of Trees in Africa

The management of invasive species is a herculean task because their transportation and dispersion are facilitated through international trade. This is further complicated by globalisation and the ease of movement by people and goods. The management of invasive species falls within international and national phytosanitary policies and legislation summarised in Fig. 11.6. At the global level, the International Plant Protection Convention (IPPC) was set to protect plants from insects and pathogens under the World Trade Organisation's agreement on the application of sanitary and phytosanitary measures (WTO-SPS). The next level of legislation is that by continental phytosanitary organisations who are responsible for enforcing the IPPC regulations.

In Africa, the African Union-Inter-African Phytosanitary Council (AU-IAPSC) is mandated to coordinate and provide support to member states on the protection of plant resources for the welfare and development of the African Union. The AU-IAPSC is specifically mandated to prevent the introduction and spread of pests which attack and damage African forests, and develop strategies against their

Table 11.5 A summary of the different sustainable methods for the control of invasive plants and pests of forestry in Africa using precision pest management approaches

Strategy	Approach/principle	Advantages	Disadvantages and challenges
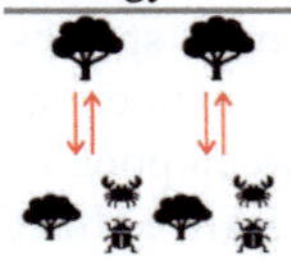 Sentinal planting	Planting of trees native to the importing country in exporting country to determine which pests and pathogens can spread to the importing countries (see Eschen et al. 2019 for more details)	Relatively inexpensive to set up, primarily foliage pests and diseases detected, possibility of planting many trees, early detection of risks. Few regulations	Not applicable to mature trees-in most cases, only young plants are traded.
 Mixed species forests	Forests and plantations should be made of more than one species. This mimics natural forests	Higher species diversity, higher productivity, low risk of biotic and abiotic disturbances, adaptation to climate change	Could be labour intensive, need proper planning and may provide diverse hosts for insects and diseases to persist all year round
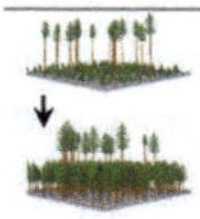 Silviculture	The practice of controlling the growth composition and structure of forests based on local/regional factors to meet needs. Usually the use of native species	Ensures sustainability and productivity at the same time. Provides un-even or even aged trees. Conserves biodiversity, soil and water. Mitigates climate change through carbon sequestration	Loss of biodiversity especially with even-aged trees, risk of pests and diseases, costs
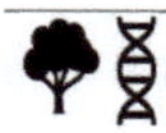 Breeding for resistance	Breeding clones and hybrids for resistance to pests and pathogens. This includes the sequencing of entire tree genomes and editing these with techniques like CRISPAR-Cas 9	Hybrids and clones that have genes resistance and tolerance for diseases, pests and changing climatic conditions	Time consuming, expensive, labour intensive. With time pests and pathogens can develop resistances

(continued)

Table 11.5 (continued)

Strategy	Approach/principle	Advantages	Disadvantages and challenges
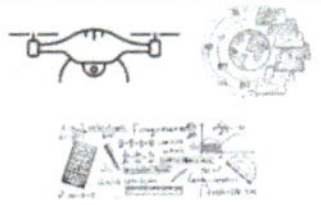 Contemporary techniques-use of drone technology, GIS, mathematical modelling etc	The use of quadcopters (unmanned aerial vehicles) to survey forests for the purpose of inspection, monitoring of tree health or the application of pesticides and herbicides Geographical information system used to manage, analyse edit and visualise geographic data in forests Data collected using the techniques above can be modelled to forecast threats by pests and diseases as well as develop management techniques	Rapid, and can be cost effective since it reduces man time in implementation. Also provides the advantage of early detection and forecast	Highly skilled, expensive due to initial investments in hardware, software equipment and personnel. Limited in its deployment to commercial plantations
 Biological control using parasitiods, entomopathogens (biopesicides).	Exploration of parasitoids/ entomopathogens of the insect pest in its native range or in the introduced range	Target and host specific, environmentally friendly. Many strategies can be deployed at the same time and no need of reapplication once the parasitoid population is established	Not rapid, time consuming, involves a lot of legislation and quarantine facilities, expensive. Requires international collaboration (see Garnas et al. 2012)
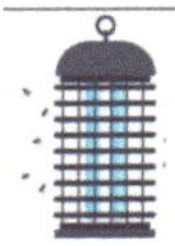 Semiochemical lures/traps	Exploring the olfactory communication of insect pests. Specifically, the odours/smell they use to locate host plants	Environmentally friendly, target specific (no non-target effect), sustainable, not labour intensive. Can be used as management or control tool	Highly skilled, interdisciplinary and can take time from exploration, testing and deployment

introduction and spread. In addition, the AU-IAPSC is also shouldered with the responsibility of enlightening member states on the implications of the WTO-SPS (AU-IAPSC 2020, www.auiapsc.org). National Plant Protection Organisations (NPPO) are responsible for regulating and reporting plant health issues to the AU-IAPSC. There are currently no regional PPOs (Fig. 11.6) on the continent which leaves a gap in monitoring and tackling invasions at the regional level. This is a gap that needs to be filled, because the economic regions are important if one

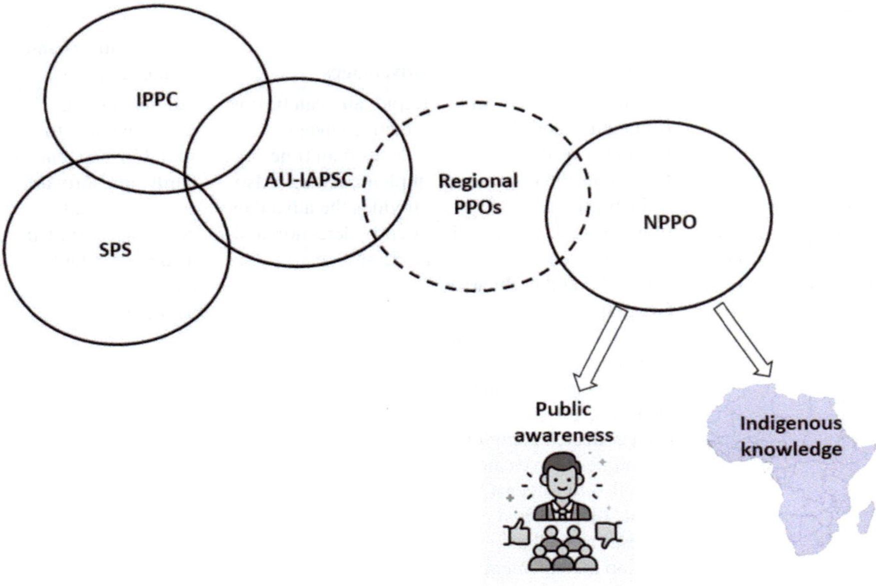

Fig. 11.6 Hierarchical structure of phytosanitary organisations including how they are nested and connected to one another. IPPC = International Plant Protection Convention set to protect plants from insects and pathogens. SPS = The agreement on the application of sanitary and phytosanitary measures sets out trade rules regulating plant health. AU-IAPSC = The African Union-Inter-African Phytosanitary Council mandated to coordinate and provide support to the protection of plant resources for the welfare and economic development in the member states of the African Union. NPPO = National Plant Protection Organisations mandated to monitor and enforce phytosanitary regulations, proposed Regional PPOs = Proposed regional plant protection organisations to coordinate activities at the different regional bodies of the continent and integration of public awareness and indigenous knowledge into national policies. (Figure modified from Lopian (2005) and Klapwijk et al. (2016))

looks at the distribution of invasive species on the continent (Tables 11.1, 11.2, 11.3, and 11.4) which takes a regional pattern before spreading to the whole of the continent.

Regional and inter-regional phytosanitary agencies will serve other purposes as eluded by Hurley et al. (2017) when advocating for the same within the South African Development Community (SADC). Regional phytosanitary organisations, will assist in consolidating efforts within the region and optimising limited resources. Other components that need to be included in the management of invasive species in Africa is the integration of public awareness campaigns on issues bothering management and monitoring. Indigenous knowledge (Fig. 11.6) that is available within communities especially those living in forests will be valuable in these efforts and give such communities a sense of belonging-the concept of 'People's forest' (Araia and Chirwa 2019.

11.8 Chapter Highlights

- Invasive species are important in African forests and multifunctional landscapes necessitating management.
- Invasive species are transported to Africa in different ways.
- Invasive pests and pathogens are established in African forests with most of the pests shifting their hosts to exotic trees used in plantation forests.
- Multifunctional forests offer a sustainable approach to forestry in Africa meeting both human and ecosystem demands
- There is the need to introduce regional phytosanitary agencies and incorporate indigenous knowledge in the management of invasive species in African multifunctional forests.

References

Adam E, Mutanga O, Ismail R (2013) Determining the susceptibility of *Eucalyptus nitens* forests to *Coryphodema tristis* (cossid moth) occurrence in Mpumalanga, South Africa. Int J Geogr Inf Sci 27(10):1924–1938. https://doi.org/10.1080/13658816.2013.772183

Adams B, Yusuf AA, Torto B, Khamis FM (2023) Non-host plant odours influence the tritrophic interaction between tomato, its foliar herbivore *Tuta absoluta* and mirid predator *Nesidiocoris tenuis*. Front Plant Sci 14:1014865. https://doi.org/10.3389/fpls.2023.1014865

Adkins S, Shabbir A (2014) Biology, ecology and management of the invasive parthenium weed (*Parthenium hysterophorus* L.). Pest Manag Sci 70(7):1023–1029. https://doi.org/10.1002/ps.3708

Araia MG, Chirwa PW (2019) Revealing the predominance of culture over the ecological abundance of resources in shaping local people's forest and tree species use behaviour: the case of the vhavenda people. South Africa Sustain 11(11):3143. https://doi.org/10.3390/su11113143

Araia MG, Chirwa PW, Syampungani S (2020) Do strictly protected areas protect vulnerable local tree species better than human land use? Disentangling conservation value from biodiversity value. J Nat Conserv 58:125919. https://doi.org/10.1016/j.jnc.2020.125919

AU-IAPSC (2020) African Union-Inter-African Phytosanitary Council. https://auiapsc.org. Accessed on 11th Nov 2023

Ayanu Y, Jentsch A, Müller-Mahn D, Rettberg S, Romankiewicz C, Koellner T (2015) Ecosystem engineer unleashed: *Prosopis juliflora* threatening ecosystem services? Reg Environ Chang 15:155–167. https://doi.org/10.1007/s10113-014-0616-x

Bajwa AA, Farooq M, Nawaz A, Yadav L, Chauhan BS, Adkins S (2019) Impact of invasive plant species on the livelihoods of farming households: evidence from *Parthenium hysterophorus* invasion in rural Punjab. Pakistan Biol Invasions 21(11):3285–3304. https://doi.org/10.1007/s10530-019-02047-0

Bellard C, Thuiller W, Leroy B, Genovesi P, Bakkenes M, Courchamp F (2013) Will climate change promote future invasions? Glob Change Biol 19(12):3740–3748. https://doi.org/10.1111/gcb.12344

Bengtsson J, Angelstam P, Elmqvist T, Emanuelsson U, Forbes C, Ihse M et al (2003) Reserves, resilience, and dynamic landscapes. Ambio 32:389–396. https://doi.org/10.1579/0044-7447-32.6.389

Bennett BM, Kruger FJ (2015) Forestry in reconstruction South Africa: imperial visions, colonial realities. Brit World 8:225–245. https://doi.org/10.3366/brw.2015.0192

Benton TG, Vickery JA, Wilson JD (2003) Farmland biodiversity: is habitat heterogeneity the key? Trends Ecol and Evol 39:273–287. https://doi.org/10.1016/S0169-5347(03)00011-9

Blackburn TM, Essl F, Evans T, Hulme PE, Jeschke JM, Kühn I, Kumschick S, Markovǻ Z, Mrugała A, Nentwig W, Pergl J (2014) A unified classification of alien species based on the magnitude of their environmental impacts. PLoS Biol 12(5):e1001850. https://doi.org/10.1371/journal.pbio.1001850

Blackburn TM, Bellard C, Ricciardi A (2019) Alien versus native species as drivers of recent extinctions. Front Ecol Environ 17(4):203–207. https://doi.org/10.1002/fee.2020

Bo Larsen J (2005) Functional forests in multifunctional landscapes-restoring the adaptive capacity of landscapes with forests and trees. In: Veltheim T and Pajari B (eds) Forests Landscape Restoration in Central and Northern Europe, EPI Proceedings, no 53

Bradley BA, Laginhas BB, Whitlock R, Allen JM, Bates AE, Bernatchez G, Sorte CJ (2019) Disentangling the abundance–impact relationship for invasive species. PNAS 116(20):9919–9924. https://doi.org/10.1073/pnas.1818081116

Brandt J, Vejre H (2004) Multifunctional landscapes – motives, concepts and perceptions. In: Brandt J and Vejre H (eds) Multifunctional Landscapes: vol 1 Theory, values and history. WIT Press, pp 3–32

Brasier CM (2008) The biosecurity threat to the UK and global environment from international trade in plants. Plant Path 57:792–808. https://doi.org/10.1111/j.1365-3059.2008.01886.x

Broennimann O, Guisan A (2008) Predicting current and future biological invasions: both native and invaded ranges matter. Biol Lett 4(5):585–589. https://doi.org/10.1098/rsbl.2008.0254

Bush SJ, Slippers B, Neser S, Harney M, Dittrich-Schroder G, Hurley BP (2016) Six recently recorded Australian insects associated with Eucalyptus in South Africa. African Entomol 24(2):539–544. https://hdl.handle.net/10520/EJC195069

CBD (2008) Ninth meeting of the Conference of the Parties to the Convention on Biological Diversity Bonn, Germany 19–30 May 2008

Coetzee MP, Wingfield BD, Harrington TC, Steimel J, Coutinho TA, Wingfield MJ (2001) The root rot fungus Armillaria mellea introduced into South Africa by early Dutch settlers. Mol Ecol 10(2):387–396. https://doi.org/10.1046/j.1365-294x.2001.01187.x

Crystal-Ornelas R, Lockwood JL (2020) The 'known unknowns' of invasive species impact measurement and the quest for comprehensive impact frameworks. Biol Invasions 22:1513–1525. https://doi.org/10.1007/s10530-020-02200-0

de Neergaard A, Saarnak C, Hill T, Khanyile M, Berzosa AM, Birch-Thomsen T (2005) Australian wattle species in the Drakensberg region of South Africa – an invasive alien or a natural resource? Agr Sys 85(3):216–233. https://doi.org/10.1016/j.agsy.2005.06.009

Didham RK, Tylianakis JM, Hutchison MA, Ewers RM, Gemmell NJ (2005) Are invasive species the drivers of ecological change? Trends Ecol Evol 20(9):470–474. https://doi.org/10.1016/j.tree.2005.07.006

Drechsler M, Touza J, White PC, Jones G (2016) Agricultural landscape structure and invasive species: the cost-effective level of crop field clustering. Food Secur 8:111–121. https://doi.org/10.1007/s12571-015-0539-5

Ehrenfeld JG (2010) Ecosystem consequences of biological invasions. Ann Rev Ecol Evol S 41:59–80. https://doi.org/10.1146/annurev-ecolsys-102209-144650

Eschen R, O'Hanlon R, Santini A, Roques A, Kirichenko N, Kenis M (2019) Safeguarding global plant health: the rise of sentinels. J Pest Sci 92:29–36. https://doi.org/10.1007/s10340-018-1041-6

FAO (2009) Global review of forest pests and diseases. Food and Agriculture Organization of the United Nations. FAO Forestry Paper, 156, 222 pp ISBN 978-92-5-106208-1

FAO (2015) Global forest resources assessment 2015: how are the world's forests changing? 2nd edn. Food and Agriculture Organization of the United Nations, Rome

FAO (2021) Global forest resources assessment 2000. Main report. FAO forestry paper 139. FAO, Rome. http://www.fao.org/forestry/fo/fra/index.jsp

Follak S, Schleicher C, Schwarz M (2018) Roads support the spread of invasive in Austria. Die Bodenkultur J Land Mgt Food Environ 69(4):257–265. https://doi.org/10.2478/boku-2018-0022

Freer-Smith PH, Webber JF (2017) Tree pests and diseases: the threat to biodiversity and the delivery of ecosystem services. Biodivers Conserv 26:3167–3181. https://doi.org/10.1007/s10531-015-1019-0

Garnas JR, Hurley BP, Slippers B, Wingfield MJ (2012) Biological control of forest plantation pests in an interconnected world requires greater international focus. Int J Pest Mange 58:211–223. https://doi.org/10.1080/09670874.2012.698764

Genovesi P, Shine C (2004) European strategy on invasive alien species: Convention on the Conservation of European Wildlife and Habitats (Bern Convention) (No. 18–137). Council of Europe

Govender P (2002) Management of insect pests: Have the goalposts changed with certification? The Southern African Forestry Journal 195(1):39–45. https://doi.org/10.1080/20702620.2002.10434602

Graziosi I, Mathias T, Kuate J, Muchugi A (2019) Pests and diseases of treat in Africa: A growing continental emergency. Plants People Planet 2(1):14–28. https://doi.org/10.1002/ppp3.31

Gullino ML, Albajes R, Aljoobry I, Angellotti F et al (2022) Climate change and pathways used by pests as challenges to plant health in agriculture and forestry. Sustain For 14(19):12421. https://doi.org/10.3390/su141912421

Gustafasson L, Susan CB, Jurgen B, William JB et al (2012) Retentions forestry to maintain multifunctional forests: a world perspective. Bioscience 62:633–645. https://doi.org/10.1525/bio.2012.62.7.6

Hirsch H, Allsopp MH, Canavan S, Cheek M, Geerts S, Geldenhuys CJ, Richardson DM (2020) *Eucalyptus camaldulensis* in South Africa–past, present, future. Trans Royal Soc South Africa 75(1):1–22. https://doi.org/10.1080/0035919X.2019.1669732

Holmes TP, Aukema JE, Von Holle B, Liebhold A, Sills E (2009) Economic impacts of invasive species in forests. Ann N Y Acad Sci 1162:18–38. https://doi.org/10.1111/j.1749-6632.2009.04446.x

Hulme PE (2009) Trade, transport and trouble: managing invasive species pathways in an era of globalization. J Appl Ecol 46(1):10–18. https://doi.org/10.1111/j.1365-2664.2008.01600.x

Hurley BP, Garnas J, Wingfield MJ, Branco M, Richardson DM, Slippers B (2016) Increasing numbers and intercontinental spread of invasive insects on eucalypts. Biol Invasions 18:921–933. https://doi.org/10.1007/s10530-016-1081-x

Hurley BP, Slippers B, Sathayapala S, Wingfield MJ (2017) Challenges to planted forests in developing economies. Biol Invasions 19:3273–3285. https://doi.org/10.1007/s10530-017-1488-z

Ilukor J, Rettberg S, Treydte A, Birner R (2016) To eradicate or not to eradicate? Recommendations on *Prosopis juliflora* management in Afar, Ethiopia, from an interdisciplinary perspective. Pastoralism 6:1–8. https://doi.org/10.1186/s13570-016-0061-1

IPCC (2021) Climate change 2021: the physical science basis. Report of working group 1, International Panel for Climate Change Assessment report. https://www.ipcc.ch/report/ar6/wg1/

Israel A, Wynberg R (2019) Multifunctional landscapes in a rural, developing country context: conflicts and synergies in Tshidzivhe, South Africa. Landscape Res. https://doi.org/10.1080/01426397.2018.1441388

IUCN (2012) International Union for the Conservation of Nature- red list for Threatened and invasive species. International Union for the Conservation of Nature

Keane RM, Crawley MJ (2002) Exotic plant invasions and the enemy release hypothesis. Trends Ecol Evol 17(4):164–170. https://doi.org/10.1016/S0169-5347(02)02499-0

Klapwijk MJ, Hopkins AJ, Eriksson L, Pettersson M, Schroeder M, Lindelöw Å et al (2016) Reducing the risk of invasive forest pests and pathogens: combining legislation, targeted management and public awareness. Ambio 45:223–234

Lenzner B, Latombe G, Schertler A, Seebens H, Yang Q, Winter M, Weigelt P, van Kleunen M, Pyšek P, Pergl J, Kreft H (2022) Naturalized alien floras still carry the legacy of European colonialism. Nat Ecol Evol 6(11):1723–1732. https://doi.org/10.1038/s41559-022-01865-1

Levine JM, D'Antonio CM (2003) Forecasting biological invasions with increasing international trade. Conserv Biol 17:322–326

Liebhold AM, Tobin PC (2008) Population ecology of insect invasions and their management. Annu Rev Entomol 53:387–408. https://doi.org/10.1146/annurev.ento.52.110405.091401

Liebhold AM, Brockerhoff EG, Garrett LJ, Parke JL, Britton KO (2012) Live plant imports: the major pathway for forest insect and pathogen invasions of the US. Front Ecol Environ 10:135–143. https://doi.org/10.1890/110198

Linders TEW, Schaffner U, Eschen R, Abebe A, Choge SK, Nigatu L, Mbaabu PR, Shiferaw H, Allan E (2019) Direct and indirect effects of invasive species: biodiversity loss is a major mechanism by which an invasive tree affects ecosystem functioning. J Ecol 107(6):2660–2672. https://doi.org/10.1111/1365-2745.13268

Lopian R (2005) Activities of the International Plant Protection Convention in regard to invasive alien species. Secretariat of the International Plant Protection Convention eng, Invasive Alien Species and the International Plant Protection Convention eng 22-26 Sep 2003 Braunschweig

Makoni M (2020) Africa's invasive species problem. Lancet Planet Health 4(8):e317–e319. https://doi.org/10.1016/S2542-5196(20)30174-1

Mazza G, Tricarico E (eds) (2018) Invasive Species and Human Health, vol 10. CAB International

McCary MA, Mores R, Farfan MA, Wise DH (2016) Invasive plants have different effects on trophic structure of green and brown food webs in terrestrial ecosystems: a meta-analysis. Ecol Lett 19(3):328–335. https://doi.org/10.1111/ele.12562

Mehari ZH (2015) The invasion of *Prosopis juliflora* and A far pastoral livelihoods in the Middle Awash area of Ethiopia. Ecol Process 4(1):13. https://doi.org/10.1186/s13717-015-0039-8

Meurisse N, Rassati D, Hurley BP, Brockerhoff EG, Haack RA (2018) Common pathways by which non-native forest insects move internationally and domestically. J Pest Sci 92:13–27. https://doi.org/10.1007/s10340-018-0990-0

Nair KSS (2001) Pest outbreaks in tropical forest plantations: is there a greater risk for exotic tree species? Centre for International Forestry Research, Kota Bogor

Nunes AL, Fill JM, Davies SJ, Louw M, Rebelo AD, Thorp CJ, Measey J (2019) A global meta-analysis of the ecological impacts of alien species on native amphibians. Proc R Soc B 286(1897):20182528. https://doi.org/10.1098/rspb.2018.2528

Paine TD, Steinbauer MJ, Lawson SA (2011) Native and exotic pests of Eucalyptus: a worldwide perspective. Annu Rev Entomol 56:181–201. https://doi.org/10.1146/annurev-ento-120709-144817

Peter E, Tamiru A, Sevgan S, Dubois T, Kelemu S, Kruger K, Torto B, Yusuf A (2023) Companion crops alter olfactory responses of the fall armyworm (*Spodoptera frugiperda*) and its larval endoparasitoid (*Cotesia icipe*). Chem Biol Tech Agric 10(1):61. https://doi.org/10.1186/s40538-023-00415-6

Pfeiffer JM, Voeks RA (2008) Biological invasions and biocultural diversity: Linking ecological and cultural systems. Environ Conserv 35(4):281–293. https://doi.org/10.1017/S0376892908005146

Pimentel D, McNair S, Janecka J, Wightman J, Simmonds C, O'connell C, Wong E, Russel L, Zern J, Aquino T, Tsomondo T (2001) Economic and environmental threats of alien plant, animal, and microbe invasions. Agric Ecosys Environ 84(1):1–20. https://doi.org/10.1016/S0167-8809(00)00178-X

Poland TM, Rassati D (2019) Improved biosecurity surveillance of non-native forest insects: a review of current methods. J Pest Sci 92:37–49. https://doi.org/10.1007/s10340-018-1004-y

Pratt CF, Constantine KL, Murphy ST (2017) Economic impacts of invasive alien species on African smallholder livelihoods. Glob Food Secur 14:31–37. https://doi.org/10.1016/j.gfs.2017.01.011

Pretty JP (2008) Agricultural Sustainability: concepts, principles, and evidence. Philos Trans R Soc B 363:447–465. https://doi.org/10.1098/rstb.2007.2163

Rai PK, Singh JS (2020) Invasive alien plant species: Their impact on environment, ecosystem services and human health. Ecol Indic 111:106020. https://doi.org/10.1016/j.ecolind.2019.106020
Rejmánek M, Richardson DM (2013) Trees and shrubs as invasive alien species – 2013 update of the global database. Diver Distr 19:1093–1094. https://doi.org/10.1111/ddi.12075
Reynolds C, Venter N, Cowie BW, Marlin D, Mayonde S, Tocco C, Byrne MJ (2020) Mapping the socio-ecological impacts of invasive plants in South Africa: are poorer households with high ecosystem service use most at risk? Ecosyst Serv 42:101075. https://doi.org/10.1016/j.ecoser.2020.101075
Ribeiro J, Reino L, Schindler S, Strubbe D, Vall-llosera M, Araújo MB, Capinha C, Carrete M, Mazzoni S, Monteiro M, Moreira F (2019) Trends in legal and illegal trade of wild birds: a global assessment based on expert knowledge. Biodivers Conserv 28:3343–3369. https://doi.org/10.1007/s10531-019-01825-5
Ricciardi A (2007) Are modern biological invasions an unprecedented form of global change? Conserv Biol 21(2):329–336. https://doi.org/10.1111/j.1523-1739.2006.00615.x
Robertson PA, Mill A, Novoa A, Jeschke JM, Essl F, Gallardo B, Geist J, Jarić I, Lambin X, Musseau C, Pergl J (2020) A proposed unified framework to describe the management of biological invasions. Biol Invasions 22:2633–2645. https://doi.org/10.1007/s10530-020-02298-2
Rogers P, Nunan F, Fentie AA (2017) Reimagining invasions: The social and cultural impacts of *Prosopis* on pastoralists in southern Afar. Ethiopia Pastoralism 7(1):1–13. https://doi.org/10.1186/s13570-017-0094-0
Roux J, Hurley BP, Wingfied MJ (2012) Diseases and pests of eucalypts, pines and wattle. In: Bredenkamp BV, Upfold SJ (eds) South African forestry handbook, 5th edn. South African Institute for Forestry, Pietermaritzburg, pp 303–336
Sertse D, Pasiecznik NM (2005) Controlling the spread of *Prosopis* in Ethiopia by its utilisation. Publication from Henry Doubleday Research Association (HDRA), Coventry
Shackleton CM, McGarry D, Fourie S, Gambiza J, Shackleton SE, Fabricius C (2007) Assessing the effects of invasive alien species on rural livelihoods: case examples and a framework from South Africa. Hum Ecol 35:113–127. https://doi.org/10.1007/s10745-006-9095-0
Shackleton RT, Witt AB, Aool W, Pratt CF (2017a) Distribution of the invasive alien weed, *Lantana camara*, and its ecological and livelihood impacts in eastern Africa. Afr J Range Forage Sci 34(1):1–11. 44. https://doi.org/10.2989/10220119.2017.1301551
Shackleton RT, Witt AB, Nunda W, Richardson DM (2017b) *Chromolaena odorata* (Siam weed) in eastern Africa: distribution and socio-ecological impacts. Biol Invasions 19(4):1285–1298. https://doi.org/10.1007/s10530-016-1338-4
Shackleton RT, Shackleton CM, Kull CA (2019) The role of invasive alien species in shaping local livelihoods and human well-being: a review. J Environ Manag 229:145–157. https://doi.org/10.1016/j.jenvman.2018.05.007
Siges TH, Hartemink AE, Hebinck P, Allen BJ (2005) The invasive shrub Piper aduncum and rural livelihoods in the Finschhafen area of Papua New Guinea. Hum Ecol 33:875–893. https://doi.org/10.1007/s10745-005-8214-7
Sileshi GW, Gebeyehu S, Mafongoya PL (2019) The threat of alien invasive insect and mite species to food security in Africa and the need for a continent-wide response. Food Secur 11:763–775. https://doi.org/10.1007/s12571-019-00930-1
Simberloff D, Martin JL, Genovesi P, Maris V, Wardle DA, Aronson J, Courchamp F, Galil B, García-Berthou E, Pascal M, Pyšek P (2013) Impacts of biological invasions: what's what and the way forward. Trends Ecol Evol 28(1):58–66
Skočajić D, Nešić M (2020) Invasive species: routes of introduction, establishment, and expansion. In: Leal Filho W, Azul A, Brandli L, Özuyar P, Wall T (eds) Life on land. Encyclopedia of the UN sustainable development goals. Springer, Cham. https://doi.org/10.1007/978-3-319-71065-5_66-1
Slippers B, Hurley BP, Allison JD (2020) Harnessing the potential of Precision Pest Management in plantation forests. South For J For Sci 82(3):197–201. https://doi.org/10.2989/20702620.2020.1813651

Smith-Ramesh LM (2017) The aliens among us: how invasive species are transforming the planet and ourselves Yale University Press, New Haven Connecticut, 2017, xii + 353 pp, ISBN 978-0-300-20890-0. Biological Invasions 1–2

Tilahun M, Birner R, Ilukor J (2017) Household-level preferences for mitigation of *Prosopis juliflora* invasion in the Afar region of Ethiopia: A contingent valuation. J Environ Plan Manag 60(2):282–308. https://doi.org/10.1080/09640568.2016.1152955

van Wilgen BW, Richardson DM, Le Maitre DC, Marais C, Magadlela D (2001) The economic consequences of alien plant invasions: Examples of impacts and approaches to sustainable management in South Africa. Environ Dev Sustain 3(2):145–168

van Wilgen BW, Witt AB, Beale T, Kiambi S (2017) A preliminary assessment of the extent and potential impacts of alien plant invasions in the Serengeti-Mara ecosystem, East Africa. Koedoe: Afr Protect Area Conserv Sci 59(1):1–16

Vilà M, Espinar JL, Hejda M, Hulme PE, Jarošík V, Maron JL, Pyšek P (2011) Ecological impacts of invasive alien plants: a meta-analysis of their effects on species, communities and ecosystems. Ecol Lett 14(7):702–708. https://doi.org/10.1111/j.1461-0248.2011.01628.x

Vimercati G, Probert AF, Volery L, Bernardo-Madrid R, Bertolino S, Céspedes V, Essl F, Evans TT, Gallardo B, Gallien L, González-Moreno P (2022) The EICAT+ framework enables classification of positive impacts of alien taxa on native biodiversity. PLoS Biol 20(8):e3001729. https://doi.org/10.1371/journal.pbio.3001729

Wakie T, Laituri M, Evangelista H (2016) Assessing the distribution and impacts of *Prosopis julifora* through participatory approaches. Appl Geogr 66:132–143. https://doi.org/10.1016/j.apgeog.2015.11.017

Westholm L, Ostwald M (2020) Food production and gender relations in multifunctional landscapes: a literature review. Agrofor Syst 94:359–374. https://doi.org/10.1007/s10457-019-00397-1

Wheeler T, Von Braun J (2013) Climate change impacts on global food security. Science 341(6145):508–513. https://doi.org/10.1126/science.1239402

Wingfield MJ, Slippers B, Hurley BP, Coutinho TA, Wingfield BD, Roux J (2008) Eucalypt pests and diseases: growing threats to plantation productivity. Southern For 70:139–144. https://doi.org/10.2989/SOUTH.FOR.2008.70.2.9.537

Wingfield MJ, Roux J, Wingfield BD (2011) Insect pests and pathogens of Australian acacias grown as non-natives—an experiment in biogeography with far-reaching consequences. Divers Distrib 17:968–977. https://doi.org/10.1111/j.1472-4642.2011.00786.x

Wise RM, Van Wilgen BW, Le Maitre DC (2012) Costs, benefits and management options for an invasive alien tree species: The case of mesquite in the Northern Cape, South Africa. J Arid Environ 84:80–90. https://doi.org/10.1016/j.jaridenv.2012.03.001

Witt A, Beale T, Van Wilgen BW (2018) An assessment of the distribution and potential ecological impacts of invasive alien plant species in eastern Africa. Trans R Soc S Afr 73(3):217–236. https://doi.org/10.1080/0035919X.2018.1529003

Zengeya T, Ivey P, Woodford DJ, Weyl O, Novoa A, Shackleton R, Richardson D, van Wilgen B (2017) Managing conflict-generating invasive species in South Africa: Challenges and tradeoffs. Bothalia-Afric Biodiver Conserv 47(2):1–11. https://doi.org/10.4102/abc.v47i2.2160

Zeray N, Legesse B, Mohamed J, Aredo M (2017) Impacts of *Prosopis* invasion on livelihoods of pastoral and agro-pastoral households of Dire Dawa Administration, Ethiopia. Pastoralism 7(1):7–14. https://doi.org/10.1186/s13570-017-0079-z

Chapter 12
Socio-ecological Management of Multifunctional Landscapes

Kouami Kokou, Stephen Syampungani, Paxie W. Chirwa, and Lucky Makhubele

Abstract This chapter explores the intricate dynamics of socio-ecological management in Sub-Saharan African multifunctional landscapes, with a specific focus on the vital role of trees. Drawing on the latest literature, the analysis reveals the multifaceted challenges posed by ecological, social, and economic factors. Advocating for a robust socio-ecological framework, the chapter emphasizes community engagement, the integration of local knowledge, and adaptive governance as fundamental pillars. The exploration extends to ecological restoration initiatives in countries like Eswatini, Madagascar, and South Africa, aligning with global commitments such as the African Forest Landscape Restoration Initiative (AFR100). Biodiversity conservation integration, recognizing Key Biodiversity Areas (KBAs) and the Southern African KBA Programme, underscores the crucial role of local knowledge in tree management. Addressing challenges in implementation, the chapter identifies adaptive strategies to navigate anthropogenic pressures, financial constraints, and conflicts of interest. It outlines opportunities and prospects, including technological innovations, community-based management, and international collaboration, while emphasizing the importance of research, education, and climate-resilient practices. A collective call to action underscores the need for collaborative efforts,

K. Kokou (✉)
Laboratoire de Recherche Forestière, Centre de Recherche sur le Changement Climatique (CRCC), Université de Lomé, BP, Lomé, Togo

S. Syampungani
Department of Plant and Soil Sciences, Faculty of Natural and Agricultural Sciences, University of Pretoria, Pretoria, South Africa

Chair of Environment and Development, Oliver R. Tambo Africa Research Chairs Initiative (ORTARChI), Copperbelt University, Kitwe, Zambia

P. W. Chirwa
Department of Plant and Soil Sciences, Faculty of Natural and Agricultural Sciences, University of Pretoria, Pretoria, South Africa

L. Makhubele
School of Agricultural Sciences, Faculty of Agriculture and Natural Sciences, University of Mpumalanga, Mbombela, South Africa

P. W. Chirwa et al. (eds.), *Trees in a Sub-Saharan Multi-functional Landscape*,
https://doi.org/10.1007/978-3-031-69812-5_12

partnerships, and knowledge-sharing among stakeholders. In conclusion, the chapter advocates for resilient, community-centric management to ensure the sustainable well-being of ecosystems and communities in Sub-Saharan African multifunctional landscapes.

Keywords Socio-ecological management · Multifunctional landscapes · Trees · Biodiversity conservation · Sub-Saharan Africa

12.1 Introduction

In the landscapes of Sub-Saharan Africa (SSA), the multifunctionality of ecosystems, particularly with respect to trees, presents a complex and crucial challenge for research, management, and policy development (Kuyah et al. 2016). As the region grapples with a myriad of socio-economic and environmental issues, the role of trees in shaping the landscape is multifaceted, impacting ecosystems, livelihoods, and biodiversity (Muthee et al. 2022). The intricate interplay between trees and their surroundings necessitates a nuanced understanding to address the simultaneous demands of agriculture, biodiversity conservation, and sustainable development. Face to climate change, population growth, and land use transformation, there is also an urgent need to explore holistic approaches that recognize the diverse functions of trees in these landscapes (Massenberg et al. 2023; Shaw et al. 2023; Djenontin et al. 2018).

Beyond their ecological contributions, trees in SSA landscapes are integral to the livelihoods and well-being of local communities. They serve as sources of food, medicine, and timber, constituting essential components of daily life and traditional practices. Non-timber forest products derived from trees, such as fruits, nuts, and resins, contribute to local economies and provide a safety net during times of agricultural uncertainty (Singh and Singh 2023). Through their root systems and canopy cover, trees form a natural defense against soil erosion and contribute to the maintenance of watershed health (Diop et al. 2022). Moreover, their ability to sequester carbon mitigates the impacts of climate change, a critical service in the face of global environmental challenges (Mori et al. 2021). The diverse tree species indigenous to SSA landscapes contribute significantly to ecological resilience by stabilizing soil, regulating water flow, and fostering biodiversity. Furthermore, the cultural significance of certain tree species is deeply embedded in the fabric of community identity, with trees often playing central roles in rituals, ceremonies, and folklore. By providing these essential ecosystem services such as carbon sequestration and water regulation to offering sustenance through non-timber forest products and supporting traditional practices deeply rooted in local communities, the multifunctionality of landscapes in SSA is both a source of resilience and vulnerability

(Song et al. 2020). Recognizing this complexity is essential for formulating effective research agendas, management strategies, and policies that harmonize ecological conservation with the diverse needs of growing populations.

In confronting the challenges posed by SSA multifunctional landscapes, the imperative for a socio-ecological management approach becomes paramount (Vlek et al. 2017). These landscapes, characterized by a complex interplay of environmental, social, and economic factors, demand a nuanced strategy that transcends conventional paradigms. The sustainability and resilience of these ecosystems hinge on an integrated management approach that acknowledges the linkages between ecological processes and the well-being of local communities.

A socio-ecological management framework recognizes that the vitality of landscapes is not solely dependent on ecological considerations but is intricately intertwined with human dynamics. This approach emphasizes the importance of community engagement, local knowledge systems, and the empowerment of stakeholders in decision-making processes. It seeks to strike a balance between conservation goals and the diverse needs of communities relying on these landscapes for their livelihoods. Winter et al. (2018) and Folke (2006) underscore the efficacy of socio-ecological systems in managing complex landscapes as an approach that embraces adaptive governance, where management strategies evolve in response to changing environmental and social conditions, combining ecological resilience with community well-being.

Drawing on a rich body of literature and empirical studies, our exploration contributes to a deeper comprehension of the challenges and opportunities posed by SSA multi-functional landscapes, fostering sustainable solutions for the benefit of both ecosystems and communities relying on them, and offering insights that bridge the gap between research, management strategies, and policy interventions.

The chapter aims to comprehensively explore the principles, approaches, and challenges associated with the socio-ecological management of trees in multifunctional landscapes within the context of SSA. The specific objectives are to:

- examine the foundational principles that underpin socio-ecological management of trees in multifunctional landscapes, considering the integrated relationship between ecological processes and human dynamics;
- investigate and analyze existing and emerging approaches to the socio-ecological management of trees, emphasizing adaptive governance, community engagement, and the integration of traditional ecological knowledge;
- explore and illuminate the rich contextual tapestry of SSA multifunctional landscapes, shedding light on their importance as nexus points for sustainable development, biodiversity conservation, and cultural heritage; and
- Identify and analyze the challenges and potential trade-offs associated with socio-ecological management, considering factors such as conflicting interests, institutional barriers, and the need for balancing conservation goals with human needs.

12.2 Understanding Dynamic Landscapes and Impact of Populations

Within the landscapes of SSA (Fig. 12.1), the complex interaction of ecological dynamics plays a crucial role in shaping the multifunctionality of ecosystems, particularly concerning trees. These dynamics encompass biodiversity, habitat changes,

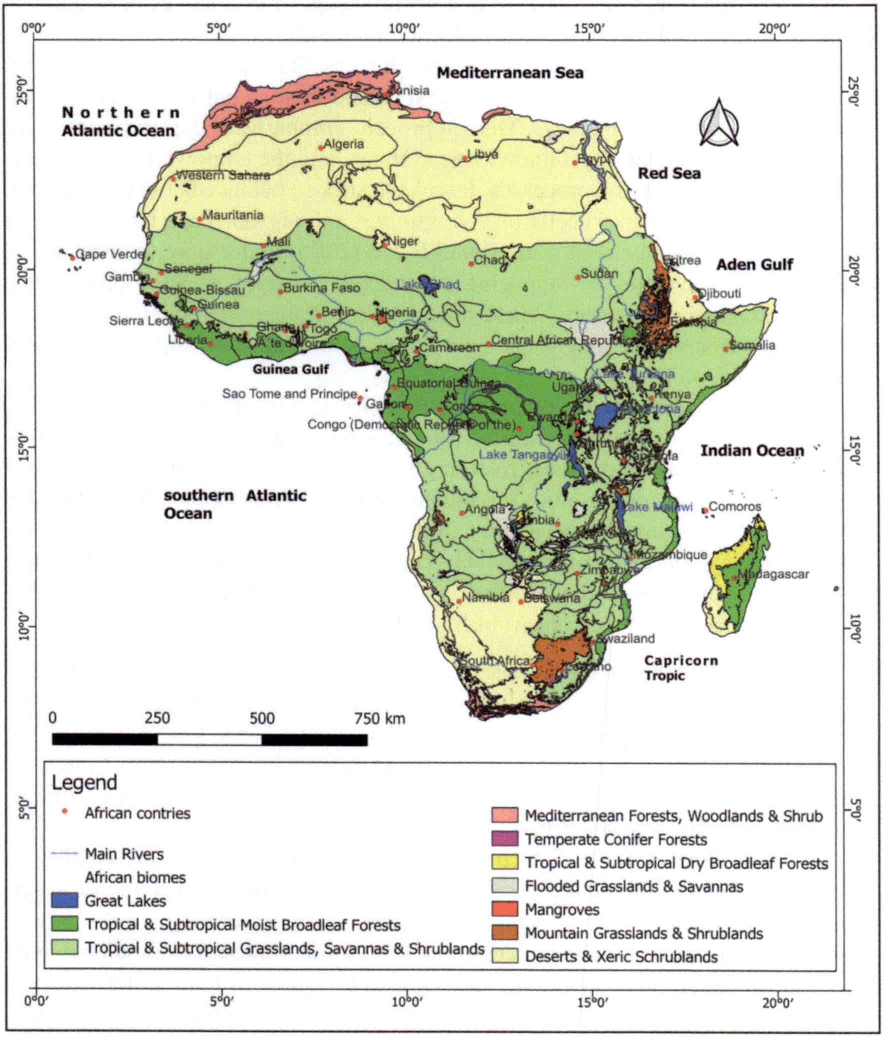

Fig. 12.1 Main biomes of Africa. (Adapted from Sahara and Sahel Observatory 2022)

habitat fragmentation, connectivity, and succession, thereby exerting a profound influence on the distribution of species and their interactions. The rich diversity of plant and animal species significantly contributes to the overall health and adaptability of these ecosystems. Landscape stability is inherently linked to habitat dynamics, whether influenced by climatic fluctuations or human activities.

Habitat fragmentation, often driven by human activities such as agriculture and urbanization, poses a major challenge. It disrupts ecological processes, impedes species movement, and leads to the formation of isolated populations (Liu et al. 2016). Preserving connectivity between different habitats is essential for maintaining biodiversity, and corridors that facilitate species movement between fragmented areas play a key role in preventing isolation and promoting genetic diversity. For instance, a need to recognize the value of small forest patches (church forest) for biodiversity conservation in Ethiopia (Abuhay et al. 2024). The ecological succession process, involving gradual and predictable changes in plant and animal communities over time, is a fundamental aspect of landscape dynamics.

Human activities, especially those related to land use, have a profound impact on landscape dynamics in SSA. Key activities, such as land use and land cover changes, driven by agriculture, urbanization, and overexploitation of natural resources significantly contribute to shaping the multifunctionality of the region's ecosystems. Agriculture, a cornerstone of human livelihoods, often leads to significant landscape changes. The expansion of agricultural areas, changes in farming practices, and the use of agrochemicals induce alterations in land cover (Cao et al. 2019). Rapid urbanization in SSA is a major driver of landscape transformation. Urban sprawl, infrastructure development, and the associated increase in impermeable surfaces modify land cover (Yiran et al. 2020). Deforestation, caused by factors like logging and land clearing for agriculture, has extensive consequences for landscapes. It contributes to changes in land cover, resulting in habitat and biodiversity loss, soil erosion, and disruptions in ecological processes (Kouassi et al. 2021). Africa had the largest net forest loss in 2010–2020, at 3.9 million ha per year, followed by South America, at 2.6 million ha. The rate of net forest loss has increased in Africa in each of the three decades since 1990 (FAO 2020). The major driver of forest loss is agricultural expansion and accounts for 73 percent of deforestation (FAO 2020).

The broader consequences of human-induced land use changes are numerous. Landscape fragmentation, arising from the conversion of natural habitats into small isolated patches, affects species mobility and genetic diversity. Soil degradation resulting from intensive agriculture and deforestation has implications for fertility and watershed health. Biodiversity loss, exacerbated by habitat destruction, and alterations in water resource management further underscore the complex interaction between human activities and landscape dynamics.

12.3 Socio-ecological Approach in the Context of Multifunctional Landscape Management

The multifunctional landscape management represents the link between nature and society in conservation, protection and utilization of natural resources (Villodre et al. 2023). Socio-ecological approach that involves integration of social and ecological objectives in a multifunctional landscape management scale is fundamental. This approach fosters a complementary and holistic approach that balances social and ecological systems. The social ecological approach is required to address the issues related to multiservice and management of multifunctional landscape. This approach is achieved through socio ecological planning that considers both social and ecological systems, decision-making and implementation (Kangas and Store 2002).

12.3.1 Innovative Socio-ecological Approaches for Sustainable Tree Management in SSA Landscapes

12.3.1.1 Agroforestry Practices and Socio-ecological Tree Management Approaches

Various agroforestry techniques have been applied successfully in SSA to improve agricultural sustainability, soil health, biodiversity, and livelihoods (Dagar et al. 2020), including shade trees in coffee-cocoa plantations, alley cropping, forest gardens, or agro-silvopastoral systems. In coffee-cocoa plantations in the forest zone of SSA, shade trees, mostly natives, enhance biodiversity and provide habitat for plant species, birds and other wildlife (Sonwa et al. 2007; Negawo and Beyene 2017). As part of eco-friendly practices, reforestation with fast-growing tree species is nowadays encouraged to replace harvested trees (Fig. 12.2).

Another most widely adopted sustainable farming system is alley cropping, where farmers plant rows of trees or shrubs in between their crop rows. These trees provide shade and windbreaks, reducing soil erosion and improving microclimates for crops. The pruned branches also serve as mulch, enhancing soil fertility. A well-known example in Nigeria is the use of *Leucaena leucocephala* in alley cropping systems (Silva-Galicia et al. 2023). Moreover forest garden systems where farmers cultivate a diverse range of food crops, fruit trees, and medicinal plants in a forest-like setting have been documented in Malawi (Ngugi 2002). By mimicking the structure of natural forests, these gardens enhance biodiversity, food security, and soil fertility. In addition, they provide a consistent source of nutritious foods for local communities, reducing dependence on single staple crops (Santos et al. 2022). In Malawi and Zambia, the EverGreen Agriculture program promotes agroforestry practices that integrate trees with annual crops. This approach helps enhance soil fertility, increase crop yields, and provide a source of firewood and timber (Mwase

Fig. 12.2 Coffee agroforest in the forest zone of Togo. (Photo Attiogbe Amen)

et al. 2015; Gello 2021). In Kenya, agro-silvopastoral systems have gained popularity. Farmers integrate trees, forage crops, and livestock in a symbiotic relationship. For instance, farmers plant fodder trees such as *Calliandra calothyrsus* Meisen. alongside Napier grass for their cattle (Franzel et al. 2014; Mohamed Nikzaad and Nusrathali 2023). The trees provide high-quality fodder, while the livestock's manure enriches the soil. This system boosts milk and meat production and conserves soil health.

Farmer-Managed Natural Regeneration (FMNR) is an innovative and low-cost approach that encourages farmers to regenerate trees from tree stumps, roots, and seeds on their land. This method is similar to the Assisted Natural Regeneration (ANR) that involves facilitating the regrowth of natural vegetation, particularly trees, in degraded or deforested areas. ANR minimizes human intervention, relying on the natural dispersal of seeds, root sprouting, and the recovery of soil fertility. The regrowth of indigenous vegetation enhances biodiversity, stabilizes soil, and improves water retention. ANR is often a cost-effective method, as it leverages natural processes, requiring minimal external inputs. Several SSA countries have embraced ANR as a sustainable restoration strategy. In Niger, the FMNR approach has led to the re-greening of vast areas, improving soil fertility and supporting agricultural productivity, and helps restore degraded landscapes and provides firewood, timber (Larwanou and Saadou 2011; Maisharou et al. 2015). Similar success stories are found in Ethiopia, where ANR contributes to landscape restoration, water regulation, and increased resilience to climate variability (Palomo et al. 2021). Similarly, area enclosures in Botswana have been reported to enhance natural regeneration and restored forest vegetation (Teketay et al. 2018).

Bamboo offers innovative solutions to the challenges of multifunctional landscape management. Ethiopia, Kenya, Nigeria and other countries are exploring the potential of bamboo agroforestry to provide income, mitigate soil erosion, and promote sustainable land use (Partey et al. 2017; Isukuru et al. 2023). Other practices and socio-ecological tree management approaches known in SSA are the community-based forest management (CBFM) initiatives. These initiatives involve local communities in the management and conservation of their forest resources. An example of CBFM is the Participatory Forest Management (PFM) program in the Kilwa District of southern Tanzania. This PFM program involves collaboration between local communities, governments, and NGOs to manage forests collectively. This program aims to balance conservation with livelihoods and sustainable resource use (Khatun et al. 2015). In addition, Payment for Ecosystem Services (PES) programs are being explored in some SSA countries to reward communities for conserving forests and promoting forest conservation and sustainable resource management (Le et al. 2023). In many countries, experiments are ongoing to develop climate-smart agroforestry and forestry practices that help mitigate climate change while enhancing agricultural and forest productivity. These approaches often involve tree planting and carbon sequestration (Ntawuruhunga et al. 2023).

12.3.1.2 Benefits of Agroforestry

Socio-ecological approaches not only contribute to environmental sustainability but also offer tangible economic and social benefits. These approaches empower local communities, improve their economic prospects, and safeguard their cultural heritage while promoting the responsible management of landscapes and natural resources. They promote gender equality by creating opportunities for women in forest and landscape management, because women often play a vital role in NTFP collection, agroforestry, and community-based conservation.

Agroforestry systems that integrate fruit trees, like citrus or guava, cocoa, coffee, avocado, banana with traditional crops provide a diversified and year-round source of food. This ensures a more stable and varied diet for communities, thereby contributing to food security. In SSA, adopting agroforestry systems in multifunctional landscapes with nitrogen-fixing trees, such as *Senegalia/Vachellia/Faidherbia* or *Leucaena*, enhances soil fertility, reduces nutrient runoff, and maintains arable land for agriculture (Sileshi et al. 2014).

Agroforestry practices incorporating high-value timber species alongside crops provide an additional income source. Communities can sell timber for construction or furniture making, diversifying their revenue streams and improving overall livelihoods (Sheppard et al. 2020a, b). Certain areas within agroforestry landscapes, such as sacred forests, preserve cultural and heritage values. These areas often host unique plant and animal species and are protected by local communities as part of their cultural traditions (Araia et al. 2019). Agroforestry systems not only contribute to biodiversity, but also support traditional healthcare practices. Cultivating and preserving medicinal herbs ensure access to traditional remedies, maintaining cultural and heritage connections.

Agroforestry systems that include fast-growing trees suitable for fuelwood, like eucalypt or *Senegalia/Vachellia/Faidherbia* species, address energy needs in rural areas. This reduces the pressure on natural forests for fuelwood, contributing to both environmental conservation and improved livelihoods. Some communities organize agroforestry festivals to celebrate the diversity of trees and crops in their landscapes. These events not only promote the cultural significance of agroforestry but also educate the younger generation on sustainable land management practices (Govind et al. 2023). Agroforestry systems that include trees with high carbon sequestration potential contribute to climate resilience (Chirwa et al. 2023). Farmers participating in climate-smart agroforestry may also benefit from carbon offset programs, providing an additional source of income. Planting flowering trees in agroforestry systems supports beekeeping. This not only enhances pollination services for crops but also provides communities with honey, contributing to both income diversification, nutrition diversity and cultural practices. Some communities practice indigenous agroforestry techniques passed down through generations. These practices often involve specific tree-crop associations that have cultural significance, emphasizing the intimate connection between agriculture and cultural heritage.

12.3.1.3 Resilience Potential of African Savannas in Developing Sustainable Silvicultural Systems

African savannas are characterized by a seasonally variable climate (Marchant 2010). Sustainable silvicultural systems in these regions consider the adaptability of tree species to drought periods, fire, and seasonal rainfall fluctuations. Tree species that are resilient to the aforementioned climatic challenges are selected to enhance the sustainability of silvicultural systems. For example, the ability of different tree species to access water resources efficiently, tolerate dry spells, and adapt to changing precipitation patterns include *Senegalia/Vachellia/Faidherbia* species, baobabs, and certain drought-resistant hardwoods (Uni et al. 2023). Tree species that have evolved fire-resistant adaptations, such as thick bark or some eucalyptus species are known for their ability to resprout and recover after fires. On the other hand, trees that could synchronize their growth and reproduction with the rainy season are promoted in silvicultural systems to respond to seasonal variations in rainfall. Hence, diverse tree species that could respond to climatic challenges or adapt in different ecological and climatic conditions are selected using site-species matching. Agroforestry systems involving tree species like *Faidherbia albida* have been promoted to improve soil fertility and crop productivity (Grovermann et al. 2023). By linking tree-based products with local markets, smallholder farmers can generate income while simultaneously restoring and improving their farmlands.

In Miombo landscape, sustainable silvicultural systems prioritize biodiversity conservation by promoting the coexistence of native species and ensuring that forest management practices do not lead to habitat loss or fragmentation (Syampungani et al. 2020). In southern Africa, including Zambia, Tanzania, Mozambique and others, selective logging is practiced where only specific tree species are harvested, and the ecological roles of these species are considered for the overall ecosystem functions, including nutrient cycling, soil stabilization, habitat provision, and support of biodiversity. After logging, or any form of harvesting such as for charcoal production or slash and burn agriculture, efforts are made to facilitate natural regeneration by maintaining the presence of important native tree species. This is premised on the potential for developing optimum management systems that provide for landscape use and a diverse range of products and services. It includes the formulation of landscape management plans, pre-harvesting resources inventory, silvicultural prescriptions for major landscape ecosystems, regulation of initial harvesting and subsequent cutting cycles and management of transition from first to second cutting cycles (in case of forests), presence of clear, official harvesting rules, monitoring of residual growing stock after logging and the extent of vegetation disturbance (Geldenhuys 1993). It calls for developing a detailed understanding of the response to various disturbances of diverse frequency, magnitude and effect. This is possible in that the African Savanna system is highly resilient due to the fact that the savanna species have a diversity of regeneration mechanisms (Fig. 12.3) that include resprouts from a persistent underground rootstock, or base as adaptation to a variety of disturbance factors.

In some cases, reforestation programs are implemented to restore harvested areas. Local and indigenous tree species are often preferred for these initiatives. Specific areas within the Miombo woodlands are designated as protected zones to safeguard critical habitats and vulnerable species. Buffer zones are established around protected areas to minimize the edge effects of human activities. Sustainable silvicultural systems including measures to manage and control invasive species to protect the integrity of the ecosystem are increasingly being considered (Dewees et al. 2011).

In the Sudanian Savanna, which spans several West African countries, shea trees (*Vitellaria paradoxa* C.F. Gaerten) are a prominent component of the landscape (Choungo Nguekeng et al. 2021) (Fig. 12.4). Shea nuts are a valuable non-timber forest product known for their use in cosmetics, food, and medicinal products. Local communities have been empowered to engage in cooperative shea nut processing and marketing, ensuring better economic returns for their efforts. This has led to increased income for rural women, who are often the primary collectors of shea nuts. The marula tree [*Sclerocarya birrea (*A. Rich. (Hochst.)] is widespread in the southern African savannas. Its fruits are harvested and processed into products like jams, juices, and liqueurs. Marula fruit processing enterprises have been developed, particularly in countries like South Africa and Namibia. The exploitation has generated income for rural communities and provided economic incentives for the conservation of marula woodlands (Murye and Pelser 2018).

Fig. 12.3 Some regeneration mechanisms of the African Savanna species. (Photo Syampungani)

Fig. 12.4 Shea agroforestry system in Northern Ghana. (Photo K. Kokou)

12.3.1.4 Modern Approaches to Tree Management in Sub-Saharan Multifunctional Landscapes

Contemporary management approaches play a crucial role in ensuring the sustainability and multifunctionality of landscapes in SSA. They often involve the use of GIS for spatial and temporal analyses and planning. GIS allows for the mapping of tree species, land use patterns, and ecological features. For instance, satellite imagery and GIS are employed to assess changes in land cover, identify deforestation hotspots, and plan afforestation or reforestation initiatives strategically. Remote sensing technologies, including satellite and drone imagery, are employed for real-time monitoring of forests. These technologies provide valuable data on forest health, changes in vegetation cover, and the impact of human activities (Hiller and Fisher 2023).

Also reduced impact logging (RIL) techniques are applied to minimize the environmental impact of logging operations, including the use of specialized machinery and methods that reduce soil disturbance, limit damage to surrounding trees, and protect wildlife habitats (Putz et al. 2008). Certification ensures that logging practices meet specific sustainability and conservation criteria (Bousfield et al. 2020; Teketay et al. 2016). It also makes the timber more attractive to international markets. Local communities actively participate in forest management and conservation; they have a say in the allocation of logging quotas, the choice of tree species to be harvested, and the use of the revenue generated from logging. This involvement creates a sense of ownership and responsibility for the forests (Teketay et al. 2016).

Contemporary approaches also include tree domestication programs, where efforts are made to cultivate and breed tree species that are well-adapted to specific environments. This helps in reforestation projects by ensuring the planting of tree varieties that can thrive under local conditions, contributing to biodiversity conservation and ecosystem resilience (Niu et al. 2023). Precision agriculture technologies, such as sensor-based monitoring and automated equipment, are increasingly applied to agroforestry systems. These technologies optimize the use of resources, enhance crop/tree yields, and minimize environmental impacts. For instance, precision irrigation systems can be integrated into agroforestry practices (Işık et al. 2017). In the face of climate change, modern tree management includes climate-smart practices involving selecting tree species resilient to changing climatic conditions, implementing agroforestry systems that enhance climate adaptation, and incorporating climate risk assessments into management strategies (Tadesse and Ahmed 2023). Online platforms and mobile applications are used to engage local communities in decision-making. These platforms facilitate communication between stakeholders, allowing communities to provide input on tree management plans, report illegal logging activities, and stay informed about sustainable practices (Gavilanes Montoya et al. 2023). Moreover, modern tree management approaches include the valuation of ecosystem services provided by forests. Economic tools and models are used to quantify the benefits of forests, including carbon sequestration, water regulation, and biodiversity conservation. This valuation informs policymakers and land managers about the broader importance of maintaining healthy tree cover (Liu et al. 2023). In many parts of SSA, charcoal is a primary source of household energy. Unsustainable and traditional methods of charcoal production have led to deforestation and environmental degradation. However, there are efficient or sustainable charcoal production systems that often use advanced kiln designs, such as retort kilns, drum kilns, and top-lit updraft (TLUD) kilns, charcoal briquettes, waste biomass from agricultural residues or prunnings (Njenga et al. 2013). These systems are increasingly utilized and aim to reduce deforestation and environmental degradation by minimizing wood consumption, utilizing agricultural residues, and improving charcoal production efficiency. The impact on forest/woodland ecosystems is positive, as these methods help mitigate the pressure on natural forests by promoting sustainable charcoal production practices and reducing reliance on unsustainable harvesting methods.

12.3.2 Ecological Restoration of Degraded Landscapes

Ecological restoration (ER) is the process of reclaiming ecological functionality and production of an ecosystem that has been degraded, destroyed and/or deforested (Dagar et al. 2020; Constant and Taylor 2020). The ER interventions counteract the effects of degradation by balancing ecosystem functionality and providing human benefits. Ecological restoration provides numerous benefits to human beings, such as provisioning ecosystem services (Besseau et al. 2018). However,. landscape degradation is attributed mainly to anthropogenic disturbances (Palmer et al. 2022). The United Nations committed and declared 2021–2030 a decade of restoring ecosystem degradation worldwide (Chazdon et al. 2021). In Africa, over 30 countries committed to restoring over 100 million hectares of degraded ecosystems by 2030 through the African Forest Landscape Restoration Initiative (AFR100) (WWF 2022; Besseau et al. 2018). According to Mansourian and Berrahmouni (2021), the Food and Agriculture Organization (FAO) estimated that over 55 million hectares of dryland need restoration in Southern Africa.

12.3.2.1 Ecological Restoration Initiatives Aimed at Rehabilitating Degraded Landscapes

To redress ecosystem biodiversity, functionality and production, professional communities embraced techniques and practices that advance the recovery of ecosystems after disturbances (Alves et al. 2022). To successfully restore ecological functionality and production, different levels of human-driven interventions are implemented, such as minimal intervention (e.g., natural regeneration) and greater intervention (e.g., tree planting) (Dagar et al. 2020; Alves et al. 2022). However, the level of interventions depends on numerous factors such as the degree of degradation, resource availability, the objective of the restoration interventions and the required time to restore the ecological functionality (Alves et al. 2022).

- **Replanting trees and re-afforestation**

The ER human-driven intervention in a given area depends on the historical use of the land and the level of its degradation. The ER's greater intervention (Replanting trees), which can also be referred to as "Assisted Natural Regeneration" (ANR) identifies and reduces impediments to natural regeneration while planting native trees (Alves et al. 2022). In natural ecosystems, the ANR is a lowercost intervention since it depends on natural reseeding than nursery-grown seedlings (Alves et al. 2022).

- **Farmer-managed natural regeneration**

Farmer-managed natural regeneration (FMNR) practice aims to address severe land degradation and deforestation (Odisi Netaji et al. 2023). The FMNR embrace sustainable farming management that restores the land functionality and production through the regeneration of shrubs and trees from existing trees and seeds (Odisi Netaji et al. 2023; Rinaudo et al. 2019), while crop production remains the main priority (Moore et al. 2020). The FMNR practice contributes to addressing the issue of declining food security, soil fertility and land production affecting the subsistence farmers in developing countries (Moore et al. 2020). The FMNR consist of three practices (1) select species and stumps, (2) prune and manage, and (3) maintain and utilize (Fig. 12.5) (Rinaudo et al. 2019).

12.3.2.2 Integration of Biodiversity Conservation in the Socio-ecological Management of Multifunctional Landscapes

Within SSA, Key Biodiversity Areas (KBAs) assume a pivotal role in the preservation of biodiversity, housing a significant concentration of species that are often unique and under threat. The incorporation of KBA protection into the framework of socio-ecological management signifies a recognition of the inherent value these areas hold. A notable example is the Southern African KBA Programme, which endeavors to systematically identify, monitor, and safeguard regions of pronounced biodiversity significance (IUCN WCPA 2018). This proactive approach, integrating KBAs into land-use planning and management strategies, reflects a deliberate commitment to harmonize human activities with the imperative to conserve these vital habitats. This concerted effort aims to strike a delicate balance between the needs of communities and the preservation of ecologically crucial environments.

Ecological restoration, as discussed in the previous section, plays a dual role in the socio-ecological management of multifunctional landscapes by addressing degradation and contributing to biodiversity conservation. The restoration of ecosystems enhances habitat quality and promotes the recovery of native plant and animal species. Initiatives such as those outlined in Table 12.1 not only focus on restoring ecosystem functionality but also emphasize the importance of restoring biodiversity as an integral part of the process. This integration ensures that ecological restoration

Fig. 12.5 Three practices of FMNR. (Source: Rinaudo et al. 2019)

efforts contribute not only to ecosystem services but also to the preservation of biodiversity.

Connectivity corridors are essential for maintaining genetic diversity and facilitating the movement of species across fragmented landscapes. These corridors serve as pathways for wildlife migration, enabling species to adapt to changing environmental conditions. In SSA, projects such as the one in Tanzania (Supporting the implementation of an integrated ecosystem management approach for landscape restoration and biodiversity conservation) highlight the importance of establishing connectivity corridors as a means of enhancing landscape resilience (Mansourian and Berrahmouni 2021). By strategically planning and implementing these corridors, socio-ecological management seeks to create a network that allows flora and fauna to thrive in interconnected ecosystems.

Successful integration of biodiversity conservation in socio-ecological management recognizes the role of local communities. Community engagement ensures that conservation strategies align with the needs and practices of local populations. Moreover, incorporating traditional ecological knowledge enhances the effectiveness of conservation efforts by tapping into the wisdom accumulated over generations. Initiatives such as the Livelihoods Vanilla Project in Madagascar (Poulos 2023) and the Smallholder Market-led Project in Eswatini (Simelane et al. 2019)

Table 12.1 Southern Africa restoration initiatives

Country	AFR100 area pledged (ha)	Example of restoration initiatives (projects)
Eswatini	500,000	**Smallholder market-led project (SMLP):** The project aims to enhance food and nutrition security and incomes among smallholder producer families through diversified agricultural production and market linkages. One of its outcomes is to ensure soil and water resources are sustainably managed for market-led smallholder agriculture in 25 chiefdoms. The project has included the rehabilitation of five wetlands through fencing to allow natural regeneration on some 50 ha of land and the rehabilitation of an area of 82 ha of heavily degraded community sites by growing various fodder and fruit tree species for the use of the communities. Importantly, the project has trained 481 farmers, many of them women, in climate-smart agricultural practices (2015–2021)
Madagascar	4,000,000	**Livelihoods Vanilla Project:** In 2017, the Livelihoods Fund for Family Farming launched a ten-year project in the vanilla-growing region of Sava in Madagascar to promote agroforestry techniques in this sector. In addition to improving practices for vanilla farming and restoring trees in the landscape (as windbreaks and firewalls), the project will enable farmers to generate higher revenues with an estimated 60 per cent of cured vanilla's value going back to farmers. The project has already helped to organize over 600 producers into associations (2017–2027) Madagascar has committed to restore four million hectares of degraded forest landscapes by 2030 in AFR100 initiative (Bamigbade 2023). The country-initiated Forest Landscape Restoration project targets the northern regions, Diana and Boeny as they are extremely degraded due to anthropogenic disturbances. The project is planned to run for 6 years, 2020–2026. Since inception, through replanting of trees, the country managed to conduct restoration measures for an area of 301 ha in DIANA and Boeny regions (Bamigbade 2023).
Mozambique	1,000,000	**Climate-resilient food security for women and men smallholders in Mozambique through integrated risk management.** The project has already started (2020–2025) The project aims (GCF 2019); Reduce vulnerability to climate risks through promotion of climate-resilient agriculture, as well as watershed restoration and enhancement, for food insecure smallholder women and men Enhance and sustain adaptive capacity of smallholder women and men through a combination of context-specific, integrated risk management tools and market-based opportunities

(continued)

Table 12.1 (continued)

Country	AFR100 area pledged (ha)	Example of restoration initiatives (projects)
		Inform adaptation planning and decision-making across smallholders, communities and national/local authorities through the generation and use of climate information **Zambezia integrated landscapes management program (ZILMP)** The project aims (World Bank 2019); A platform that promotes an enabling environment for forest conservation Complementary investments to scale up the tackling of deforestation and forest degradation
South Africa	3,600,000	**Working for ecosystems:** This programme aims to reverse environmental degradation through ecological restoration and maintenance programmes. It aims to regain natural ecosystem composition, structure and function and enhance ecosystem services. At the same time, the aim is to improve livelihood security and the productive potential of land, improve natural species diversity, and promote the development of a market for ecosystem services and pro-poor economic development and empowerment in rural areas. It follows from the subtropical thicket restoration project started in 2004 (2004–ongoing) Ecological restoration initiatives such as ANR through tree planting have been implemented in several parts of the country (Moyo et al. 2021). These initiatives aimed to restore the ecosystem's biodiversity, functionality, goods, and services to improve the livelihoods of the local communities (Moyo et al. 2021). To successfully restore degraded ecosystems, these initiatives involved different stakeholders such as government, communities, non-government organizations, university academics and economists (Moyo et al. 2021). Amongst other initiatives, South Africa implements ANR through planting in projects such as the Subtropical Thicket Restoration Project in Eastern and Western Capes, Restoration of Coastal Dunes and Buffelsdraai Landfill Site Community Reforestation Project in KwaZulu Natal (Moyo et al. 2021)
Tanzania	5,200,000	**Supporting the implementation of an integrated ecosystem management approach for landscape restoration and biodiversity conservation in Tanzania:** The objective of the project is to strengthen integrated natural resources management and restoration of degraded landscapes for building resilient socioecological systems in Tanzania. It has three components: (1) establishing national landscape restoration governance and regulatory structures, and mainstreaming landscape restoration and sustainable land management into policies, regulations and strategies; (2) engaging communities and local authorities in identifying and implementing feasible restoration options at landscape level; (3) putting in place effective M&E and data management systems that would enable the project to gather gender-disaggregated data, disseminate lessons and facilitate learning, and scale up good practices (2018–2023)

(continued)

Table 12.1 (continued)

Country	AFR100 area pledged (ha)	Example of restoration initiatives (projects)
Zambia	2,000,000	**Strengthening climate resilience of agricultural livelihoods in Agro-Ecological Regions I and II in Zambia.** The project has already started (2018–2025). The project targeted interventions to capitalize on opportunities to strengthen and promote viable climate-resilient value chains relating to smallholder agriculture in the target regions, specifically targeting value chains that are gender sensitive and provide viable economic opportunities for women. This includes three interrelated outputs: 1) strengthening capacity of farmers to plan for climate risk; 2) strengthening resilient agricultural production and diversification practices (for both food security and income generation); and 3) strengthening farmers' access to markets and commercialization of introduced resilient agricultural commodities (GCF 2018)
Zimbabwe	2,000,000	**Building Climate Resilience of Vulnerable Agricultural Livelihoods in Southern Zimbabwe.** The project has already started (2018–2025). The project aimed to enhance the water security for smallholder farmers in light of evolving climate risks by enabling revitalization and climate-proofing of irrigation schemes and improving water use efficiency and enhancing soil moisture management on rain-fed lands. And strengthen the capacities of vulnerable smallholder farmers through Farmer Field Schools and peer-to-peer support to scale up climate-resilient agriculture, with access to resilient inputs, markets, and actionable climate information. The project empowers vulnerable smallholders through multi-stakeholder Innovation Platforms for climate-resilient agriculture—including value-chain actors and financial intermediaries—to make a transformative shift away from subsistence livelihoods to climate-resilient, market-oriented agricultural livelihoods. It leverages government budget on input supply to smallholder farmers in the three provinces so that it is directed towards climate-resilient varieties. The project yields significant environmental, social and economic co-benefits, including climate risk-informed, sustainable land management, strengthened gender norms and women's empowerment, private sector engagement, and increased income and food security including income and productivity benefits over the project's lifetime (GCF 2020)

(continued)

Table 12.1 (continued)

Country	AFR100 area pledged (ha)	Example of restoration initiatives (projects)
Mauritius	4,000,000 (Bamigbade 2023)	**Mainstreaming biodiversity into the management of the coastal zone in the Republic of Mauritius:** The project's objective is to mainstream the conservation and sustainable use of biodiversity and ecosystem services in coastal zone management and the operations and policies of the tourism and physical development sectors in the Republic of Mauritius through a 'land- and seascape-wide' integrated management approach. It includes an ecosystem restoration component to reduce erosion and soil loss in 200 ha of erosion-prone watersheds and restore ecosystem services on 15.4 ha of freshwater wetlands and 23.9 ha of associated buffer. Restoration includes the removal of invasive exotic species such as the Chinese guava *(Psidium cattleianum)* plant which outcompetes the growth and regeneration of native plants (2018–2021)

Adopted: Mansourian and Berrahmouni (2021)

demonstrate how community-focused projects can simultaneously address biodiversity conservation and improve the well-being of local communities.

While the integration of biodiversity conservation in socio-ecological management offers significant benefits, challenges such as conflicting interests, limited resources, and institutional barriers need careful consideration. Striking a balance between conservation goals and human needs requires adaptive governance and collaboration among stakeholders. Opportunities for synergy between ecological conservation and community well-being emerge when conservation initiatives align with sustainable development goals, fostering a holistic approach to landscape management.

The integration of biodiversity conservation within the socio-ecological management of multifunctional landscapes in SSA is a dynamic and intricate process. Through the protection of key biodiversity areas, ecological restoration initiatives, creation of connectivity corridors, and meaningful community engagement, there exists a pathway to achieve both conservation objectives and the well-being of local communities. This holistic approach recognizes the interconnectedness of ecological and human dynamics, paving the way for resilient and sustainable landscapes in the face of ongoing environmental challenges.

12.4 Integrating Socio-ecological Approach in the Management of Biodiversity at the Landscape Level

12.4.1 Importance of Integrating Social, Economic and Ecological Dimensions in the Management of Trees in Multifunctional Landscapes

In the management of trees within multifunctional landscapes, recognizing and integrating social, economic, and ecological dimensions is paramount for fostering sustainable and resilient ecosystems. This holistic approach acknowledges the interplay between natural processes, human dynamics, and economic considerations. Social considerations are fundamental in understanding the role of trees within communities. Trees often hold cultural significance, playing central roles in local traditions, ceremonies, and daily practices (Fig. 12.6). Moreover, the well-being of communities is intricately tied to the services trees provide, such as sources of food, medicine, and timber. Integrating social dimensions into tree management involves engaging local communities, respecting traditional knowledge systems. In addition, trees in multifunctional landscapes often hold deep cultural significance within local communities (Atindehou et al. 2022). Ethnobotanical knowledge, which encompasses the traditional uses of plants, plays a crucial role in sustainable tree management. Understanding which tree species are used for medicine, rituals, or construction provides insights into the intricate relationship between communities and their natural surroundings. Integrating cultural and ecological dimensions into management strategies ensures that conservation efforts align with local values and customs and enhance the health and functionality of landscapes.

Fig. 12.6 Sacred *Antiaris africana* in a convent at Aného, Togo (Photo Akoété)

12.4.1.1 Local Knowledge and Practices of Tree Management in Multifunctional Landscapes

The integration of local and scientific knowledge in decision-making and planning at the landscape level is essential for developing holistic and culturally sensitive approaches to multifunctional landscape management. Local communities possess valuable traditional knowledge about the landscape and have a deep understanding of their natural environments, and their wisdom in tree management is invaluable for promoting sustainable practices (Kanene 2016). Participatory mapping, combined with Geographic Information Systems (GIS), is a powerful tool for integrating this local knowledge with scientific data. Communities can contribute information about resource locations, land use patterns, and cultural significance, which, when mapped, provides a comprehensive understanding of the landscape. Therefore, empowering local communities through active participation in tree management fosters a sense of ownership and responsibility. Community engagement allows for the exchange of knowledge between formal conservation practices and traditional wisdom. Participatory approaches ensure that management strategies are not only effective in ecological terms but also resonate with the needs and aspirations of the people who inhabit these landscapes.

Local communities in multifunctional landscapes have developed a profound understanding of the ecological dynamics of their surroundings. Traditional Ecological Knowledge (TEK) in Agroforestry is within the capacity local farmers have through generations of experience in selecting tree species that are well-adapted to specific soil conditions and climate. Traditional knowledge includes methods of seed collection, germination, and tree propagation, contributing to the preservation of critical species. TEK includes the medicinal and cultural uses of trees; local communities have a deep understanding of the healing properties of various tree species and their cultural significance. It encompasses insights into seasonal patterns, plant-animal interactions, and the sustainable use of natural resources. By integrating this wisdom into tree management strategies, there is an opportunity to leverage centuries-old practices that promote biodiversity conservation, soil health, and overall ecosystem resilience (Rossier and Lake 2014).

Local practices often reflect adaptive strategies that have evolved over time in response to the dynamic nature of landscapes. These practices may include specific methods of agroforestry, rotational tree harvesting, or community-led initiatives for sustainable resource use. Incorporating these adaptive tree management practices into formal management plans enhances the resilience of ecosystems, allowing them to adapt to changing environmental conditions, climate variability, and other external pressures. In most countries in West Africa, traditional practices frequently involve the protection of sacred forests or areas of high biodiversity (Adjonou et al. 2021; Hounkpati et al. 2022). The preservation of sacred forests reflects the deep connection between local communities, their cultural heritage, and the environment. It is a testament to the role of traditional knowledge in fostering conservation and maintaining the ecological balance (Kokou et al. 2008). Similar approaches have been reported in some parts of South Africa (Araia and Chirwa 2019).

Indigenous communities have traditional practices for managing wildfires, rotational grazing, water management systems that have sustained them for generations. Traditional beekeeping practices in agroforestry systems enhances biodiversity through pollination services. Simultaneously, farmers can harvest honey as a marketable product, contributing to income diversification. Integrating local knowledge with hydrological studies can inform modern water management practices.

12.4.2 Socio-ecological Value of Trees in Multifunctional Landscapes of Sub-Saharan Africa

12.4.2.1 Multiple Services of Trees in These Landscapes

In the multifunctional landscapes of SSA, trees play a pivotal role in providing a diverse array of essential services (Sheppard et al. 2020a, b; Le et al. 2023), encompassing climate regulation, biodiversity conservation, natural resource production, and food security. The socio-ecological value of trees extends far beyond their individual components, contributing significantly to the overall health and resilience of these dynamic landscapes. These services align with established classifications for ecosystem services, including provisioning, regulating, cultural, and supporting services, thereby highlighting the diverse and interconnected roles that trees play in sustaining both ecological and human well-being in SSA landscapes. These include the following:

- **Climate regulation:** Trees act as crucial regulators of climate within multifunctional landscapes. Through the process of carbon sequestration, they play a vital role in mitigating climate change by absorbing and storing carbon dioxide. Additionally, trees contribute to the local climate by providing shade, reducing temperatures, and influencing precipitation patterns. Their intricate role in climate regulation is fundamental to maintaining ecological balance and supporting agricultural practices in the region.
- **Biodiversity conservation:** The presence of trees is synonymous with biodiversity conservation in SSA landscapes. Trees provide habitats for a myriad of plant and animal species, fostering biodiversity hotspots. They serve as corridors for wildlife movement, enabling genetic diversity and supporting various ecological niches. The conservation of indigenous tree species is particularly critical, as these trees often play host to endemic flora and fauna, contributing to the overall richness and resilience of the ecosystem.
- **Natural resource production:** Trees in multifunctional landscapes are prolific producers of various natural resources essential for local communities. Timber, non-timber forest products, medicinal plants, and fruits are among the valuable resources derived from trees. Sustainable management practices ensure a continuous supply of these resources, supporting local economies, and meeting the diverse needs of communities while simultaneously preserving the integrity of the landscape.

- **Food security:** The role of trees in ensuring food security is multifaceted. Fruit-bearing trees provide a direct source of nutrition, contributing to the dietary diversity of local communities. Agroforestry practices, integrating trees with agricultural crops, enhance soil fertility, water retention, and create microclimates conducive to crop growth. This symbiotic relationship between trees and agriculture promotes sustainable food production, resilience to climate variability, and the maintenance of livelihoods.
- **Livelihood support:** Trees are lifelines for local communities in terms of livelihood sustenance. They provide a multitude of resources, including timber, non-timber forest products, and medicinal plants, which form the economic backbone of many communities. Timber from trees serves as a primary construction material for housing, while non-timber products such as fruits, nuts, and resins contribute to income generation and act as safety nets during periods of agricultural uncertainty. Sustainable tree management practices ensure the continued availability of these resources, securing the economic stability of local populations.
- **Cultural significance:** Trees play a central role in the cultural fabric of communities across SSA. Many tree species are imbued with cultural symbolism, featuring prominently in rituals, ceremonies, and folklore. Certain trees are considered sacred, becoming focal points for community gatherings and spiritual practices. The longevity and resilience of trees often mirror the endurance and continuity of cultural traditions, creating a deep connection between the people and the trees that surround them.
- **Heritage preservation:** The landscapes adorned with diverse tree species become living repositories of cultural heritage. Local knowledge, traditions, and stories passed down through generations find tangible expression in the trees that have witnessed the evolution of communities. Ancient or particularly significant trees are often designated as cultural heritage sites, representing a tangible link to the past and an anchor for future generations. Preserving these trees becomes synonymous with safeguarding the cultural identity of the community (Akpovo and Fandohan 2021; Ganka et al. 2022).
- **Community well-being and identity:** Trees contribute to the overall well-being of communities by providing aesthetic beauty, shade, and spaces for communal activities. The presence of green spaces and diverse tree cover fosters a sense of environmental harmony and mental well-being. Additionally, the conservation of indigenous tree species becomes integral to maintaining community identity, as these trees often hold unique ecological and cultural significance specific to the region.

Importantly, the services provided by trees in SSA landscapes are interconnected, creating a synergistic web of ecological functions. For example, the shade provided by trees not only regulates local temperatures but also creates microenvironments conducive to diverse plant and animal life. Agroforestry initiatives not only enhance food security but also contribute to carbon sequestration and biodiversity conservation. Recognizing the importance of trees to local communities calls for management practices that are rooted in sustainability and inclusivity. Engaging communities in decision-making processes, incorporating traditional ecological knowledge, and implementing adaptive governance models ensure that tree management aligns with the needs and aspirations of the people. This approach not only safeguards the ecological health of landscapes but also reinforces the resilience of local communities.

12.5 Challenges and Prospects for the Socio-ecological Management of Trees in Multifunctional Landscapes

12.5.1 Agro-ecosystem Management

The challenges and prospects of agro-ecosystem management in SSA landscapes are multifaceted, demanding comprehensive strategies to address various key aspects. Landscape planning faces difficulties due to competing land uses like urbanization and agricultural expansion, leading to habitat fragmentation. However, integrated planning approaches and conservation corridors offer promising solutions, emphasizing ecological connectivity and engaging local communities in participatory planning. Farming systems, influenced by intensive agriculture and monoculture, contribute to deforestation and biodiversity loss. The adoption of sustainable practices like agroforestry presents a hopeful prospect by integrating trees into agricultural landscapes, enhancing biodiversity, improving soil health, and providing additional income. Semi-natural habitats encounter threats from degradation, invasive species, and climate change impacts, requiring investments in restoration, protected areas, and community-based conservation initiatives. Understanding ecosystem structure and processes is hindered by insufficient knowledge and climate change uncertainties, but investing in research, monitoring, and integrating traditional ecological knowledge can contribute to adaptive management strategies (Table 12.2).

12.5.2 Social Systems in the Context of Africa

Navigating the socio-economic, political and ecological management of trees in multifunctional landscapes in Africa requires a nuanced understanding of these systems, encapsulated in challenges and prospects (Table 12.3). The intricate structure of social systems, influenced by socio-cultural, economic, and political factors, poses challenges related to resource distribution, access disparities, and external influences. However, promising prospects emerge through inclusive initiatives promoting equity, community empowerment, and collaborative decision-making. Dynamic social processes, including demographic shifts and cultural transformations, challenge the sustainable management of trees. Adaptive governance

Table 12.2 Challenges and prospects of agro-ecosystem management in SSA

Key aspect	Challenges	Prospects
Landscape planning and management	Rapid urbanization, expansion of agricultural frontiers, and infrastructure development often encroach upon natural habitats. Fragmentation of landscapes disrupts ecological connectivity, leading to biodiversity loss and compromising ecosystem services.	Designing and implementing conservation corridors that facilitate the movement of species across fragmented landscapes enhances biodiversity conservation. Engaging local communities and stakeholders in participatory landscape planning ensures that conservation goals align with socio-economic needs, fostering a harmonious coexistence between humans and nature
Farming systems	Intensive agriculture, monoculture, and unsustainable land management practices contribute to deforestation, soil degradation, and loss of biodiversity. The overreliance on certain cash crops may lead to the neglect of indigenous tree species, impacting ecosystem resilience	Sustainable farming systems, such as agroforestry, offer promising solutions. Agroforestry integrates trees into agricultural landscapes, promoting biodiversity, improving soil health, and providing additional income streams for farmers. Promoting agroecological practices that prioritize conservation alongside food production contributes to the long-term sustainability of farming systems
Semi-natural habitats	The preservation of semi-natural habitats faces challenges from habitat degradation, invasive species, and the impacts of climate change. These threats compromise the ability of semi-natural habitats to support biodiversity and ecosystem functions. The loss of native vegetation in semi-natural habitats reduces the availability of resources for local communities and disrupts ecological balance	Investing in habitat restoration and the establishment of protected areas enhances the prospects for preserving semi-natural habitats. Implementing sustainable land management practices, controlling invasive species, and restoring degraded areas contribute to the resilience of semi-natural habitats. Community-based conservation initiatives that involve local residents in habitat restoration efforts can lead to more effective outcomes
Ecosystem structure and processes	Understanding and managing ecosystem structure and processes are hindered by inadequate scientific knowledge, insufficient monitoring, and a lack of data on biodiversity and ecosystem functions. Climate change further exacerbates challenges, affecting the distribution of species and altering ecosystem dynamics. Unpredictable changes in precipitation patterns and temperature extremes impact tree growth, survival, and reproductive processes	Investing in ecological research and monitoring programs enhances the understanding of ecosystem structure and processes. This knowledge is crucial for developing adaptive management strategies that account for climate change impacts. Integrating traditional ecological knowledge with scientific research contributes to a more comprehensive understanding of local ecosystems and fosters community engagement in sustainable management practices

Table 12.3 Social economic and political systems' challenges and prospects

Key aspect	Challenges	Prospects
Social system structure	One of the key challenges in social system structure lies in the often intricate web of socio-cultural, economic, and political factors that shape decision-making. Inequitable distribution of resources, unequal access to decision-making processes, and historical disparities can create barriers to effective collaboration. Additionally, the influence of external forces, such as global market dynamics and international policies, can impact local social structures, sometimes leading to disempowerment and marginalization	Promising prospects arise from initiatives that promote inclusivity, equity, and community empowerment. Strengthening local governance structures, ensuring representation of diverse voices, and fostering community-led decision-making processes enhance the resilience of social systems. Collaborative platforms that bridge traditional knowledge with modern insights create a foundation for sustainable socio-ecological management
Social system processes	Dynamic social processes, such as demographic shifts, cultural transformations, and changes in land tenure systems, pose challenges to the sustainable management of trees. Urbanization, migration, and evolving economic landscapes can alter traditional relationships among communities and their environments. Moreover, the erosion of traditional knowledge systems and practices may lead to a disconnect between communities and their ecological heritage	Addressing these challenges requires adaptive governance mechanisms that accommodate evolving social processes. Engaging communities in the co-creation of policies, incorporating traditional knowledge into management strategies, and fostering education and awareness initiatives contribute to more resilient social systems. Recognizing the adaptive capacity of communities to navigate change is central to the prospects of sustainable socio-ecological management
Integration of social systems with ecological dynamics	One of the central challenges is the potential disconnect between social systems and ecological dynamics. Imbalances in resource use, conflicting interests, and insufficient communication channels between communities and environmental scientists can hinder effective collaboration. Failure to integrate local knowledge with scientific insights may lead to management strategies that are disconnected from the realities faced by communities	Promoting interdisciplinary collaboration and fostering a dialogue between traditional knowledge holders and scientific experts can bridge the gap between social systems and ecological dynamics. Integrated research approaches that recognize the value of both local wisdom and scientific rigor contribute to holistic socio-ecological management. Community-based monitoring and participatory research initiatives empower local communities to actively engage in the conservation and sustainable use of trees

(continued)

Table 12.3 (continued)

Key aspect	Challenges	Prospects
Cultural resilience and identity	Cultural resilience faces threats from globalization, cultural homogenization, and the erosion of traditional practices. As societies undergo rapid changes, there is a risk of losing the intricate connections between cultural identity and environmental stewardship. The commodification of cultural practices and the neglect of indigenous knowledge systems further challenge cultural resilience	Safeguarding cultural resilience involves revitalizing and preserving traditional knowledge, practices, and languages. Supporting cultural initiatives that celebrate the relationship between communities and their landscapes fosters a sense of pride and identity. Recognizing the cultural value of trees in rituals, ceremonies, and folklore reinforces the importance of preserving both cultural heritage and ecological integrity
Economic	Disparities in resource access, limited economic opportunities, and reliance on unsustainable practices hindering sustainable land management	Fostering inclusive economic models prioritizing community ownership, promoting sustainable livelihoods, strengthening market access, implementing payment for ecosystem services, and enhancing supply chains for sustainably sourced products
Political	Governance deficits, institutional weaknesses, and lax enforcement of environmental regulations leading to ineffective resource management	Strengthening governance, enhancing transparency, enforcing environmental laws, fostering political commitment to sustainability, empowering communities, and promoting regional cooperation for addressing transboundary challenges

mechanisms (e.g. decentralizing decision-making authority, involving internal and external stakeholders), community engagement in policy creation, and the integration of traditional knowledge address these challenges, contributing to resilient social systems. Bridging the potential disconnect between social systems and ecological dynamics is a central challenge, with interdisciplinary collaboration and community-based monitoring offering solutions for holistic socio-ecological management. Cultural resilience faces threats from globalization, but preserving traditional knowledge, supporting cultural initiatives, and recognizing the cultural value of trees contribute to safeguarding cultural identity and fostering pride.

12.5.3 Challenges Related to the Implementation of Socio-ecological Management

The implementation of socio-ecological management for trees in multifunctional landscapes faces a web of challenges, each intricately connected and posing distinct threats. Anthropogenic pressures from population growth and unsustainable practices lead to habitat destruction and biodiversity loss. Insufficient financial resources hinder comprehensive restoration and conservation efforts, particularly in socio-economically challenged regions. Continuous ecosystem degradation, exacerbated by climate change, weakens landscapes' capacity to provide essential services. Deforestation, driven by timber extraction and, fires, agriculture, directly threatens tree populations and disrupts ecological balance. Institutional constraints, economic pressures prioritizing short-term gains, conflicts of interest, and limited community involvement impede effective management. Additionally, underinvestment in education and awareness programs hinders stakeholders' understanding of the importance of sustainable practices (Table 12.4).

Table 12.4 Challenges in the implementation of socio-ecological management

Key aspect	Challenges	Implications
Anthropogenic pressures	Human activities, driven by population growth and unsustainable practices, including land use and land cover changes (LULCC) exert immense pressure on multifunctional landscapes. Agricultural expansion, urbanization, and resource extraction often lead to habitat destruction, fragmentation, and a loss of biodiversity. Anthropogenic pressures jeopardize the ecological integrity of landscapes, making it challenging to sustainably manage trees and their ecosystems	The encroachment of human activities into natural areas disrupts wildlife habitats, contributes to the decline of native species, and compromises the overall resilience of ecosystems. Balancing the needs of growing populations with environmental conservation becomes a complex task in the face of increasing anthropogenic pressures
Lack of financial resources	Insufficient financial resources represent a significant hurdle in implementing effective socio-ecological management. Limited funding hampers the execution of comprehensive restoration, conservation, and community engagement initiatives. This financial gap is particularly pronounced in regions facing socio-economic challenges, hindering the adoption of sustainable practices	The lack of financial resources impedes the implementation of critical interventions, including afforestation projects, ecosystem restoration, and community capacity building. It also undermines the establishment of monitoring and enforcement mechanisms, making it difficult to address issues such as illegal logging or land degradation
Continuous degradation of ecosystems	Ongoing degradation of ecosystems, fueled by factors such as unsustainable land use, LULCC, overexploitation of natural resources, and pollution, poses a persistent challenge. Degradation diminishes the capacity of ecosystems to provide essential services, hinders natural regeneration processes, and exacerbates the vulnerability of landscapes to external stressors	Continuous ecosystem degradation compromises the ability of trees to fulfill their multifunctional roles. Soil fertility declines, water sources and biodiversity are compromised, and the overall health of landscapes deteriorates. Effective socio-ecological management requires addressing the root causes of degradation to ensure the long-term sustainability of ecosystems

(continued)

Table 12.4 (continued)

Key aspect	Challenges	Implications
Climate change	The impacts of climate change, including altered precipitation patterns, rising temperatures, and extreme weather events, present formidable challenges to socio-ecological management. These changes disrupt ecological processes, affect the distribution of plant species, and introduce uncertainties that complicate long-term planning and conservation efforts	Climate change exacerbates existing challenges and introduces new dynamics that demand adaptive strategies (e.g. climate-smart agriculture). Trees may face increased stress, altering their growth patterns and distribution. Mitigating and adapting to climate change becomes integral to ensuring the resilience of both ecosystems and the communities dependent on them
Deforestation	The persistent issue of deforestation, driven by timber extraction, agricultural expansion, and infrastructure development, poses a direct threat to tree populations and the ecosystems they support. Deforestation disrupts ecological balance, reduces carbon sequestration capacity, and diminishes the overall functionality of landscapes	The loss of forests contributes to biodiversity decline, soil erosion, disrupts water cycles, and releases stored carbon into the atmosphere, exacerbating climate change. Efforts to implement socio-ecological management must address the drivers of deforestation, promote sustainable land-use practices, and encourage reforestation initiatives
Institutional constraints	Inadequate institutional frameworks and governance structures present a significant impediment to successful socio-ecological management. Weak enforcement mechanisms, lack of coordination among relevant agencies, and ambiguous land tenure systems can undermine the implementation of sustainable practices	Institutional constraints contribute to a lack of clarity in property rights, making it difficult to regulate land use and resource extraction. This ambiguity may lead to uncontrolled deforestation, unsustainable agricultural practices, and a failure to enforce conservation measures, hampering the overall goals of socio-ecological management
Economic pressures	Economic considerations often prioritize short-term gains over long-term sustainability. Economic pressures, driven by the demand for natural resources, can incentivize practices that degrade landscapes. Unsustainable logging, conversion of natural areas for agriculture, and extractive industries may prevail due to economic motivations	The pursuit undesirable LULCC FOR of immediate economic benefits may lead to the overexploitation of trees and ecosystems, undermining their resilience and long-term viability. Economic pressures contribute to a mindset that prioritizes profit without adequately accounting for the ecological consequences, posing a substantial challenge to the principles of sustainable socio-ecological management

(continued)

Table 12.4 (continued)

Key aspect	Challenges	Implications
Conflicts of interest	Divergent interests among stakeholders, including local communities, commercial enterprises, and conservation organizations, create conflicts that hinder effective socio-ecological management. Discrepancies in priorities, land use preferences, and profit motives may lead to disputes that compromise the implementation of sustainable practices	Conflicts of interest can result in a lack of consensus on land use planning, resource allocation, and conservation strategies. Competing agendas may undermine collaborative efforts and perpetuate unsustainable practices, exacerbating the challenges faced by both ecosystems and communities relying on them
Limited community involvement	Insufficient inclusion of local communities in decision-making processes can be a barrier to successful socio-ecological management. Failure to recognize and incorporate traditional knowledge and practices may lead to strategies that are disconnected from the needs and realities of the communities residing in multifunctional landscapes	Excluding local communities from decision-making processes can result in a lack of ownership and commitment to management initiatives. This disconnect may lead to resistance, non-compliance, or unintended negative consequences, including ecosystem degradation. Successful socio-ecological management requires active community engagement to ensure the effectiveness and sustainability of implemented strategies.
Underinvestment in education and awareness	Limited investment in education and awareness programs regarding the importance of socio-ecological management can hinder its successful implementation. Lack of understanding among stakeholders, including local communities and decision-makers, about the value of preserving ecosystems and sustainable practices may impede progress	Without a comprehensive understanding of the ecological and socio-economic benefits of sustainable management, stakeholders may prioritize short-term gains over long-term sustainability. Promoting awareness and education is crucial for fostering a collective commitment to socio-ecological principles and overcoming resistance to change

12.5.4 Future Prospects for Effective Socio-ecological Management

The future prospects for effective socio-ecological management of trees in the multifunctional landscapes of Sub-Saharan Africa are promising, offering opportunities for innovative approaches, collaborative partnerships, and sustainable solutions. The exploration of these opportunities is essential for enhancing the resilience of ecosystems and improving the well-being of communities dependent on these landscapes. These future prospects are:

1. Advancements in technology provide a unique opportunity to integrate traditional ecological knowledge with modern scientific insights. Remote sensing, Geographic Information Systems (GIS), and mobile applications can facilitate data collection, monitoring, and assessment of tree ecosystems. This integration supports informed decision-making that respects local wisdom while leveraging technological tools for effective management.
2. Empowering local communities as active participants in natural resource management is a key opportunity for improving socio-ecological management. Community-based initiatives, supported by inclusive decision-making processes, can harness traditional knowledge and practices. This approach fosters a sense of ownership and responsibility among communities, ensuring the sustainability of management strategies.
3. Designing and implementing incentive mechanisms that reward sustainable practices can encourage positive contributions to socio-ecological management. Financial incentives, certification programs for eco-friendly products (Poratelli et al. 2022), and recognition for conservation efforts create motivations for individuals, communities, and businesses to adopt and maintain sustainable practices.
4. Effective socio-ecological management necessitates collaboration across diverse sectors, including agriculture, forestry, conservation, and community development. Encouraging partnerships among governmental agencies, non-governmental organizations, academia, and private enterprises facilitates a holistic approach. Integrated efforts that address the interconnected challenges faced by multifunctional landscapes enhance the overall effectiveness of management strategies.
5. Increased investment in research is crucial for building a knowledge base that informs sustainable management practices. Research initiatives focused on the ecological, social, and economic dimensions of multifunctional landscapes contribute to evidence-based decision-making.
6. Given the challenges posed by climate change, adopting climate-smart and resilient practices is a forward-looking opportunity. This involves promoting agroforestry systems, selecting climate-adapted tree species, and implementing landscape restoration projects that enhance resilience to changing climatic conditions. Climate-smart approaches contribute to the long-term sustainability of socio-ecological management.
7. Innovative policy frameworks that recognize the multifunctionality of landscapes and integrate conservation with development goals present an opportunity for effective socio-ecological management. Policy coherence, clear guidelines for sustainable resource use, and incentives for conservation efforts create an enabling environment for responsible management practices.
8. Investing in the capacity building of local communities, governmental institutions, and non-governmental organizations is a fundamental opportunity. Empowering individuals with the knowledge and skills necessary for sustainable

land management enhances the ability of communities to actively contribute to and benefit from socio-ecological initiatives.

9. International collaboration and funding initiatives, driven by global commitments to environmental sustainability, offer a significant opportunity. Collaborative projects, knowledge and innovative technology exchange, and financial support from international organizations contribute to the scaling up of successful socio-ecological management models, especially in regions facing resource constraints.
10. Environmental awareness and education enhance ecological literacy among communities, decision-makers, and stakeholders. Understanding the intricate relationships within ecosystems, the role of trees in biodiversity conservation, and the importance of maintaining ecosystem services empowers individuals to make informed choices that contribute to the health and resilience of landscapes. By promoting an understanding of the interconnectedness of environmental, social, and economic systems, education contributes to the achievement of objectives such as those outlined in the United Nations Sustainable Development Goals (SDGs). This includes goals related to biodiversity conservation, climate action, and the well-being of communities. Educational programs raise awareness and empower stakeholders with the understanding needed for informed and responsible management.

12.6 Chapter Highlights

The exploration of the socio-ecological management of trees in Sub-Saharan African (SSA) multifunctional landscapes has uncovered a complex interplay of ecological, social, and economic dimensions. The multifunctionality of these landscapes, particularly about trees, emerges as a critical challenge necessitating nuanced approaches for research, management, and policy development. The significance of SSA multifunctional landscapes lies in their pivotal role in the socio-economic and ecological fabric of the region. Trees, as key components of these landscapes, offer essential ecosystem services, contribute to livelihoods, and hold cultural significance. From carbon sequestration to providing sustenance through non-timber forest products, trees serve as both sources of resilience and vulnerability in the face of diverse challenges, including climate change and population growth.

The chapter underscores:

- **socio-ecological framework** advocating for a holistic approach that recognizes intricate linkages between ecological processes and human dynamics. This framework promotes active community engagement, integration of local knowledge, and stakeholder empowerment through adaptive governance, aiming to balance conservation goals with the diverse needs of communities for sustainable multifunctional landscapes.

- **ecological restoration initiatives** highlighting commitments in countries like Eswatini, Madagascar, and South Africa and their alignment with the African Forest Landscape Restoration Initiative (AFR100). Emphasizing the significance of landscape restoration for ecological health and its contribution to broader international goals for sustainable environmental management.
- **biodiversity conservation integration** stressing the importance of Key Biodiversity Areas (KBAs) and initiatives like the Southern African KBA Programme. Recognizing the crucial role of local knowledge in tree management and advocating for the integration of traditional practices in effective and culturally sensitive conservation strategies.
- **challenges in implementation** acknowledging diverse challenges from anthropogenic pressures and financial constraints to climate change and conflicts of interest. Emphasizing the need for adaptive strategies to address multifaceted challenges and promote sustainable tree management in multifunctional landscapes.
- **opportunities and prospects i**dentifying avenues for improvement, including education and technological innovations, community-based management, and cross-sectoral collaboration. Stressing the importance of investment in research, education, climate-resilient practices, and policy innovation for sustainable socio-ecological management.
- **collective action** advocating for collaborative efforts, partnerships, knowledge and technology sharing, and the inclusion of diverse perspectives. Forging resilient, community-centric management strategies to ensure the sustainable well-being of ecosystems and the communities dependent on them

References

Abuhay A, Kassaw E, Masresha G, Asefa M (2024) Small forest patches in Ethiopian highlands uniquely support high plant biodiversity. Biodivers Conserv 33:1711–1727

Adjonou K, Kpeli Poukpezi M, Segla KN, Kokou K (2021) Impacts of traditional practices on biodiversity and structural characteristics of sacred groves in northern Togo. West Africa Acta Oecologica 110:103680. https://doi.org/10.1016/j.actao.2020.103680

Akpovo AH, Fandohan AB (2021) Usages, distribution des connaissances traditionnelles et valeur économique de *Ricinodendron heudelotii* au Bénin. Revue Marocaine des Sciences Agronomiques et Vétérinaires 9(2):276–287

Alves J, Oliveira M, Chazdon RL, Calmon M, Pinto A, Darvin E, Pereira B (2022) The role of assisted natural regeneration in accelerating forest and landscape restoration: practical experiences from the field. Practice note. WRI Brasil, São Paulo

Araia MG, Chirwa PW (2019) Revealing the predominance of culture over the ecological abundance of resources in shaping local People's Forest and tree species use behavior: the case of the Vhavenda people, South Africa. Sustainability 11(11):3143. https://doi.org/10.3390/su11113143

Araia ZZ, Gebregziabher NK, Mesfun AB (2019) Self-medication practice and associated factors among students of Asmara College of Health Sciences, Eritrea: a cross sectional study. J Pharm Policy Pract 12:3. https://doi.org/10.1186/s40545-019-0165-2

Atindehou MML, Avakoudjo HGG, Idohou R, Azihou FA, Assogbadjo AE, Adomou AC, Sinsin B (2022) Old sacred trees as memories of the cultural landscapes of southern Benin (West Africa). Land 11:478. https://doi.org/10.3390/land11040478

Bamigbade P (2023) Giving Forests a future. Forests4Future Madagascar, Deutsche Gesellschaft für Internationale Zusammenarbeit (GIZ) GmbH. www.giz.de/de/weltweit/85060.html

Besseau P, Graham S, Christophersen T (2018) Restoring forests and landscapes: the key to a sustainable future. Global Partnership on Forest Landscape Restoration, Vienna, Austria

Bousfield CG, Cerullo GR, Massam MR, Edwards DP (2020) Protecting environmental and socio-economic values of selectively logged tropical forests in the Anthropocene. In: Dumbrell AJ, Turner EC, Fayle TM (eds) Advances in ecological research, vol 62. ElsevierLtd, pp 1–52

Cao M, Zhu Y, Lü G, Chen M, Qiao W (2019) Spatial distribution of global cultivated land and its variation between 2000 and 2010, from both agro-ecological and geopolitical perspectives. Sustain For 11:1242. https://doi.org/10.3390/su11051242

Chazdon RL, Falk DA, Banin LF, Wagner M, Wilson SJ, Grabowski RC, Suding KN (2021) The intervention continuum in restoration ecology: rethinking the active-passive dichotomy. Restor Ecol:1–13. https://doi.org/10.1111/rec.13535

Chirwa PW, Musokwa M, Mwale SE, Handavu F, Nyamadzawo G (2023) Agroforestry systems for mitigating climate change and reducing carbon footprints of land-use systems in southern Africa. Carbon Footprints 2(1):1. https://doi.org/10.20517/cf.2022.081601

Choungo Nguekeng PB, Hendre P, Tchoundjeu Z, Kalousová M, Tchanou Tchapda AV, Kyereh D, Masters E, Lojka B (2021) The current state of knowledge of Shea butter tree (*Vitellaria paradoxa* C.F.Gaertner.) for nutritional value and tree improvement in west and Central Africa. Forests 12:1740. https://doi.org/10.3390/f12121740

Constant NL, Taylor PJ (2020) Restoring the forest revives our culture: ecosystem services and values for ecological restoration across the rural-urban nexus in South Africa. Forest Policy Econ 118:102222

Dagar JC, Sileshi GW, Akinnifes FK (2020) Agroforestry to enhance livelihood security in Africa: research trends and emerging challenges. In: Dagar JC, Gupta SR, Teketay D (eds) Agroforestry for degraded landscapes: recent advances and emerging challenges, vol 1. Springer, Singapore, pp 71–134. https://doi.org/10.1007/978-981-15-4136-0_3

Dewees P, Campbell B, Katerere Y, Sitoe A, Cunningham AB, Angelsen A, Wunder S (2011) Managing the Miombo woodlands of southern Africa: policies, incentives, and options for the rural poor, Washington, DC. Program on Forests (PROFOR)

Diop M, Chirinda N, Beniaich A, El Gharous M, El Mejahed K (2022) Soil and water conservation in Africa: state of play and potential role in tackling soil degradation and building soil health in agricultural lands. Sustain For 14(20):13425

Djenontin INS, Foli S, Zulu LC (2018) Revisiting the factors shaping outcomes for Forest and landscape restoration in sub-Saharan Africa: a way forward for policy. Pract Res Sustain 10:906. https://doi.org/10.3390/su10040906

FAO (2020) Global Forest resources assessment 2020: main report. Rome. https://doi.org/10.4060/ca9825en

Folke C (2006) Resilience: the emergence of a perspective for social–ecological systems analyses. Glob Environ Chang 16(3):253–267. https://doi.org/10.1016/j.gloenvcha.2006.04.002

Franzel S, Carsan S, Lukuyu B, Sinja J, Wambugu C (2014) Fodder trees for improving livestock productivity and smallholder livelihoods in Africa. Curr Opin Environ Sustain 6:98–103. https://doi.org/10.1016/j.cosust.2013.11.008

Ganka G, Salako KV, Fandohan AB (2022) Importance des cultes dans la préservation des espèces d'arbre, le cas du samba (*Triplochiton scleroxylon* K. Schum.) au Bénin. Bois et Forêts des Tropiques 351:53–65

Gavilanes Montoya AV, Castillo Vizuete DD, Marcu MV (2023) Exploring the role of ICTs and communication flows in the forest sector. Sustain For 15:10973. https://doi.org/10.3390/su151410973

Geldenhuys CJ (1993) The use of diameter distributions in sustained-use management of forests: examples from southern Africa. In: Piearce GD, Gumbo DJ (eds) The ecology and management of indigenous forests in southern Africa: proceedings of an international symposium, Victoria Falls, Zimbabwe, 27–29 July 1992. Zimbabwe Forestry Commission & SAREC, pp 154–167

Gello J (2021) Adoption of agroforestry technology in Malawi: hindrances experienced by farmers in Nsanje District. GRÓ land restoration training Programme [final project]. https://www.grocentre.is/static/gro/publication/828/document/gello2021.pdf

Govind SB, Choudhary R, Deshmukh PP (2023) Social and cultural aspect of agroforestry. In: Banoo R, Rajput NB, Abbas G, Packialakshmi M, Keprate A (eds) Advances in forestry and agro-forestry, vol 1. Stella International Publication, pp 209–226

Green Climate Fund (2018) FP072: strengthening climate resilience of agricultural livelihoods in AgroEcological regions I and II in Zambia. Zambia I United Nations Development Programme (UNDP) I B.19/12

Green Climate Fund (2019) SAP011: climate-resilient food security for women and men smallholders in Mozambique through integrated risk management. Mozambique I World Food Programme (WFP)I Decision B.24/094

Green Climate Fund (2020) FP127: building climate resilience of vulnerable agricultural livelihoods in Southern Zimbabwe Zimbabwe I United Nations Development Programme (UNDP) I Decision B.25/04

Grovermann C, Rees C, Beye A, Wossen T, Abdoulaye T, Cicek H (2023) Uptake of agroforestry-based crop management in the semi-arid Sahel – analysis of joint decisions and adoption determinants. Front Sustain Food Syst 7:1042551. https://www.frontiersin.org/articles/10.3389/fsufs.2023.1042551

Hiller B, Fisher J (2023) A multifunctional 'scape approach for sustainable management of intact ecosystems—a review of tropical peatlands. Sustain For 15:2484. https://doi.org/10.3390/su15032484

Hounkpati K, Adjonou K, Kokou K (2022) Distribution and cultural identity of sacred groves in Togo. Int For Rev 24(2):163–174

Işık MF, Sönmez Y, Yılmaz C, Özdemir V, Yılmaz EN (2017) Precision irrigation system (PIS) using sensor network technology integrated with IOS/android application. Appl Sci 7:891. https://doi.org/10.3390/app7090891

Isukuru EJ, Ogunkeyede AO, Adebayo AA, Uruejoma MF (2023) Potentials of bamboo and its ecological benefits in Nigeria. Adv Bamboo Sci 4:100032. https://doi.org/10.1016/j.bamboo.2023.100032

IUCN WCPA (2018) PARKS. The International Journal of Protected Areas and Conservation, Volume 24 special issue. IUCN, Gland

Kanene KM (2016) Indigenous practices of environmental sustainability in the Tonga community of southern Zambia. Jàmbá J Disast Risk Stud 8(1):a331. https://doi.org/10.4102/jamba.v8i1.331

Kangas J, Store R (2002) Socioecological landscape planning: an approach to multi-functional forest management. Silva Fennica 36(4). https://doi.org/10.14214/sf.527

Khatun K, Gross-Camp N, Corbera E, Martin A, Ball S, Massao G (2015) When participatory forest management makes money: insights from Tanzania on governance, benefit sharing, and implications for REDD+. Environ Plan A Econ Space 47:2097–2112. https://doi.org/10.1177/0308518X15595899

Kokou K, Adjossou K, Kokutse AD (2008) Considering sacred and riverside forests in criteria and indicators of forest management in low wood producing countries: the case of Togo. Ecol Indic 8:158–169

Kouassi JL, Gyau A, Diby L, Bene Y, Kouamé C (2021) Assessing land use and land cover change and farmers' perceptions of deforestation and land degradation in south-west Côte d'Ivoire. West Africa Land 10:429. https://doi.org/10.3390/land10040429

Kuyah S, Öborn I, Jonsson M, Dahlin AS, Barrios E, Muthuri C, Malmer A, Nyaga J, Magaju C, Namirembe S, Nyberg Y, Sinclair LF (2016) Trees in agricultural landscapes enhance

provision of ecosystem services in sub-Saharan Africa. Intl J Biodiv Sci Ecosyst Serv Manage 12(4):255–273
Larwanou M, Saadou M (2011) The role of human interventions in tree dynamics and environmental rehabilitation in the Sahel zone of Niger. J Arid Environ 75:194–200. https://doi.org/10.1016/j.jaridenv.2010.09.016
Liu Z, He C, Wu J (2016) The relationship between habitat loss and fragmentation during urbanization: an empirical evaluation from 16 world cities. PLoS One 11(4):e0154613. https://doi.org/10.1371/journal.pone.0154613. PMID: 27124180; PMCID: PMC4849762
Liu S, Dong Y, Liu H, Wang F, Yu L (2023) Review of valuation of forest ecosystem services and realization approaches in China. Land 12:1102. https://doi.org/10.3390/land12051102
Maisharou A, Chirwa PW, Larwanou M, Babalola FD, Ofoegbu C (2015) Sustainable land management practices in the Sahel: review of practices, techniques and Technologies for Land Restoration and Strategy for up-scaling. Int For Rev 17(S3):1–19
Mansourian S, Berrahmouni N (2021) Review of forest and landscape restoration in Africa. FAO and AUDA-NEPAD, Accra. https://doi.org/10.4060/cb6111en
Marchant R (2010) Understanding complexity in savannas: climate, biodiversity and people. Curr Opin Environ Sustain 2(1–2):101–108. https://doi.org/10.1016/j.cosust.2010.03.001
Massenberg JR, Schiller J, Schröter-Schlaack C (2023) Towards a holistic approach to rewilding in cultural landscapes. People Nat 5(1):45–56
Mohamed Nikzaad R, Nusrathali N (2023) Integrating livestock and crop Systems for Enhanced Productivity and Grassland Conservation in developing countries [internet]. Grasslands – conservation and development [working title]. IntechOpen; Available from: https://doi.org/10.5772/intechopen.113109
Moore E, van Dijk T, Asenga A, Bongers F, Sambalino F, Veenendaal E, Lohbeck M (2020) Species selection and management under farmer managed natural regeneration in Dodoma, Tanzania. Front For Global Change 3:563364
Mori AS, Dee LE, Gonzalez A, Ohashi H, Cowles J, Wright AJ, Loreau M, Hautier Y, Newbold T, Reich PB, Matsui T (2021) Biodiversity–productivity relationships are key to nature-based climate solutions. Nat Clim Chang 11(6):543–550
Moyo H, Slotow R, Rouget M, Mugwedi L, Douwes E, Tsvuura Z, Tshabalala T (2021) Adaptive management in restoration initiatives: lessons learned from some of South Africa's projects. S Afr J Bot 139:352–361
Murye AF, Pelser AJ (2018) Commercial Harvesting of Marula (*Sclerocarya Birrea*) in Swaziland: a quest for sustainability. Available online: https://cdn.intechopen.com/pdfs/60753.pdf. Accessed on 26 Nov 2018
Muthee K, Duguma L, Majale C, Mucheru-Muna M, Wainaina P, Minang P (2022) A quantitative appraisal of selected agroforestry studies in the sub-Saharan Africa. Heliyon 8(9):e10670. https://doi.org/10.1016/j.heliyon.2022.e10670. PMID: 36164539; PMCID: PMC9508564
Mwase W, Sefasi A, Njoloma J, Nyoka BI, Manduwa D, Nyaika J (2015) Factors affecting adoption of agroforestry and evergreen agriculture in southern Africa. Environ Nat Resour Res 5(2). https://doi.org/10.5539/enrr.v5n2p148
Negawo WJ, Beyene DN (2017) The role of coffee based agroforestry system in tree diversity conservation in eastern Uganda. J Landsc Ecol 10(2):1–18. https://doi.org/10.1515/jlecol-2017-0001
Ngugi ND (2002) Agroforestry study visit to Malawi and Zambia. CTA working document number 8028. https://core.ac.uk/download/pdf/132673942.pdf
Niu S, Ding J, Xu C, Wang J (2023) Modern and future forestry based on biotechnology. Mod Agric 1(1):27–33. https://doi.org/10.1002/moda.3
Njenga M, Karanja N, Munster C, Iiyama M, Neufeldt H, Kithinji J, Jamnadass R (2013) Charcoal production and strategies to enhance its sustainability in Kenya. Dev Pract 23(3):359–371. http://www.jstor.org/stable/23469243

Ntawuruhunga D, Ngowi EE, Mangi HO, Salanga RJ, Shikuku KM (2023) Climate-smart agroforestry systems and practices: a systematic review of what works, what doesn't work, and why. Forest Policy Econ 150:102937. https://doi.org/10.1016/j.forpol.2023.102937

Odisi Netaji MV, Lakshmi DS, Sahoo U (2023) Farmer managed natural regeneration (FMNR): a key to environmental sustainability. In: Kushwaha S, Kamalvanshi V (eds) Climate change and agriculture, vol 8. AkiNik Publications, pp 85–98. https://doi.org/10.22271/ed.book.2264

Palmer CG, Fry A, Libala N, Ralekhetla M, Mtati N, Weaver M, Mtintsilana Z, Scherman PA (2022) Engaging society and building participatory governance in a rural landscape restoration context. Anthropocene 37:100320

Palomo I, Locatelli B, Otero I, Colloff M, Crouzat E, Cuni-Sanchez A, Gómez-Baggethun E, González-García A, Grêt-Regamey A, Jiménez-Aceituno A, Martín-López B, Pascual U, Zafra-Calvo N, Bruley E, Fischborn M, Metz R, Lavorel S (2021) Assessing nature-based solutions for transformative change. One Earth 4(5):730–741. https://doi.org/10.1016/j.oneear.2021.04.013

Partey ST, Sarfo DA, Frith O, Kwaku MY, Thevathasan NV (2017) Potentials of bamboo-based agroforestry for sustainable development in sub-Saharan Africa: a review. Agric Res 6(1):22–32. https://doi.org/10.1007/s40003-017-0244-z

Poratelli F, Blanc S, Pippinato L, Zanchini R, Bruzzese S, Brun F (2022) Willingness to pay for certified wooden products: a critical literature review. For Stud 76(1):46–63

Poulos M (2023) Supporting women vanilla farmers in Madagascar: the promise of VSLAs and alternative livelihoods. Master's project, Duke University. Retrieved from https://hdl.handle.net/10161/27369

Putz FE, Sist P, Fredericksen T, Dykstra D (2008) Reduced-impact logging: challenges and opportunities. For Ecol Manag 256:1427–1433

Rinaudo T, Muller A, Morris M (2019) Farmer Managed Natural Regeneration (FMNR) manual: a resource for project managers, practitioners and all who are interested in better understanding and supporting the FMNR movement. Melbourne: WVA. https://fmnrhub.com.au/wp-content/uploads/2019/03/FMNR-Field-Manual_DIGITAL_FA.pdf

Rossier C, Lake F (2014) Indigenous traditional ecological knowledge in agroforestry. Agrofor Notes 44

Sahara and Sahel Observatory (2022) African ecosystems: between degradation and restoration. Sahara and Sahel Observatory:160 p. http://www.oss-online.org/sites/default/files/2022-06/OSS-LivreEcosystemesAfrique.pdf

Santos M, Moreira H, Cabral JA, Gabriel R, Teixeira A, Bastos R, Aires A (2022) Contribution of home gardens to sustainable development: perspectives from a supported opinion essay. Int J Environ Res Public Health 21(20):13715. https://doi.org/10.3390/ijerph192013715

Shaw RE, Spencer PB, Gibson LA, Dunlop JA, Kinloch JE, Mokany K, Byrne M, Moritz C, Davie H, Travouillon KJ, Ottewell KM (2023) Linking life history to landscape for threatened species conservation in a multiuse region. Conserv Biol 37(1):e13989

Sheppard JP, Chamberlain J, Agúndez D, Bhattacharya P, Chirwa PW, Gontcharov A, Sagona WCJ, Shen H, Tadesse W, Mutke S (2020a) Sustainable Forest management beyond the timber-oriented status quo: transitioning to co-production of timber and non-wood Forest products—a global perspective. Curr For Rep. https://doi.org/10.1007/s40725-019-00107-1

Sheppard JP, Bohn Reckziegel R, Borrass L, Chirwa PW, Cuaranhua CJ, Hassler SK, Hoffmeister S, Kestel F, Maier R, Mälicke M, Morhart C, Ndlovu NP, Veste M, Funk R, Lang F, Seifert T, du Toit B, Kahle HP (2020b) Agroforestry: an appropriate and sustainable response to a changing climate in southern Africa? Sustain For 12(17):6796. https://doi.org/10.3390/su12176796

Sileshi GW, Mafongoya PL, Akinnifesi FK, Phiri E, Chirwa P, Beedy T, Makumba W, Nyamadzawo G, Njoloma J, Wuta M, Nyamugafata P, Jiri O (2014) Agroforestry: fertilizer trees. In: van Alfen N (ed) Encyclopedia of agriculture and food systems, vol 1. Elsevier, pp 222–234

Silva-Galicia A, Larsen J, Álvarez-Espino R, Ceccon E (2023) Agroecological and agroforestry strategies to improve organic hibiscus productivity in an indigenous non-governmental organization from Mexico. Renew Agricult Food Syst 38(e3):1–10. https://doi.org/10.1017/S1742170522000369

Simelane SM, Terblanche SE, Masarirambi MT (2019) Collective action for access to inputs, finance, markets and extension for smallholder farmers in Eswatini. S Afr J Agric 47(2):21–36. https://doi.org/10.17159/2413-3221/2019/v47n2a500

Singh S, Singh G (2023) Agroforestry for sustainable development: assessing frameworks to drive agricultural sector growth. Environ Dev Sustain 1-37. https://doi.org/10.1007/s10668-023-03551-z

Song B, Robinson GM, Bardsley DK (2020) Measuring multifunctional agricultural landscapes. Land 9:260. https://doi.org/10.3390/land9080260

Sonwa D, Nkongmeneck BA, Weise S, Tchatat M, Adesina AA, Janssens M (2007) Diversity of plants in cocoa agroforests in the humid forest zone of southern Cameroon. Biodivers Conserv. http://www.springerlink.com/content/52m3727651133vp7/fulltext.html

Syampungani S, Chirwa PW, Geldenhuys CJ, Handavu F, Chishaleshale M, Rija AA, Mbanze AA, Ribeiro NS (2020) Managing Miombo: ecological and Silvicultural options for sustainable socio-economic benefits. In: Ribeiro NS, Katerere Y, Chirwa PW, Grundy IM (eds) Miombo woodlands in a changing environment: securing the resilience and sustainability of people and woodlands. Springer, Cham, pp 101–137. https://doi.org/10.1007/978-3-030-50104-4_4

Tadesse B, Ahmed M (2023) Impact of adoption of climate smart agricultural practices to minimize production risk in Ethiopia: a systematic review. J Agricult Food Res 13:100655. https://doi.org/10.1016/j.jafr.2023.100655

Teketay D, Mbolo AMM, Kalonga SK, Ahimin O (2016) Forest certification in Africa: achievements, challenges and opportunities. African Forest Forum, Nairobi. https://afforum.org/oldaff/sites/default/files/English/English_13.pdf

Teketay D, Kashe K, Madome J, Kabelo M, Neelo J, Mmusi M, Masamba W (2018) Enhancement of diversity, stand structure and regeneration of woody species through area exclosure: the case of a mopane woodland in northern Botswana. Ecol Process 7:1–15

Le TT, Vodden K, Wu J, Bullock R, Sabau G (2023) Payments for ecosystem services programs: a global review of contributions towards sustainability. Heliyon 10(1):1–32. https://doi.org/10.1016/j.heliyon.2023.e22361

Uni D, Sheffer E, Klein T, Shem-Tov R, Segev N, Winters G (2023) Responses of two *Acacia* species to drought suggest different water-use strategies, reflecting their topographic distribution. Front Plant Sci 14:1154223. https://doi.org/10.3389/fpls.2023.1154223s

Villodre M, Arnaiz-Schmitz C, Schmitz MF (2023) Landscape conservation in the natural-rural interface. A social-ecological approach in natural Parks of Andalusia (Spain). Landsc Ecol 38(12):3517–3535

Vlek PLG, Khamzina A, Azadi H, Bhaduri A, Bharati L, Braimoh A, Martius C, Sunderland T, Taheri F (2017) Trade-offs in multi-purpose land use under land degradation. Sustain For 9:2196. https://doi.org/10.3390/su9122196

Winter KB, Lincoln NK, Berkes F (2018) The social-ecological keystone concept: a quantifiable metaphor for understanding the structure, function, and resilience of a biocultural system. Sustain For 10:3294. https://doi.org/10.3390/su10093294

World Bank (2019) Zambezia Integrated Landscapes Management Program (ZILMP). National Sustainable Development Fund (FNDS) (Fundo Nacional de Desenvolvimento Sustentável) Rua Joe Slovo, 21, Cidade de Maputo

World Wide Fund for Nature (WWF) (2022) Forest Landscape Restoration (FLR) In Africa Initiative. FY22 Annual Report

Yiran GAB, Ablo AD, Asem FE, Owusu G (2020) Urban sprawl in sub-Saharan Africa: a review of the literature in selected countries. Ghana J Geogr 12(1):1–28. https://doi.org/10.4314/gjg.v12i1.1

Chapter 13
Socio-economic Factors, Policy and Governance Systems Influencing Multifunctional Landscapes

Paxie W. Chirwa, Witness Kozanayi, Amani J. Uisso, Ratsodo P. Tshidzumba, Folaranmi D. Babalola, and Tajudeen O. Amusa

Abstract Multifunctional landscapes, characterised by their ability to simultaneously fulfil diverse societal needs ranging from agricultural production to biodiversity conservation and cultural heritage preservation, represent complex socio-ecological systems at the nexus of human and natural interactions. Drawing on insights from several countries in sub-Saharan Africa, this chapter looked at the role of trees in sustainable livelihoods within multifunctional landscapes through the myriad of their ecosystem services. The role of socio-economic drivers such as demographics, in particular population growth, economic and market forces, as well as cultural aspirations on land use decisions and resource allocation within

P. W. Chirwa (✉)
Department of Plant and Soil Sciences, Faculty of Natural and Agricultural Sciences, University of Pretoria, Pretoria, South Africa
e-mail: paxie.chirwa@up.ac.za

W. Kozanayi
Department of Wildlife and Fisheries Management, Marondera University of Agricultural Sciences and Technology, Marondera, Zimbabwe

A. J. Uisso
Department of Forest Production Research, Tanzania Forestry Research Institute, Morogoro, Tanzania

R. P. Tshidzumba
Department of Plant and Soil Sciences, Faculty of Natural and Agricultural Sciences, University of Pretoria, Pretoria, South Africa

School of Agricultural Sciences, Faculty of Agriculture and Natural Sciences, University of Mpumalanga, Mbombela, South Africa

F. D. Babalola
Department of Forest Resources Management, Faculty of Agriculture, University of Ilorin, Ilorin, Nigeria

National Centre for Food Manufacturing, University of Lincoln, Holbeach, UK

T. O. Amusa
Department of Forest Resources Management, Faculty of Agriculture, University of Ilorin, Ilorin, Nigeria

P. W. Chirwa et al. (eds.), *Trees in a Sub-Saharan Multi-functional Landscape*, https://doi.org/10.1007/978-3-031-69812-5_13

multifunctional landscapes were also examined. The chapter further scrutinized the role of policy frameworks in guiding the trajectory of multifunctional landscapes toward sustainable outcomes. The efficacy of diverse policy interventions, ranging from land-use zoning regulations to incentive-based schemes and participatory decision-making processes was also highlighted. Emerging trends and challenges on the horizon of multifunctional landscape management including landscape complexity, conflicting interests, fragmented governance arrangements, and insufficient data were also identified. It is believed that by embracing adaptive strategies that integrate socio-economic considerations, policy imperatives, and governance principles, societies can unlock the full potential of multifunctional landscapes to meet the evolving needs of present and future generations.

Keywords Ecosystem services · Socio-ecological systems · Sustainable livelihoods · Landscape complexity · Trees

13.1 Introduction

The African continent is endowed with abundant forest resources providing valuable ecosystem services to sustain livelihoods of the poor and wealthier population in both rural and urban communities (Oljirra 2019) and also contributing to national gross domestic products (FAO 2022). Globally, over one million animal and plant species are threatened with extinction largely due to over-harvesting (IPBES 2019). Simultaneously, the African forest cover has experienced an immense decline in the past two decades resulting from various socio-economic pressures (Xiao et al. 2022). Thus, both the demand and unsustainable use of forest resources due to dire socio-economic status and emergence of markets for some of the forest products (Kozanayi 2018) in many African countries continue to threaten and put immense pressure on the generative and provisioning capacity of forests.

Furthermore, the need to continue satisfying other land uses with the aim to reduce poverty and improve economic status of the rural dwellers, have also perpetuated an increase in deforestation and forest degradation in Africa (Kouadio and Singh 2021). Deforestation has caused a need for upscaling of effective and sustainable measures to protect and implement environmental and biodiversity conservation efforts to safeguard the provisioning, regulating, cultural and supporting ecosystem services of the forest (Chirwa et al. 2015). To that end, various initiatives aimed at forest restorations have been adopted in Africa, including the African Union inspired Great Green Wall (GGW) which aims to plant 100 million hectares of forests (https://www.unccd.int/our-work/ggwi), as an approach to restock the deforested and degraded forests, up-scaling its functionality and contribution to mitigate climate change through carbon sequestration and/or storage capacity. A similar initiative is the African Forest Landscape Restoration Initiative (AFR-100) which is a country-led effort to bring 100 million hectares of land in Africa into restoration (Mansourian and Berrahmouni 2021). Nevertheless, the uptake of these efforts and/or implementation varies across countries due to different societal and economic levels as well as their potential to protect forest resources (Xiao et al. 2022).

Multifunctional landscapes offer the capacity to supply a high diversity and quantity of ecosystem services within a specified region, hence satisfying many social, cultural, ecological, and economic needs (Stürck and Verburg 2017). A comprehensive valuation of the global ecosystem functions by Constanza et al. (1997), revealed that ecosystems provide about $15 trillion dollars per year. The number of people deriving direct and indirect benefits from forests and trees—in the form of employment, forest products, and contributions to livelihoods and incomes—is estimated at 1.3 billion (FAO 2014). The forest sector contributes more than USD 1.52 trillion to world gross domestic product and employs 33 million people (FAO 2022). According to Kato and Ahern (2009), multifunctional landscapes effectively employ time and space while resolving disputes and maximising the benefits of competing stakeholders (Huber et al. 2020). To fulfil the demands of future generations and to endure the difficulties of global change, multifunctionality is crucial for the development of resilient and sustainable landscapes (Kato and Ahern 2009; Reyers et al. 2012). Multifunctionality with a strong emphasis on agricultural land use systems has gained prominence in land use strategies (Maier and Shōbayashi 2001; Wiggering et al. 2003; Tait 2016).

For the forest sector, the benefits of trees within the multifunctional landscape extend far beyond mere ecological considerations but encompass social, cultural, and even spiritual dimensions. One such aspect is the cultural significance attributed to trees by local communities, often manifested in the establishment of sacred landscapes that form an integral part of rural livelihoods (Byers et al. 2001; Babalola et al. 2014) and some sacred tree species. In this regard, traditional land management practices are mostly based on cultural norms and religious beliefs, and they have proved effective in sustainable land management (Berkes et al. 2000; Colding and Folke 2001; Kayambazinthu et al. 2003; Mgumia and Oba 2003; Soutter et al. 2003; Moller et al. 2004; Saj et al. 2006; Banadda 2010; Uisso et al. 2019). Historically, the most effective traditional land management practices in forest conservation in sub-Saharan Africa include tree planting; beliefs in sacredness of trees; beliefs in sacred forests; respect of cultural trees and forests; protection of plants at the burial sites; selective harvesting; secrecy; collection of deadwood for firewood, and use of energy-saving traditional stoves (Msuya and Kideghesho 2009; Banadda 2010). Sacred landscapes, deeply rooted in indigenous knowledge and cultural beliefs, are governed by rules established by local communities and spiritual leaders (Liu 2006; Louman et al. 2009; Adesiji and Babalola 2012; Jimoh et al. 2012). Moreover, sacred forests are known for their rich plant diversity, serving as important sites for the collection of plants used in traditional herbal medicine (Bhagwat and Rutte 2006). Beyond their ecological value, sacred forests play a crucial role in preserving traditional religious practices (Lawrence 2003; Salick et al. 2007; Jepson and Canney 2003), indigenous local knowledge systems, and cultural beliefs, thereby bridging the gap between biodiversity conservation and cultural heritage (Byers et al. 2001; Hens 2006; Babalola 2009).

Taboos and sacred places have also been used to protect and preserve the environment, a fact that was noted by early anthropologists (Schoffeleers 1979). Water sources too, have been kept clean and protected by adhering to taboos regarding their use (Chemhuru and Masaka 2010). Apart from the use of taboos and sacred places, indigenous people have other ways of conserving the environment. Nomads practise nomadic herding to allow vegetation to regenerate. There are also traditional silvicultural practices that local people use, for example farmer managed natural regeneration, assisted natural regeneration to enrich biodiversity.

Key features that characterize the traditional forest resources management practices include social organisations that control access to resources within the community; customary norms and procedure for control, acquisition, maintenance and transfer of the resources; and traditional utilisation and conservation practices (Kayambazinthu et al. 2003). According to Abesha et al. (2023), recognition and incorporation of local people's or traditional ecological knowledge is fundamental in promoting environmental and landscape conservation. However, despite the traditional land management practices and knowledge, land ownership or tenure have proven to be a serious barrier affecting effective sustainable land management practices and/or generation of reasonable multifunctional landscape benefits. In this case, most people are exposed to the highest level of land resource exploitation over established foreign investors that often prioritise their profit over the rural population. In this case, the lack of local people's negotiation power due to poor scientific knowledge and capacity often compromise their contribution towards effective resource management (Djenontin et al. 2018). As such, the need to consolidate tenure rights and redistribution of power across all stages of decision making remain an important strategy (Bansard and Schroder 2021).

While secure tenure is conceptualised in the global North as a panacea to problems related to resource access, access to resources on the landscape across the African landscape is regulated by a gamut of complex approaches that are usually juxtaposed to sometimes conflicting statutory systems. Ribot and Peluso (2003), provide a more nuanced notion of access, that argues that access is not just about rights- rather is it about the ability to gain access to a resource. That ability is derived from many sources including power and adroit use of wealth, access to labour, use of legitimate and illicit power, corruption, and drawing from social capital. Yet another complication with the governance of resources on the African landscape is that there is a duality of rules governing access- comprising statutory forms of governance, overlaid on customary forms of governance (Kozanayi 2018). In fact, much of today's forest tenure combines elements of traditional forest management with government-controlled conservation (FAO 2008; Gilmour 2016). Facilitating hybridisation of customary and statutory forms of governance for effective governance of resources on the landscape has remained elusive.

Practically, the landscape level approach has become widely used to inform decision-makers, policymakers, academicians, researchers, planners and land-use managers of the links between landscape-level and sustainable natural resources governance. Landscape level approach is a prerequisite for biodiversity conservation and the functioning of ecosystems. A heterogeneous landscape with multiple land uses can serve as a buffer zone to adapt to environmental and economic challenges. Thus, a landscape approach is critical to effectively manage biodiversity and ecosystem goods and services (Ibid). Landscapes, be it in the form of agriculture, heterogeneous plantations, agroforestry (plantation or indigenous), settlement and forestry, are multifunctional as they support mosaics of biodiversity, and a wide range of ecosystem goods and services especially, those supporting livelihoods of local communities (Rantala et al. 2011; Kremen and Merenlender 2018). However, the intensity and diversity of support to biodiversity (Birkhofer et al. 2018) and livelihood (Malmborg et al. 2018) vary across the host landscapes. For example, plantations and built-up areas tend to support less biodiversity while pristine miombo woodlands would support richer forms of biodiversity. In this context, governance of the landscape becomes vital. Without prescribing or proscribing, the principle of subsidiarity (Ribot 2002), which postulates nesting of powers in any governance hierarchy so that problems and issues are best addressed and discussed at the level at which they occur, appears as guiding star. Thus, while there is need to conceptualise and frame governance at the landscape level, local level peculiarities need to be accorded due attention. In sub-Saharan Africa, the concept of landscape level management has been embraced in the natural resources management and conservation sectors (Rantala et al. 2011; Nindi et al. 2019). It is now gaining momentum and widely accepted in addressing multiple land uses in a particular wider landscape.

Worldwide, and across Africa, multifunctional landscapes are linked to climate change and biodiversity loss. An intricate link between climate change mitigation and biodiversity conservation exists. Changes in biodiversity exacerbates the impact of climate change on people. On the other hand, climate change has directly and indirectly altered biodiversity globally and particularly in Africa (Muluneh 2021; Shivanna 2022). Although climate change has negative impacts on biodiversity loss including loss of tree species, trees within the landscapes have the potential to offset fossil-fuel emissions and have become the centre of the carbon market debate. Trees are the main sinks of atmospheric carbon dioxide and thus contribute to climate change mitigation (Muthuri et al. 2023; Kirschbaum et al. 2024). Forests also nourish local populations (WRI 2005), and their use is dependent on how the solutions to climate change are framed. The World Bank (2007) estimated that forest resources directly contribute to the livelihoods of 90% of the 1.2 billion people living in extreme poverty in the developing world. Over the years, due to the debilitating effects of climate change and poor governance by governments, more people have been pushed to mining the forest to eke out a living (Campbell et al. 2002; Ribot 2022). Thus, about 13 million hectares of tropical forests are cleared every year, predominantly through conversion into other land uses (World Bank 2007). There is also a growing concern about the role of trees in climate change mitigation and the tree-based carbon market especially a widespread criticism of voluntary carbon markets and forest carbon credits (Aryal et al. 2024). The process by which the mechanism of carbon funding is done is shrouded in secrecy and benefits to ordinary farmers remain elusive (Mashingaidze et al. 2021). Thus, biodiversity climate change issues have to be addressed concurrently.

Against this background, this chapter highlights how sustainable livelihoods are derived from trees, the socio-economic factors, policy frameworks and forms of governance shaping the multifunctionality of landscapes in sub-Saharan Africa.

13.2 Sustainable Livelihoods Derived from Trees on the Landscape

Over the past decades, landscape conservation efforts have been of increasing concern. Landscapes have been greatly recognized as key arenas for sustainable development. The livelihoods and well-being of local communities are predominantly based on landscape through agriculture (cropping and livestock), logging, non-timber forest products (NTFPs) and fisheries. Better management of the landscapes is needed to meet the challenges of increasing population, climate change, and sustainable livelihoods. While livelihood refers to the way people live (their activities, capabilities and assets) and earn income (Chambers and Conway 1991; DFID 1999), sustainable livelihoods entail the long-term process of pursuing livelihoods. In this context, landscapes are gaining policy attention as a means to address the multiple aspects of the environment and associated stressors. In one case, there is a push for trees on the landscapes in the form of agricultural/agroforestry and forest (natural/plantation) landscapes (Minang et al. 2015).

Trees on the landscape provide opportunities for addressing sustainable livelihoods in several ways from social-ecological, to cultural or economic aspects. This means trees on the landscapes can achieve a "multiple win-win scenario" in the context of livelihoods. Trees in the landscapes play an important role in protecting the environment through ecosystem services such as clean air, stormwater mitigation, and provision of habitat and microclimate regulation (Pataki et al. 2021). Trees in the landscapes regulate the quality of surface water in bodies. They enhance water infiltration and minimise soil erosion in agricultural landscapes. They also supply and enhance nutrients for agricultural production and fodder for crops, thus ensuring food supply and income. These potential ecosystem goods and services are categorised into four groups namely: provisional, supporting, regulating, and cultural (Table 13.1).

Further, trees are linked to livelihood capitals, namely natural, financial, physical, human and social. The different capital bases are the building blocks to pursue a range of livelihood strategies for positive livelihood outcomes. The livelihood capitals derived from the trees on the landscapes are described in Table 13.2.

However, despite having multiple wins, trees on the landscape have been threatened and degraded, linked with unsustainable utilisation, thus undermining their potential contribution to livelihoods. According to the Millennium Ecosystem

Table 13.1 Tree based landscape ecosystem services for livelihoods

S/ no	Ecosystem category	Example of service/goods to support livelihoods
1.	Provisioning	Fish, insects, crops, wild foods, honey, fresh water, nuts, fibre, fuel, medicinal plants, fodder and forage, firewood, raw materials for building, timber, poles and logs
2.	Supporting	Biomass production, production of atmospheric oxygen, soil formation and retention, nutrient cycling, water cycling, and habitat provision
3.	Regulating	Pollination, biodiversity habitat, water filtration/ purification, erosion and flood control and carbon sequestration/ storage, disease regulation and decomposition
4.	Cultural	Recreation and tourism, education opportunities, preservation of fauna and flora species and ecosystems, opportunities for cultural and spiritual connection and practice, heritage protection

Source: Barrow and Mlenge (2003), Dawson et al. (2014), and Pataki et al. (2021)

Table 13.2 Livelihood capital category and examples of their actual livelihood benefits

S/ no	Livelihood capital category	Examples of livelihood capital
1.	Natural	Building materials (timber, poles, logs), improve microclimate, fodder for livestock, bees habitat, food (fruits, vegetables), water cycling, bio oil/diesel, fibre
2.	Physical	Infrastructure development (transport, communication, housing)
3.	Financial	Employment opportunity, source of income from sales of timber, poles and logs, access to loan, membership to financial institutions
4.	Human	Medicine for human health care, human skills and knowledge
5.	Social	Ritual practices, social cohesion/relations

Source: Colfer et al. (2008), Akrofi-Atitianti et al. (2018), and Muthuri et al. (2023)

Assessment (2005), forest landscapes are subjected to environmental stress from agricultural production thus undermining their capacity to provide goods and services. Several interventions have been implemented to improve trees on the landscapes by local, national and international communities. These include among others, Payment for Ecosystem Services (PES), Reduced Emission from Deforestation and Forest Degradation and sustainable management of the forests and conservation and enhancement of forest carbon stock (REDD+), land use planning, forest certification, landscape restoration, integrated landscape management and participatory forest management (PFM). These initiatives among others, aim to improve natural resources management, conservation and community livelihoods. For instance, in Tanzania, land use planning, PFM and REDD+ have been implemented for the management and conservation of forest landscapes at the same time to improve livelihoods.

13.3 Socioeconomic Factors Shaping Multifunctionality of Landscapes

Landscape multifunctionality is described as achieving multiple objectives or functions simultaneously in a certain landscape through reducing trade-offs and promoting synergies (Mastrangelo et al. 2014). It is also considered as the capacity of a landscape to simultaneously support multiple ecosystem goods and services or benefits to people from its interacting ecosystems within a landscape (Mastrangelo et al. 2014, Stürck and Verburg 2017, Hölting et al. 2019a). Overall, these ecosystem goods and services have the potential to achieve multiple social, cultural and economic needs of people. According to Reyers et al. (2012), multifunctionality of landscapes enables the development of sustainable and resilient landscapes to respond to climate change and related disasters and meet the needs of future generations. However, literature suggests that multifunctional landscapes have been subjected to fundamental structural changes due to socio-economic factors such as demographic and population, economic and market forces, and social-cultural spaces (Hölting et al. 2019a, 2019b; Peterson et al. 2014, Sunderland et al. 2017). Hölting et al. (2019b), noted that worldwide there are no landscapes left that are pristine and have not been shaped by humans. The next section explores the understanding of how socio-economic factors reshape the landscape's multifunctionality. Understanding and considering these socio-economic factors is critical when developing sustainable multifunctional landscape policies and strategies.

13.3.1 Demographics and Population Dynamics

Globally, and in Africa in particular, changes in landscape multifunctionality can be explained by changes in demographic and population dynamics. Demographic and population growth has been reported to influence the multifunctionality of landscapes, negatively affecting ecosystem services provision in the main. Demographic elements such as gender, age, education, wealth, household size and residence status are significant determinants of forest product use and hence landscape change. Villamor et al. (2014), noted that a multifunctional landscape perspective should include gender-specific factors that may contribute to the pattern of landscape multifunctionality. Population growth is a significant determinant of changes in land use and land cover thus affecting the multifunctionality of landscapes. Population growth creates a high level of dependence on forest land and products including for agriculture, settlement, mining and grazing among communities. For example, in the African context, more than 60% of the people depend on agriculture as their economic activity (Odusola 2021), thus forested landscapes have been transformed into human-induced landscapes (agricultural landscapes). Agriculture supports 80% of people and agriculture expansion explains the main driver of deforestation and forest degradation in forested landscapes (Doggart et al. 2020). Thus, approaches to meeting human population growth needs while maintaining the multifunctionality landscapes are critical.

Further, the African population is fairly young, with two thirds of the population being youths. This dynamic is important for two reasons that have a direct bearing on the governance of landscape. First, across much of sub-Saharan Africa, unemployment is rife and to eke out a living, the youths harvest forest resources, including in some cases, settling or farming in fragile lands. Second, the young generation do not respect governance systems (both customary and statutory) in place to regulate access to resources on the landscape (Kozanayi 2018).

Related to population growth is the issue of urbanisation. Urbanisation is growing fast across Africa. Urbanisation has two main impacts on the landscape, first-forests are converted into settlements for the increasing urban population. Second, urbanisation results in more demand for natural resources to support the increasing urban population. For example, due to rapid urbanisation, demand for sand to construct buildings is expected to increase from about 40 billion tons to 60 billion tons per year by 2030 (UNEP 2019).

13.3.2 Economic Drivers and Market Forces

Across Africa, multifunctional landscapes are underpinned by global and local political, economic and market forces (Peterson et al. 2014). For example, agriculture as an economic activity linked to food production, is largely driven by the demand for food to feed and increasing human population. Agriculture seems to be critical in an African landscape and as noted above, agricultural production employs over 60% of African people (Ba 2016; Waddington and White 2014; Odusola 2021). Global market forces have shifted agricultural production from subsistence to commercially intensive endeavours across African countries. Land grabbing is a common feature in an African landscape as there are emerging investors opening up large-scale agricultural production. Generally, there is a notable shift from a forest-dominated landscape to an agricultural landscape which is a human-dominated landscape, thus reshaping the multifunctionality of the landscape. Energy production such as charcoal production is another leading socio-economic activity shaping multifunctional landscapes in sub-Saharan Africa (Liyama et al. 2017). According to Mwampamba et al. (2013) and Liyama et al. (2017), charcoal is an important energy source for the majority of the population in urban areas of sub-Saharan Africa. Although worldwide, the African continent especially sub-Saharan Africa is more reliant on charcoal (60%) (Rose et al. 2022), within Africa, charcoal production by regions is disproportionately high. Nyarko et al. (2021) reported the regional production pattern as East Africa (32 million tonnes, 43.2%), West Africa (23 million tonnes, 32.1%), North Africa (8.6 million tonnes, 11.6%), Middle Africa (8.5 tonnes million, 11.5%) and South Africa (1.2 million tonnes, 1.6%). This means East Africa and West Africa are the major producers of charcoal in Africa. Rose et al. (2022) estimated that 195 and 200 million people in sub-Saharan Africa rely on charcoal as their primary and secondary cooking fuel respectively. It is an affordable energy source and generates employment and hence incomes. However, charcoal production is the major source of biodiversity loss and climate change through deforestation and forest degradation (Ellegård and Nordström 2013). There are also human health risks associated with the use of charcoal including the respiratory illness of women and children and indoor air pollution (Amugune 2020). This has implications for the increasing burden of human health care. The fact that agriculture and charcoal cannot be readily substituted in sub-Saharan Africa, there must be practical policies and strategies to support sustainable management of these economic drivers in multifunctional landscapes in Africa. Commercialisation of NTFPs in the region is also gaining momentum (Shackleton 2015). A number of NTFPs are being commercialised in sub–Saharan Africa to meet demand for novel foods in the global North. Such commercialisation has brought in a raft of customary systems of governance characterised by complex permit systems in the governance of the targeted NTFPs (Wynberg et al. 2015) (see Chaps. 8 and 9).

Transfrontier national parks are yet another emerging form of landscape use and governance. These transfrontier parks in Southern Africa, of which the Kavango-Zambezi (KAZA), is one of the biggest conservation areas in the world covering 5 countries namely Angola, Zimbabwe, Namibia, Zambia and Botswana, are promoted for double dividend of biodiversity conservation and economic benefits through tourism (Stoldt et al. 2020). Thus, these transfrontier parks have potential to improve the livelihoods of communities living near or around the designated parks. On the other hand, the creation of these transfrontier parks can result in the displacement of indigenous people and in some cases heightened human-wildlife conflicts.

13.3.3 Social Dynamics and Cultural Influences

In an African context, changes in multifunctionality are echoed in social dynamics and cultural influences. Socially, gender appreciation of trees can influence landscape multifunctionality. This is reflected in gender-differentiated access, use and decision-making activities related to trees in the landscape including participation for decision-making in the management of multifunctional landscapes at large (Bourne et al. 2015). Culturally, across Africa, the prevailing traditional values and belief systems that consider some landscapes preserved as sacred forests for prayers, rituals, sacrifices and burials are common and have positively influenced the management and conservation of multifunctional landscapes. The culturally protected landscapes employ the system of taboos and traditional rules which has been instrumental in preserving the landscapes (UNU-IAS and IR3S/UTIAS 2016). Such systems of culturally protected multifunctional forest landscapes are found in several regions across Africa, for example, the Mijikenda Kaya forests in the coastal area of

Kenya, Chome forests in North East Tanzania, Muzarabani Communal Forest Lands of Zimbabwe (Zambezi valley), Licuati forest and Chao Highlands in the south and Manica Province respectively in Mozambique, Ankodida forests in Madagascar, Accra plains in Ghana, and Loma area dominated by the Upper Guinea forests in Liberia (UNU-IAS and IR3S/UTIAS 2016; Ekblom et al. 2019). All these culturally protected forests have positively influenced the conservation of landscapes (Ekblom et al. 2019). According to UNU-IAS and IR3S/UTIAS (2016), the culturally protected multifunctional landscapes are important for biodiversity conservation, livelihoods and well-being of communities and play critical roles in local adaptation to climate change by strengthening the adaptive capacity of local communities.

13.4 Policy Frameworks for Multifunctional Landscapes: Forms of Governance in Sub-Saharan Africa

Landscapes are multifunctional and contain a mosaic of land uses and land cover. Each mosaic has a unique form of governance that regulates its use (Colfer and Pfund 2011). Landscape restoration is being done in different parts of the world and in sub-Saharan Africa in particular with different implementation frameworks (Minang et al. 2015). Worldwide before the 1980s, the governance of natural resources relied on a "centralised management". This approach followed a "top-down approach" and people were considered as objects. During this period, planning and decision-making on natural resource governance was done by the government without consideration of the local community.

Having realised that the centralized management approach to natural resource governance including landscape governance did not work to the desired extent, a new approach commonly known as "decentralised management" was proposed and adopted in the 1980s. The centralised approach was not effective and thus didn't yield the intended management and conservation outcomes. Under the decentralised approach, people have to be fully involved in the governance process, especially in the planning and implementation of natural resources management and conservation initiatives. This is one way that the power of planning rests on people who could value and feel a sense of ownership and responsibility for the initiatives (Maier et al. 2014). Following the decentralised approach to natural resources governance, many countries in the world including sub -Saharan Africa adopted the same.

13.4.1 Sectoral Policies on the Governance of Landscapes in Sub–Saharan Africa

A number of countries have revised their policies to include the involvement of communities in the management of natural resources. More importantly, the policies have recognised a human rights-based approach to integrating international human rights standards and principles in natural resources management including a gender-responsive approach. This comes after the realisation of the need to put people at the centre of planning and implementation of sustainable land management approaches and practices. Further, sustainable management of natural resources outcomes requires transformative shifts in gender norms. In addition, this is driven by the need to address some emerging issues such as climate change. This calls for approaches that help communities to either adapt or mitigate to factors of climate change such as forest or farm tree-based adaptation and mitigation interventions, identification of sustainable livelihood mechanisms to food production such as conservation agriculture, climate-smart/resilient agriculture, ecological agriculture, agroforestry, community-based natural resources management (CBNRM), among others.

Landscape governance requires a policy framework that guides landscape management and conservation practices. Good governance practices support successful sustainable landscape management (Nzyoka et al. 2021). The management of multifunctional landscapes is dependent on several policy formulations including cultural customs at a local level and land tenure rights at national government level. Some countries implemented policies that promoted multifunctional landscape governance while others had different policies that marred the development of agroforestry systems or land use options. Furthermore, some countries like South Africa have developed policies to promote agroforestry with limited implementation. Also, the land tenure policy has not been fully implemented since the dawn of democracy leaving the utilization of land for multifunction production difficult and challenging. Yet, it is significant to note that promoting the inclusion of local people in forest resource management using landscape governance approaches that foster incentives for the participation in the ecosystem and landscape conservation. Hedden-Dunkhorst and Schmitt (2020) emphasised that stakeholder participation and cooperation are critical conditions for implementation of integrated landscape approaches. However, van der Duim and Pellis (2022), suggested that partnerships play a significant and/or substantial role in biodiversity management and protection as a landscape governance approach, while at the same time, serving as a policy vehicle that promotes negotiations between various interested stakeholders including community, private and public. Similarly, the report by Loza (2023) emphasized the need for the community and government to work together in the management and conservation of natural forest resources initiatives in order to achieve sustainability.

However, the main challenges with this partnership are poor community involvement, unequal power (power inequalities) and unclear benefit-sharing mechanisms which need to be addressed. (Mbaiwa 2017; Shereni and Saarinen 2021; Kengoum et al. 2024). Generally, several factors including poverty, apportioning of land amongst various stakeholders and/or impact of climate change in earmarked landscapes remain less influenced by partnership efforts. It is, therefore, important to note that without government support, partnerships would not effectively achieve landscape governance roles. In support of this, the international partnership has strengthened the capacity and steered successful outcomes of conservation and development in various landscapes (Hedden-Dunkhorst and Schmitt 2020).

Table 13.3 shows policy and legislation factors that contribute to the design and implementation of forest-based mitigation and adaptation actions, which could be directly linked to trees on a multifunctional landscape. The success of implementing forest-based mitigation activities seems to fundamentally rely on good governance and security of tenure and access to the resources. In fact, like many other tree-based activities, if there is no secure land tenure, communities are not willing to participate in such activities. For example, in Tanzania, the village governments that have promoted secure land tenure have resulted in effective implementation of REDD+ and other forest resource management (Uisso et al. 2018; Uisso 2022). Most forest policies in Anglophone countries have mainstreamed involvement of communities in management of forest resources, but with no devolvement on benefits sharing. In a study conducted in six countries, governance in the form of tenure resource rights and political will was ranked as the most important policy and legislation factor contributing to design and implementation of forest-based mitigation and adaptation actions (Chirwa 2020; Table 13.3).

Table 13.3 Policy and legislation factors contributing to design and implementation of forest-based mitigation and adaptation actions

Policy and legislation factors	Zambia	Ghana	Mozambique	Sierra Leone	South Africa	Average
Governance—tenure and resource rights, political will etc.	3.0	3.0	3.0	2.0	3.0	2.8
Technical capacity to develop bankable projects and robust MRV	2.0	3.0	2.0	3.0	2.0	2.4
Economic incentives for project proponents	3.0	3.0	2.0	2.0	2.0	2.4
Availability of sufficient and sustainable financing	1.0	1.0	1.0	3.0	3.0	1.6

Source: Chirwa (2020)
Rating scale: 1= Strongly disagree, 2 = Agree, 3 = Strongly agree

13.5 Challenges and Opportunities for Operating at Scale/Landscape Level

13.5.1 Challenges

The multifunctionality of landscapes in Africa is challenged by various factors that hinder effective management and conservation efforts. In the first place, the complexity and interconnectedness inherent in landscapes pose significant hurdles. These systems comprise diverse ecological, social, and economic elements, making it challenging to understand and navigate their dynamic nature. Managing such complexity requires a nuanced approach to address the intricate web of interactions between these elements (Rantala et al. 2011). Furthermore, conflicting interests and priorities among stakeholders within landscapes exacerbate the challenge. Divergent interests often lead to conflicts over land use, resource allocation, and conservation measures. Achieving a balance between these competing interests while ensuring the sustainability of landscapes remains a persistent challenge for policymakers and practitioners (See Chap. 2).

Furthermore, effective governance structures and coordination mechanisms are essential for managing landscapes sustainably. However, governance arrangements may be fragmented across multiple jurisdictions and sectors, resulting in institutional barriers and coordination challenges. This fragmentation hampers efforts to address multifaceted landscape issues effectively (Rantala et al. 2011). Additionally, the limited availability of comprehensive data and information on landscapes inhibits informed decision-making and planning processes. Addressing data gaps and improving information sharing among stakeholders are critical for evidence-based landscape management. Without adequate data, management actions risk being based on incomplete or inaccurate information, undermining their effectiveness (Ibid).

In addition, most African countries have set up the Environmental Social Safeguards (ESS) designed to prevent and mitigate undue harm to people and their environment but are not adhered to. There is also a lack of resource mechanisms for local communities that are marginalised (Kinsella and Brehony 2009).

Moreover, landscape approaches are not explicitly pronounced in policies and laws in many African countries (Kimengsi et al. 2022), such as Tanzania, which affects the integration of landscape-level forest conservation efforts. Inadequate funding and capacity for institutions coordinating landscape natural resource management further compound the challenge (Rantala et al. 2011). The multiple and interconnected nature of landscapes requires a comprehensive approach to tackle them effectively (Sitas et al. 2014). However, landscape-level planning has not been widely implemented beyond conservation corridors, limiting the scope of management interventions (Nindi et al. 2019). Additionally, increasing human population pressure exacerbates demands on natural resources and the environment in rural landscapes (Yawson et al. 2021; Singh et al. 2023), further straining their resilience and sustainability (Rantala et al. 2011). Addressing these multifaceted challenges requires a holistic and collaborative approach, involving stakeholders at multiple levels and sectors to ensure the long-term viability of landscapes and resilience of people in Africa (Van Oosten et al. 2018).

13.5.2 Opportunities

The multifunctionality of landscapes in Africa presents numerous opportunities for holistic management and sustainable development. Firstly, adopting integrated landscape approaches promote comprehensive management that considers ecological, social, and economic objectives simultaneously (Neyret et al. 2023). By fostering collaboration among diverse stakeholders (Ros-Cuéllar et al. 2019), these approaches promote synergies and minimise trade-offs in land use decision-making, thereby enhancing the resilience and sustainability of landscapes (Nindi et al. 2019).

Secondly, participatory governance presents a valuable opportunity for engaging local communities, indigenous peoples, and other stakeholders in decision-making processes. By involving these groups, decision-making becomes more inclusive and participatory, empowering local communities to contribute their knowledge and perspectives to management initiatives (Wheeler and Root-Bernstein 2020). This approach fosters ownership and legitimacy of management interventions, leading to more effective and sustainable outcomes (Nindi et al. 2019).

Furthermore, exploring innovative financing mechanisms such as payments for ecosystem services, biodiversity offsets, and green bonds can provide financial incentives for sustainable landscape management practices (Yasué and Kirkpatrick 2020). These mechanisms not only generate revenue for conservation efforts but also incentivize the adoption of practices that enhance ecosystem services and biodiversity conservation. By aligning financial incentives with conservation objectives, these mechanisms contribute to the long-term sustainability of landscapes (Nindi et al. 2019). Notwithstanding, increasingly, there is another school of thought that posits that offsets can be dangerous whether carbon or biodiversity. It is argued that one cannot destroy an ecosystem and recreate the same elsewhere. Besides, green bonds have not been taken up like a 'wildfire' considering when they were first issued in 2007. Green bonds, it is argued, are another form of loans and generate huge amounts of debt, and therefore not conducive to debt ridden Africa. In fact, the mechanism favours private companies who benefit the most from all these credit systems. There is, thus, a need to re-look at the 'polluter pays principle' as it has no mechanism on the sharing of benefits to the communities. In fact, both carbon and biodiversity offset projects do not generate conservation finance but instead, generate profits for the private companies and some small amounts of money that go to the communities. This only encourages the wealthier nations to continue business as usual and claim they are offsetting their bad behaviour through offset projects.

Additionally, advances in technology offer promising opportunities for enhancing landscape management. Technologies such as remote sensing, geographic information systems (GIS), drones, and participatory mapping enable enhanced monitoring, assessment, and evaluation of landscapes (Hodbod et al. 2019). By providing real-time data and spatial analysis capabilities, these tools improve decision-making processes and facilitate more informed and evidence-based management interventions. Harnessing the power of technology can significantly enhance the effectiveness and efficiency of landscape management efforts in Africa (Nindi et al. 2019).

Moreover, the presence of supportive policies and initiatives such as REDD+, PFM, Village Land Use Planning, and AFR100 provide a conducive environment for landscape-level conservation and planning. These initiatives, coupled with the willingness of local and international organisations to promote and invest in landscape-level approaches, create favourable conditions for implementing integrated and sustainable landscape management strategies (Nindi et al. 2019). By leveraging these opportunities, stakeholders can work together to unlock the full potential of landscapes in Africa for the benefit of both current and future generations. Nonetheless, there are challenges associated with these initiatives for example REDD+ has become controversial in terms of forced removals, lack of safeguards, the unclear benefit-sharing mechanism (Birhan et al. 2021) and voluntary carbon market (Aryal et al. 2024). In the case of AFR 100, major challenges relate to restoration costs as compared to preventing forest degradation (Mansourian and Berrahmouni 2021). Overcoming these constraints and others could provide a pathway to accelerating the implementation of conservation initiatives at a large scale.

13.6 Conclusion

The concept of multifunctional landscapes has emerged as a promising approach to effectively manage natural resources, maximising ecosystem benefits, and addressing competing land uses in Africa. Landscape multifunctionality in Africa holds significant potential for achieving multiple ecological, social, cultural, and economic objectives simultaneously, thereby fostering sustainable and resilient landscapes capable of responding to climate change and meeting the needs of future generations. Also, the multifunctionality of landscapes is increasingly shaped by various socio-economic factors, including demographic and population dynamics, economic drivers and market forces, and social dynamics, cultural influences and governance structures. Efforts to ensure sustainable livelihoods require better landscape management strategies to address the challenges posed by increasing population pressures and climate change. Trees on landscapes play a pivotal role in this regard, offering opportunities for social-ecological, cultural, and economic benefits, thereby contributing to a "multiple win-win scenario" for livelihoods. Trees provide essential ecosystem services such as clean air, water regulation, soil fertility enhancement, and habitat provision, which are vital for supporting agricultural production and income generation.

13.7 Chapter Highlights

This chapter has examined the nuanced dynamics that shape multifunctional landscapes, acknowledging the pivotal role of socioeconomic context, policy directives, and governance structures.

The socioeconomic fabric weaves a complex narrative, influencing land use decisions, resource allocation, and community dynamics within multifunctional landscapes. Factors such as population growth, urbanization trends, income disparities, and cultural heritage aspirations interplay to shape the landscape mosaic. These drivers manifest on the ground, shaping the distribution of functions and the resilience of these landscapes in the face of socioeconomic pressures.

Meanwhile, policy interventions serve as the guiding compass, steering the trajectory of multifunctional landscapes towards sustainable outcomes. This factor highlights the importance of adaptive governance approaches that can navigate the complexities of multifunctional landscapes while fostering synergies among competing interests.

Embedded within the institutional architecture, governance systems also wield significant influence over the management and evolution of multifunctional landscapes. From local grassroots initiatives to national-level strategies, the role of governance structures in mediating conflicts, fostering collaboration, and ensuring equitable access to resources is underscored. It further underscores the need for inclusive governance mechanisms that empower stakeholders across diverse sectors and scales to co-create resilient landscapes that transcend narrow sectoral interests.

As the global landscape continues to evolve in response to rapid urbanization, climate change, and shifting socioeconomic dynamics, the chapter identifies emerging trends and looming challenges. It explores innovative approaches such as landscape-level planning, ecosystem services valuation, and digital technologies that hold promise in navigating the complexities of multifunctional landscapes in the twenty-first century.

In conclusion, this chapter underscores the intricate interplay of socio-economic factors, policy frameworks, and governance systems in shaping multifunctional landscapes. By embracing adaptive approaches that foster collaboration, equity, and resilience, societies can harness the full potential of these landscapes to meet the evolving needs of present and future generations.

References

Abesha N, Assefa E, Petrova MA, Seid S (2023) Farmers' perceptions of large-scale agricultural investment and its impacts on traditional ecological knowledge in Gololcha District, Southeastern Ethiopia. Geo J 88(1):985–999

Adesiji GB, Babalola FD (2012) Sacred forests: indigenous knowledge for conservation of forest at Ifo local government area of Ondo state. Nigeria Afr J Sustain Develop 2(1):117–134

Akrofi-Atitianti F, Speranza CI, Bockel L, Asare R (2018) Assessing climate smart agriculture and its determinants of practice in Ghana: a case of the cocoa production system. Land 7:30. https://doi.org/10.3390/land7010030

Amugune IM (2020) Technological factors influencing the quality and quantity of charcoal produced in Western Mau Forest, Kericho County, Kenya, Africa Nazarene University, Nairobi

Aryal K, Masaren T, Subedi PB, Laudari KH, Ghimire LP, Khanal CS, Zhang H, Timilsina R (2024) REDD+ at risk: emerging ten questions that REDD+ must answer. Environ Sci Policy 156:103744

Ba MN (2016) Analysis of agricultural commodities value chains and greenhouse gas emission in rice and maize in West Africa: impact on food security. Agric Sci 7:457–468

Babalola FD (2009) Modern religions versus traditional beliefs: threat to wilderness conservation in Nigeria. In: Alan W, Joaquin M-S, Brooke M (eds). Proceedings of world wilderness congress symposium: 125-129

Babalola FD, Lawal I, Opii EE, Oso AO (2014) Roles of and threats to indigenous cultural beliefs in sacred forests of Southwest Nigeria. Albanian J Agricult Sci 13(2):41–50

Banadda N (2010) Gaps, barriers and bottlenecks to sustainable land management (SLM) adoption in Uganda. Afr J Agric Res 5(25):3571–3580

Bansard J, Schroder M (2021) The sustainable use of natural resources: the governance challenge, international institute for sustainable development. https://www.iisd.org/articles/deep-dive/sustainable-use-natural-resources-governance-challenge Accessed on 17 Mar 2024

Barrow E, Mlenge W (2003) Trees as key to pastoralist risk management in semi-arid landscapes in Shinyanga, Tanzania and Turkana, Kenya. In: Proceedings of the international conference on rural livelihoods, forests and biodiversity, Bonn, Germany, 19–23 May 2003, pp 19–23

Berkes F, Gardner JS, Sinclair AJ (2000) Comparative aspects of mountain land resources management and sustainability: case studies from India and Canada. Int J Sust Dev World 7(4):375–390

Bhagwat SA, Rutte C (2006) Sacred groves: potential for biodiversity management. Front Ecol Environ 4(10):519–524

Birhan E, Assefa E, Petrova MA (2021) Challenges of forest governance in addressing redd+: status, effects and prospects. The case of bale eco-region, Oromia regional state, Ethiopia. Geogr Environ Sustain 14(1):185–195. https://doi.org/10.24057/2071-9388-2020-108

Birkhofer K, Andersson GKS, Bengtsson J, BommarcoR, Dänhardt, J Ekbom B, Ekroos J, Hahn T, Hedlund K, Jönsson AM et al. (2018) Relationships between multiple biodiversity components and ecosystem services along a landscape complexity gradient. Biol Conserv 218: 247–253

Bourne M, Kimaiyo J, Tanui J, Catacutan D, Otiende V (2015) Can gender appreciation of trees enhance landscape multifunctionality? A case of smallholder farming systems on Mount Elgon. Int For Rev 17(4):33–45

Byers BA, Cunliffe RN, Huda AT (2001) Linking the conservation of culture and nature: a case study of sacred forests in Zimbabwe. Human Ecol 29:187–218

Campbell BM, Luckert M, Mutamba M, Zindi C, Kozanayi W (2002) Household livelihoods in semi-arid regions: options and constraints. CIFOR, Bogor

Chambers R, Conway G (1991) Sustainable rural livelihoods: practical concepts for the 21st century. In: IDS discussion paper 296. IDS, Brighton, UK

Chemhuru M, Masaka D (2010) Taboos as sources of Shona people's environmental ethics. J Sustain Develop Afr 12(7):121–133

Chirwa PW (2020) Strengthening capacity of African stakeholders to integrate and uptake forest and farm tree-based adaptation and mitigation options in response to climate change in Anglophone and Lusophone Africa. AFF Report, Nairobi

Chirwa PW, Larwanou M, Syampungani S, Babalola FD (2015) Management and restoration practices in degraded landscapes of eastern Africa and requirements for up-scaling. Int For Rev 17(3):20–30

Colding J, Folke C (2001) Social taboos: invisible systems of local resource management and biological conservation. Ecol Appl 11(2):584–600

Colfer CJP, Pfund JL (2011) Collaborative governance of tropical landscapes. Routledge, 302 p

Colfer CJP, Sheil D, Kishi M (2008) Human health and forests: a global overview of issues. Practice and policy. Earthscan, London

Constanza R, d'Arge R, de Groot RS, Farber S, Grasso M, Hannon B, Limburg K, Naeem S, O'Neill RV, Paruelo J, Raskin RG, Sutton P, van den Belt M (1997) The value of the world's ecosystem services and natural capital. Nature 387:253–260

Dawson IK, Leakey R, Clement CR, Weber JC, Cornelius JP, Roshetko JM et al (2014) The management of tree genetic resources and the livelihoods of rural communities in the tropics: non-timber forest products, smallholder agroforestry practices and tree commodity crops. Forest Ecol Manage 333:9–21. https://doi.org/10.1016/j.foreco.2014.01.021

Department for International Development (DFID) (1999) Sustainable livelihoods framework: guidance sheets, Department for International Development (DFID), London, UK

Djenontin INS, Foli S, Zulu LC (2018) Revisiting the factors shaping outcomes for forest and landscape restoration in sub-Saharan Africa: a way forward for policy, practice and research. Sustain For 10(4):906. https://doi.org/10.3390/su10040906

Doggart N, Morgan-Brown T, Lyimo E, Mbilinyi B, Meshack KC, Sallu SM, Spracklen DV (2020) Agriculture is the main driver of deforestation in Tanzania. Environ Res Lett 15:1–9

Ekblom A, Shoemaker A, Gillson L, Lane P, Lindholm K (2019) Conservation through biocultural heritage—examples from sub-Saharan Africa. Land 8(1):5. https://doi.org/10.3390/land8010005

Ellegård A, Nordström M (2013) Deforestation for the poor? Renew Energy Develop 16(2):4–6

FAO (2008) Tenure security for better forestry: understanding forest tenure in Africa. FAO, Rome. http://www.fao.org/docrep/011/i0352e/i0352e00.htm

FAO (2014) State of the world's forests 2014: enhancing the socioeconomic benefits from forests. FAO, Rome. http://www.fao.org/3/i3710e/i3710e.pdf

FAO (2022) The state of the World's forests 2022: Forest pathways for green recovery and building inclusive, resilient and sustainable economies. FAO, Rome. https://www.fao.org/3/cb9360en/cb9360en.pdf

Gilmour D (2016) Forty years of community-based forestry: a review of its extent and effectiveness. FAO, Rome. http://www.fao.org/3/bi5415e.pdf

Hedden-Dunkhorst B, Schmitt F (2020) Exploring the potential and contribution of UNESCO biosphere reserves for landscape governance and management in Africa. Land 9(8):237. https://doi.org/10.3390/land9080237

Hens L (2006) Indigenous knowledge and biodiversity conservation and management in Ghana. J Hum Ecol 20(1):21–30

Hodbod J, Tebbs E, Chan K, Sharma S (2019) Integrating participatory methods and remote sensing to enhance understanding of ecosystem service dynamics across scales. Land 8(9):132

Hölting L, Beckmann M, Volk M, Cord AF (2019a) Multifunctionality assessments—more than assessing multiple ecosystem functions and services? A quantitative literature review. Ecol Indic 103:226–235. https://doi.org/10.1016/j.ecolind.2019.04.009

Hölting L, Felipe-Lucia R, Cord AF (2019b) Multifunctional landscapes. In: Goldstein MI, DellaSala DA (eds). Elsevier, Encyclopedia of the world's biomes, pp 128–134. https://doi.org/10.1016/B978-0-12-409548-9.12098-6

Huber L, Schirpke U, Marsoner T, Tasser E, Leitinger G (2020) Does socioeconomic diversification enhance multifunctionality of mountain landscapes? Ecosyst Serv 44:101122. https://doi.org/10.1016/j.ecoser.2020.101122

Iiyama M, Neufeldt H, Njenga M, Derero A, Ndegwa GM, Mukuralinda A, Dobie P, Jamnadass R, Mowo J (2017) Conceptual analysis: the charcoal-agriculture nexus to understand the socio-ecological contexts underlying varied sustainability outcomes in African landscapes. Front Environ Sci 5:31. https://doi.org/10.3389/fenvs.2017.00031

IPBES (2019) Media Release: nature's dangerous decline 'unprecedented'; Species Extinction Rates 'Accelerating': Media release. 7 May, 2019. https://ipbes.net/news/Media-Release-Global-Assessment

Jepson P, Canney S (2003) Values-led conservation. Glob Ecol Biogeogr 12:271–274

Jimoh SO, Ikyaagba ET, Alarap AA, Obioha EE, Adeyemi AA (2012) The role of traditional Laws and Taboos in wildlife conservation in the Oban Hill sector of Cross River National Park (CRNP). Nigeria J Human Ecol 39(3):209–219

Kato S, Ahern J (2009) Multifunctional landscapes as a basis for sustainable landscape development. J Japanese Instit Lands Architect 72:799–804. https://doi.org/10.5632/jila.72.799

Kayambazinthu D, Matose F, Kajembe G, Nemarundwe N (2003) Institutional arrangements governing natural resource management of the miombo woodland. In: Campbell BM, Sumaila UR (eds) Kowero G. Policies and governance structures in woodlands of Southern Africa, CIFOR, pp 45–64

Kengoum F, Pham TT, Ntirumenyerwa MBP (2024) REDD+ benefit sharing mechanisms in The Democratic Republic of the Congo: legal and institutional frameworks, policy implementation, and project experiences. Working paper 27. CIFOR/Nairobi, Bogor/Kenya

Kimengsi JN, Owusu R, Balgah RA (2022) Nexus approach and environmental resource governance in sub-Saharan Africa: a systematic review. Sustain Sci 17(3):1091–1108

Kinsella J, Brehony E (2009) Are current aid strategies marginalising the already marginalised? Cases Tanzania Develop Pract 19(1):51–60. http://www.jstor.org/stable/27752009

Kirschbaum MU, Cowie AL, Peñuelas J, Smith P, Conant T, Sage RF, Brandão M, Cotrufo MF, Luo Y, Way DA, Robinson SA (2024) Is tree planting an effective strategy for climate change mitigation? Sci Total Environ 909:168479. https://doi.org/10.1016/j.scitotenv.2023.168479

Kouadio I, Singh R (2021) Deforestation and threat to biodiversity in developing countries: case of Ivory Coast. Wesleyan J Res 14(07):32–43

Kozanayi W (2018) Influences of customary and statutory governance on sustainable use and livelihoods: the case of baobab, Chimanimani District, Zimbabwe. Unpublished PhD thesis, University of Cape Town, South Africa

Kremen C, Merenlender AM (2018) Landscapes that work for biodiversity and people. Science 362(6412):eaau6020. https://doi.org/10.1126/science.aau6020

Lawrence SH (2003) Forest and tree conservation through metaphysical constraints. Forest Tree Conserv 19(3):57–78

Liu JL (2006) Traditional knowledge and implication to development from anthropological perspective. Acad J China's Agricult Univ 2:133–141

Liyama M, Neufeldt H, Njenga M, Derero A, Ndegwa GM, Mukuralinda A, Dobie P, Jamnadass R, Mowo J (2017) Conceptual analysis: the charcoal-agriculture nexus to understand the socio-ecological contexts underlying varied sustainability outcomes in African landscapes. Front Environ Sci 5:31. https://doi.org/10.3389/fenvs.2017.00031

Louman B, Fischlin A, Peter G et al (2009) Forest ecosystem services: a cornerstone for human Well-being. In: Risto S, Alexander B, Pia K (eds) Adaptation of forests and people to climate change. A global assessment report, vol 22. IUFRO. World Series, Helsinki, pp 15–27

Loza J (2023) Engaging local communities in conservation and governance of the world heritage in a transnational context: an example from the maloti-Drakensberg Park world heritage site (Lesotho/South Africa). In: Houehounha D, Moukala E (eds) Managing transnational UNESCO world heritage sites in Africa. Springer, Cham, pp 47–56. https://doi.org/10.1007/978-3-030-80910-2_5

Maier L, Shōbayashi M (2001) Multifunctionality: towards an analytical framework. Organisation for Economic Co-operation and Development (OECD) Publications, Paris, p 159p

Maier C, Lindner T, Winkel G (2014) Stakeholders' perceptions of participation in forest policy: a case study from Baden-Württemberg. Land Use Policy 39:166–176

Malmborg K, Sinare H, Kautsky EE, Ouedraogo I, Gordon LJ (2018) Mapping regional livelihood benefits from local ecosystem services assessments in rural Sahel. PLoS One 13(2):e0192019. https://doi.org/10.1371/journal.pone.0192019

Mansourian S, Berrahmouni N (2021) Review of forest and landscape restoration in Africa. FAO and AUDA-NEPAD, Accra. https://doi.org/10.4060/cb6111en

Mashingaidze N, Chirisa I, Mutambisi T, Matamanda AR (2021) Paradoxes surrounding carbon credits and local area development: the case of Mbire District. Zimbabwe Local Environment 26(10):1175–1185. https://doi.org/10.1080/13549839.2021.1967899

Mastrangelo M, Weyland EF, Villarino SH, Barral MP, Nahuelhual L, Laterra P (2014) Concepts and methods for landscape multifunctionality and a unifying framework based on ecosystem services. Landsc Ecol 29(2):345–358. https://doi.org/10.1007/s10980-013-9959-9

Mbaiwa JE (2017) Poverty or riches: who benefits from the booming tourism industry in Botswana? J Contemp Afr Stud 35(1):93–112. https://doi.org/10.1080/02589001.2016.1270424

Mgumia F H Oba G(2003) Potential role of sacred groves in biodiversity conservation in Tanzania. Environ Conserv 30:259–265. https://doi.org/10.1017/S037689

Millennium Ecosystem Assessment (2005) Ecosystems and human well-being: synthesis. Island Press, Washington, DC

Minang PA, Van NM, Freeman OE, Mbow C, de Leeuw J, Catacuta D (2015) Climate-smart landscapes: multifunctionality in practice world. Agroforestry Centre (ICRAF), Nairobi

Moller H, Berkes F, Lyver PO, Kislalioglu M (2004) Combining science and traditional ecological knowledge: monitoring populations for co-management. Ecol Soc 9(3):2. http://www.ecologyandsociety.org/vol9/iss3/art2/

Msuya TS, Kideghesho JR (2009) The role of traditional management practices in enhancing sustainable use and conservation of medicinal plants in West Usambara Mountains. Tanzania Trop Conserv Sci 2(1):88–105

Muluneh MG (2021) Impact of climate change on biodiversity and food security: a global perspective – a review article. Agricult Food Secur 10:36

Muthuri CW, Kuyah S, Njenga M, Kuria A, Öborn I, Van Noordwijk M (2023) Agroforestry's contribution to livelihoods and carbon sequestration in East Africa: a systematic review. Trees Forests People 14:100432. https://doi.org/10.1016/j.tfp.2023.100432

Mwampamba TH, Ghilardi A, Sander K, Chaix KJ (2013) Dispelling common misconceptions to improve attitudes and policy outlook on charcoal in developing countries. Energy Sustain Dev 17:158–170. https://doi.org/10.1016/j.esd.2013.01.001

Neyret M, Peter S, Le Provost G, Boch S, Boesing AL, Bullock JM, Hölzel N, Klaus VH, Kleinebecker T, Krauss J, Müller J (2023) Landscape management strategies for multifunctionality and social equity. Nature Sustain 6(4):391–403

Nindi S, Flintan F, Kalenzi D, Mwita V, Luambano I (2019) Landscape approach in addressing land use and tenure arrangements among pastoral communities in Eastern Africa: The case of Tanzania. National Land Use Planning Commission (NLUPC), International Livestock Research Institute (ILRI), Ministry of Livestock and Fisheries (MOLF), Tanzania, Paper prepared for presentation at the 2018 World bank conference on land and poverty. The World Bank – Washington DC, March 25–29, 2019

Nyarko I, Nwaogu C, Miroslav H, Peseu PO (2021) Socio-economic analysis of wood charcoal production as a significant output of forest bioeconomy in Africa. Forests 12(5):568. https://doi.org/10.3390/f12050568

Nzyoka J, Minan PA, Wainaina P, Duguma L, Manda L, Temu E (2021) Landscape governance and sustainable land restoration: evidence from Shinyanga. Tanzania Sustain 13:7730. https://doi.org/10.3390/su13147730

Odusola A (2021) Agriculture as the fulcrum of inclusive development in Africa. In: Odusola A (ed) Africa's agricultural renaissance: from paradox to powerhouse. Palgrave Macmillan, Cham, pp 15–54. https://doi.org/10.1007/978-3-030-65748-2_2

Oljirra A (2019) The causes, consequences and remedies of deforestation in Ethiopia. J Degraded Mining lands Manage 6(3):1747–1754

Pataki DE, Alberti M, Cadenasso ML, Felson AJ, McDonnell MJ, Pincetl S, Whitlow TH (2021) The benefits and limits of urban tree planting for environmental and human health. Front Ecol Evol 9:603757. https://doi.org/10.3389/fevo.2021.603757

Peterson JM, Caldas MM, Bergtold JS, Sturm BS, Graves RW, Earnhart D, Hanley EA, Brown JC (2014) Economic linkages to changing landscapes. Environ Manag 53:55–66

Rantala S, Lyimo E, Powell B, Kitalyi A, Vihemäki H (2011) Challenges and opportunities for collaborative landscape governance in the east Usambara Mountains, Tanzania. ICRAF working paper no. 127 and CIFOR working paper no 61. Nairobi Bogor

Reyers B, O'Farrell PJ, Nel JL, Wilson K (2012) Expanding the conservation toolbox: conservation planning of multifunctional landscapes. Landsc Ecol 27:1121–1134. https://doi.org/10.1007/s10980-012-9761-0

Ribot JC (2002) African decentralization: local actors, powers and accountability. UNRISD Programme on democracy, governance and human rights paper no 8. UNRISD and IDRC, Geneva

Ribot J (2022) Violent silence: framing out social causes of climate-related crises. J Peasant Stud 49(4):683–712. https://doi.org/10.1080/03066150.2022.2069016

Ribot J, Peluso LN (2003) A theory of access. Rural Sociol 68(2):153–181

Ros-Cuéllar J, Porter-Bolland L, Bonilla-Moheno M (2019) Can multifunctional landscapes become effective conservation strategies? Challenges and opportunities from a Mexican case study. Land 8(1):6

Rose J, Bensch G, Munyehirwe A, Peters J (2022) The forgotten coal: charcoal demand in sub-Saharan Africa. World Development Perspectives 25:100401. https://doi.org/10.1016/j.wdp.2022.100401

Saj TL, Mather C, Sicotte P (2006) Traditional taboos in biological conservation: the case of Colobus vellerosus at the Boabeng-Fiema monkey sanctuary. Central Ghana Soc Sci Inform 45(2):285–310

Salick J, Amend A, Anderson D et al (2007) Tibetan sacred sites conserve old growth trees and cover in the eastern Himalayas. Biodivers Conserv 16:693–706

Schoffeleers JM (1979) Territorial cults in Central Africa. In: Schoffeleers JM (ed) Guardians of the land: essays on central African territorial cults. Mambo Press, Gweru

Shackleton CM (2015) Non-timber forest products in livelihoods. In: Shackleton CM, Pandey AK, Ticktin T (eds) Ecological sustainability for non-timber forest products: dynamics and case studies of harvesting. Earthscan from Routledge, London, pp 10–29

Shereni NC, Saarinen J (2021) Community perceptions on the benefits and challenges of community-based natural resources management in Zimbabwe. Dev South Afr 38(6):879–895. https://doi.org/10.1080/0376835X.2020.1796599

Shivanna KR (2022) Climate change and its impact on biodiversity and human welfare. Proceedings of the Indian National Science Academy. Part A Phys Sci 88(2): 160–171. https://doi.org/10.1007/s43538-022-00073-6

Singh S, Sharma GD, Radulescu M, Balsalobre-Lorente D, Bansal P (2023) Do natural resources impact economic growth: an investigation of P5+ 1 countries under sustainable management. Geosci Front 15(3):101595

Sitas N, Prozesky HE, Esler KJ, Reyers B (2014) Opportunities and challenges for mainstreaming ecosystem services in development planning: perspectives from a landscape level. Landsc Ecol 29:1315–1331

Soutter R, Ntiamoa-Baidu Y, Smith J, Rana D (2003) Recognising the contribution of sacred natural sites for biodiversity conservation. World Parks Congress, Durban

Stoldt M, Göttert T, Mann C, Zeller U (2020) Transfrontier conservation areas and human-wildlife conflict: the case of the Namibian component of the Kavango-Zambezi (KAZA) TFCA. Sci Rep 10:7964. https://doi.org/10.1038/s41598-020-64537-9

Stürck J, Verburg PH (2017) Multifunctionality at what scale? A landscape multifunctionality assessment for the European Union under conditions of land use change. Landsc Ecol 32:481–500. https://doi.org/10.1007/s10980-016-0459-6

Sunderland T, Abdoulaye R, Ahamma R, Asah S, Baudron F, Deakin E, Duriauxs J, Eddy I, Foli S, Gumbo D, Khatun K, Kondwani M, Kshatriya M, Leonald L, Rowland D, Stacey N, Tomscha S, Yan K, Gergel S, Vianen JV (2017) A methodological approach for assessing cross-site landscape change: understanding socio-ecological systems. Forest Policy Econ 84:83–91

Tait J (2016) Science, governance and multifunctionality of European agriculture. Outlook Agr 30:91–95. https://doi.org/10.5367/000000001101293517

Uisso AJ (2022) Nexus of village land use planning, village land forest reserves and REDD+ in Tanzania: a synthesis of concepts. SN Soc Sci 2(5):70. https://doi.org/10.1007/s43545-022-00376-x

Uisso AJ, Chirwa PW, Ackerman PA, Mbwambo L (2018) The role of village land Forest reserves in the implementation of land use plans: experience from the REDDþInitiative in Tanzania. Int For Rev 20(2):236–249. https://doi.org/10.1505/146554818823767591

Uisso AJ, Chirwa PW, Ackerman PA, Mbwambo L (2019) Forest management and conservation before and after the introduction of village participatory land use plans in the Kilosa district REDD+ initiative. Tanzania J Sustain Forest 38(2):97–115

UNEP (2019) Sand and sustainability: finding new solutions for environmental governance of global sand resources. GRID-Geneva, United Nations Environment Programme, Geneva

UNU-IAS, IR3S/UTIAS (2016) Socio-ecological production landscapes and seascapes (SEPLS) in Africa. United Nations University Institute for the Advanced Study of Sustainability, Tokyo

Van der Duim R, Pellis A (2022) Landscape governance in sub-Saharan Africa. In: Stone LS, Stone MT, Mogomotsi PK, Mogomotsi GEJ (eds) Protected areas and tourism in southern Africa, 1st edn. Routledge, pp 73–86

Van Oosten C, Uzamukunda A, Runhaar H (2018) Strategies for achieving environmental policy integration at the landscape level. A framework illustrated with an analysis of landscape governance in Rwanda. Environ Sci Policy 83:63–70

Villamor GB, Palomo I, Santiago CAL, Oteros-Rozas E, Hill J (2014) Assessing stakeholders' perceptions and values towards social-ecological systems using participatory methods. Ecol Process 3:1–12

Waddington H, White H (2014) Farmer field schools: from agricultural extension to adult education. Systematic review summary 1. International Initiative for Impact Evaluation (3ie), London. https://3ieimpact.org/evidence-hub/publications/systematic-review-summaries/farmer-field-schools-agricultural-extension

Wheeler HC, Root-Bernstein M (2020) Informing decision-making with indigenous and local knowledge and science. J Appl Ecol 57(9):1634–1643

Wiggering H, Müller K, Werner A, Helming K (2003) The concept of multifunctionality in sustainable land development Helming K. In: Wiggering H (ed) Sustainable development of multifunctional landscapes. Springer, Berlin/Heidelberg, pp 3–18. https://doi.org/10.1007/978-3-662-05240-2_1

World Bank (2007) At loggerheads: agricultural expansion, poverty reduction, and environment in the tropical forests. A world bank policy research report. World Bank Group, Washington, DC

WRI (2005) World resources 2005. The wealth of the poor: managing ecosystems to fight poverty. UNEP/UNDP/WRI? World Bank. https://files.wri.org/d8/s3fs-public/pdf/wrr05_full_hires.pdf. Accessed 22 Apr 24

Wynberg R, Laird S, Van Nieker J, Kozanayi W (2015) Formalisation for natural product trade in Southern Africa: unintended consequences and policy blurring in bio trade and bioprospecting. Soc Nat Resour J 28(5):559–574

Xiao H, Liu J, He G, Zhang X, Wang H, Long T, Zhang Z, Wang W, Yin R, Guo Y, Cheng B (2022) Data-driven forest cover change and its driving factors analysis in Africa. Front Environ Sci 9:780069

Yasué M, Kirkpatrick JB (2020) Do financial incentives motivate conservation on private land? Oryx 54(4):499–510

Yawson DO, Adu MO, Asare PA, Armah FA (2021) Multifunctional landscape transformation of urban idle spaces for climate resilience in sub-Saharan Africa. In: Oguge N, Ayal D, Adeleke L, da Silva I (eds) African handbook of climate change adaptation. Springer, Cham, pp 2193–2219. https://doi.org/10.1007/978-3-030-45106-6_214

Chapter 14
Securing Multifunctional Landscapes in Africa: What Is the Future Policy, Research and Development Direction?

Paxie W. Chirwa, Stephen Syampungani, Theodore M. Mwamba, Kokou Kouami, Tajudeen O. Amusa, Jonathan M. Kamwi, Folaranmi D. Babalola, Simon A. Mng'omba, Abdullahi A. Yusuf, Gillian Kabwe, Eméline S. P. Assèdé, Gudeta W. Sileshi, Paul Thangata, and Lackson Chama

Abstract Rural populations in sub-Saharan Africa (SSA) face livelihood challenges due to limited opportunities, and increasing dependency on vulnerable landscapes. These challenges are exacerbated by population growth and the consequent ecological footprint, primarily driven by agricultural expansion, leading to significant landscape changes thus affecting tropical forests and semi-arid environments. This book explores the concept of multifunctional landscapes and tree-based

P. W. Chirwa (✉)
Department of Plant and Soil Sciences, Faculty of Natural and Agricultural Sciences, University of Pretoria, Pretoria, South Africa
e-mail: paxie.chirwa@up.ac.za

S. Syampungani
Department of Plant and Soil Sciences, Faculty of Natural and Agricultural Sciences, University of Pretoria, Pretoria, South Africa

Chair of Environment and Development, Oliver R. Tambo Africa Research Chairs Initiative (ORTARChI), Copperbelt University, Kitwe, Zambia

T. M. Mwamba
Chair of Environment and Development, Oliver R. Tambo Africa Research Chairs Initiative (ORTARChI), Copperbelt University, Kitwe, Zambia

Faculty of Agriculture, University of Lubumbashi, Lubumbashi, Democratic Republic of the Congo

K. Kouami
Laboratoire de Recherche Forestière, Centre de Recherche sur le Changement Climatique (CRCC), Université de Lomé, BP, Lomé, Togo

T. O. Amusa · F. D. Babalola
Department of Forest Resources Management, Faculty of Agriculture, University of Ilorin, Ilorin, Nigeria

P. W. Chirwa et al. (eds.), *Trees in a Sub-Saharan Multi-functional Landscape*,
https://doi.org/10.1007/978-3-031-69812-5_14

ecosystem approaches, highlighting the need for employing sustainable landscape management practices to maximize ecosystem services and ensure equitable benefit distribution.

The book details how the integration of traditional and contemporary agroforestry practices enhances local food security and nutrition, health care provision, and income, identifying a diversity of tree species/taxonomic groups of interest. It emphasizes the role of trees in biodiversity conservation, climate change mitigation and restoration of degraded landscapes while addressing the socio-economic challenges of landscape management. Key tree-based ecosystem management practices such as tree domestication, natural regeneration of trees, assisted or enrichment planting, fire management, sustainable harvesting of non-timber forest products (NTFPs) and agroforestry systems and technologies, are discussed to provide a basis for innovation in integrating trees in multifunctional landscapes in different biophysical contexts across SSA. Ways of overcoming various ecological and socio-economic challenges of landscape management are also discussed, including bio-control strategies against invasive species and approaches for quality tree germplasms provisioning. The book underscores the need to integrate scientific and indigenous knowledge, and develop adaptive policy and governance to support sustainable landscape management while fostering cultural resilience. It emphasizes the critical

J. M. Kamwi
Department of Natural Resource Sciences, Namibia University of Science and Technology, Windhoek, Namibia

S. A. Mng'omba
Lilongwe University of Agriculture and Natural Resources (LUANAR), Lilongwe, Malawi

A. A. Yusuf
Department of Zoology and Entomology, University of Pretoria, Pretoria, South Africa

Forestry and Agricultural Biotechnology Institute, University of Pretoria, Pretoria, South Africa

G. Kabwe
School of Natural Resources, Plant and Environmental Sciences Department, Copperbelt University, Kitwe, Zambia

E. S. P. Assèdé
Department of Plant and Soil Sciences, Faculty of Natural and Agricultural Sciences, University of Pretoria, Pretoria, South Africa

Département Aménagement et Gestion des Ressources Naturelles, Faculté d'Agronomie, Université de Parakou, Parakou, Bénin

G. W. Sileshi
Department of Plant Biology and Biodiversity Management, Addis Ababa University, Addis Ababa, Ethiopia

P. Thangata
AGRA, West End Towers, Nairobi, Kenya

L. Chama
Department of Zoology and Aquatic Sciences, School of Natural Resources, Copperbelt University, Kitwe, Zambia

role of landscape-scale interventions and large-scale stakeholders' collaboration through the Community-based natural resource management approach. Notably, the book recognizes the importance of establishing incentives to bolster community engagement, essential to achieving sustainable outcomes.

Future research directions identified include (i) deepening the understanding of ecosystem services provided by trees and the underlying functional traits, (ii) developing innovative practices for sustainable agroforestry practices, (iii) promoting climate-smart agriculture, and (iv) developing more rigorous approaches in assessing both the socio-economic benefits of trees in SSA and the efficacy of policy interventions for sustainable landscape management. By prioritizing holistic and collaborative approaches, stakeholders can ensure the resilience and vitality of SSA's landscapes, thereby supporting rural livelihoods and biodiversity conservation for future generations.

Keywords Sub-Saharan Africa · Rural livelihoods · Landscape multifunctionality · Tree-based ecosystem approaches · Policy development

14.1 Introduction

Rural populations in sub-Saharan Africa (SSA), like those of other rural areas in the developing world, have limited livelihood opportunities (Kraybill 2013). This is generally regarded as one of the major causes of poverty in the region. The direct link of people's livelihoods to landscapes in which they live, implies the vulnerability of rural communities to changes in these landscapes (Suich et al. 2015). The high rate of population growth, associated with the need for expansion of agricultural activities to feed the growing population has been identified as the key driver of landscape change in the region, with a massive footprint on tropical forest and semi-arid environments recorded across the region. Though the majority of population in SSA is engaged in agriculture, this region remains one of the poorest in the world (Yengoh 2020; Baah and Lakner 2023). In recent years, livelihood diversification has emerged as a promising approach towards reducing poverty in SSA (Loison 2015). Therefore, sustainable landscape management to maximize the provisioning of different ecosystem services while maintaining equitable distribution of benefits across different stakeholders (Neyret et al. 2023) is necessary.

The book begins with the provision of the definition and concept of multifunctional landscapes in Africa. In the process, it highlights the merits of the different landscape approaches with a recognition that no approach can address the needs of all existing African landscapes (Chap. 2). Different tree-based ecosystem approaches and technologies are presented in which trees provide a broad range of ecosystem services (Chap. 3) that often go beyond their specific habitat, e.g., habitat for pollinators, corridors for wildlife mobility and contribution to the local and global climate systems. Traditional tree-based ecosystem management practices across different African landscapes such as the tropical forests, savanna woodlands and

Sahel are highlighted. The selected practices presented are neither exhaustive nor meant to be replicated or taken as a blueprint for successful multifunctional landscape approaches, but rather intended to provide a basis for innovation to integrate trees in multifunctional landscapes. These different approaches and systems cover practices that provide benefits in terms of biodiversity conservation (Chaps. 3 and 12), climate change mitigation and adaptation (Chaps. 3, 5, and 12), rehabilitation of degraded lands (Chaps. 5, 6, and 12) and management of biological invasions (Chap. 11). Approaches for sustainable harvesting of forest resources including timber and non-timber products are also discussed (Chaps. 4, 8, and 9). Alternative approach involving domestication of indigenous trees and associated resources are also highlighted (Chaps. 8, 9, and 10). The book also focuses on the implications of socio-ecological factors (Chap. 12) and socioeconomic drivers (Chap. 13) on multifunctional landscapes, in relation to shifts in land use and land cover (Chap. 7), which shape the classification, systems, structure, and functionality of trees within landscapes, affecting their ability to provide ecosystem services and support sustainable livelihoods.

14.2 What About Trees in a Multifunctional Landscape in Sub-Saharan Africa?

The concept of multifunctional landscapes in the African context, draws from scientific literature to emphasize the need for diverse approaches necessary for addressing the varied needs of African ecosystems. As presented in Chap. 2, it underscores the importance of a people-centred approach, aligning with principles advocated in scholarly works such as "Multifunctional Landscapes: Interconnecting Sustainable Agriculture, Rural Development, and Ecosystem Services" by Scherr and McNeely (2008). This approach aims to achieve biodiversity conservation, climate change mitigation, food security, and poverty alleviation, reflecting the findings of studies like "Linking Sustainable Agriculture and Biodiversity: Implications for Multifunctional Landscape Management" by Tscharntke et al. (2012). The prospects of managing landscapes to serve multiple functions while prioritizing the needs of local communities is emphasized, aligning with recommendations put forth in scholarly works like "People and Trees: The Role of Social Forestry in Sustainable Development" by Arnold and Dewees (1995). In areas where communities are faced with various emerging environmental challenges, the health and prosperity of people are greatly impacted. To address this, the potential application of nature-based solutions is gaining wide recognition at global level (SETAC 2019). Trees are considered as an essential component for their contribution towards creating landscapes that provide multiple services and functions, improving the health and livelihoods of communities (Konijnendijk 2020). As countries in SSA work towards achieving sustainable development, it is important to realize the role of trees in ensuring that the landscape contributes towards rural development. SSA is associated with a diversity of land uses including forests, woodlands, agricultural and rocky outcrops

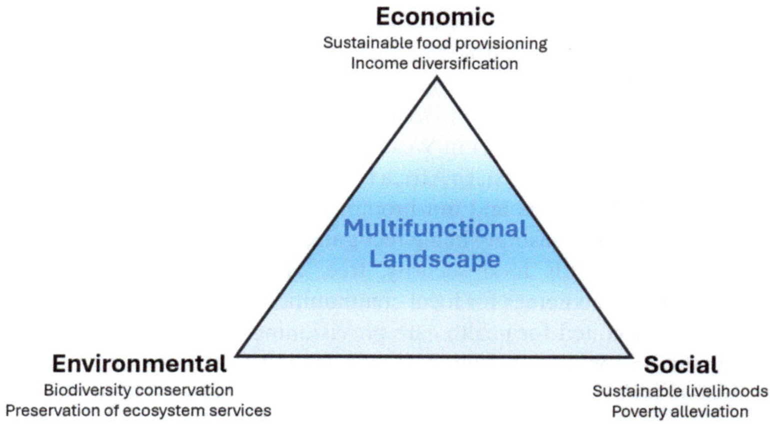

Fig. 14.1 Multifunctional landscape framework showing the links between sustainable food provisioning, income diversification (economic), biodiversity conservation, preservation of ecosystem services (environmental), and sustainable livelihoods, poverty alleviation (social) dimensions

landscapes, with each of these contributing towards sustainable livelihoods provisioning.

From the different chapters presented in the book, we categorize these within the three pillars of sustainability as well encapsulated in the framework expounded by Lovell and Johnston (Lovell and Johnston 2009, Fig. 14.1). The framework for transforming landscapes to improve performance is thus achieved by integrating ecological principles into landscape design focusing on the development of multifunctional landscapes, guided by the rapidly growing knowledge base of ecosystem services provided by landscape features.

14.2.1 Socioeconomic: Productive Aspects of Trees in the African Multifunctional Landscape

A diversity of tree species has been recorded for enhancing local household nutrition in SSA. As described in Chaps. 3, 5, 6, 8, 9 and 12, trees and associated nontimber forest products (NTFPs) offer a vital source of incomes and livelihood for local communities, supporting millions of people in SSA. In particular, exploitation of wild fruits and edible roots have been observed as a subsistence strategy towards ensuring food security, especially in dry season, for a large proportion of resource-poor populations in rural areas (Chaps. 5 and 9). They are basically used as a source of nutrients (essential minerals, vitamins, antioxidants) to balance the dietary requirements that are mostly dominated by nutrient-deficient cereal-based diets (Akinnifesi et al. 2006; Sileshi et al. 2023a). Thus, livelihoods are improved at local household level through income diversification by exploiting landscape resources beyond cereal-based agriculture. For example, on a larger scale, NTFPs sector contribute significantly to increased revenue of rural communities through enhanced employment opportunities arising from trade in indigenous fruit products such as Marula (*Sclerocary birrea (A.Rich.) Hochst.*) fruit in Southern Africa, Shea butter (*Vitellaria paradoxa C.F.Gaertn.*) in West Africa, and Allan blackia (*Allanblackia floribunda* Oliv.) products in Eastern Africa (Chap. 9). Likewise, the socio-economic contribution of edible insects and mushrooms to both the national and local economic well-being are increasingly being recognized in the region (Chap. 8). Besides their contribution towards food security, tree and tree-associated products are important sources of medicines for local communities (Chaps. 3 and 8). Various tree parts have been exploited for health care provisioning for a diversity of infections such as stimulating appetite, treating digestive, intestinal, urinary respiratory, reproductive system skin infections and malaria. Similarly, therapeutic properties have also been observed in African mushrooms and other non-timber forest products (Bastos et al. 2023, Chap. 8). There is promising medicinal potential for African mushrooms and other non-timber forest products as these have been observed to be valuable assets in therapeutic applications (Sileshi et al. 2023b). Recently, the use of edible insects for therapeutic purposes has also been revealed (Tanga and Ekesi 2024). Additionally, SSA landscapes are important for provisioning of other non-timber forest products, such as construction material and utensils e.g., from the use of bamboo (*Bambusa spp*), jute (*Corchorus spp.*), *Ischyrolepis eleocharis* (Mast.) H.P.Linder (Chap. 8). Likewise, a specific plant genus (drosera) from rocky outcrop landscapes is put forth as a source of materials for research on adhesive mucilages and for the textile industry (Chap. 4).

While the socio-economic benefits of trees and NTFPs are largely well- recognized, the evidence of these benefits is inadequately quantified (Malkamäki et al. 2018). This makes it difficult to demonstrate the socio-economic benefits of tree and forest resources in the SSA. As such, there is urgent need for long-term assessment of the socio-economic benefits of trees in the SSA landscapes (Mensah et al. 2024).

14.2.2 Environmental Benefits of Trees in African Multifunctional Landscapes

14.2.2.1 Conservation of Biodiversity

One of the key aspects of environmental benefits of trees in multifunctional landscapes is biodiversity conservation. This is because landscapes offer habitats for both faunal and floral species, and therefore provide conservational opportunities. The multifunctional landscapes in Africa are considered as biodiversity hubs (Imarhiagbe et al. 2022), with agricultural landscapes increasing biodiversity through agroforestry practices. Chapters 3 and 12 provide examples of evidence of agroforestry practices enhancing biodiversity, notably in coffee agroforests in Ethiopia (Buechley et al. 2015; Martens et al. 2020) and cacao agroforests in west

African countries (Clough et al. 2009; Maney et al. 2022), which present a wider range of floral and faunal species than nearby natural forest or monoculture plantations. Similarly, from Chap. 4, we showed that even African rocky outcrop landscapes play an important role in biodiversity conservation, as they harbour a rich floristic diversity with unique carnivores, geophytes, graminoids, resurrection, succulents and therophytes. They present rare micro-climates that offer specialized habitats for floral and faunal species, with some being centres of endemism. As such, they contribute towards widening the diversity of genetic resources in SSA. It is therefore important that management of multifunctional landscape in SSA consider different rocky outcrops as part of the major landscapes in the region (Allen 2020).

14.2.2.2 Climate Change Mitigation, Adaptation and Resilience

The nexus between global change and land-use dynamics is characterized by unprecedented population growth (2.5% per annum on average) in most SSA countries. SSA is among the most vulnerable regions to the impacts of climate change. As such, climate change adaptation and mitigation in SSA remain crucial in ensuring sustainable rural livelihoods. It is apparent from Chaps 5 and 12 that the most common mitigation and adaptation practices include climate-smart agriculture, woodlot establishment for energy and poles, different agroforestry systems and technologies, reforestation, and restoration. All these land use practices are prevalent in east and southern Africa and the Sahel. Their application mainly focuses on promoting sustainable land management (Maisharou et al. 2015; Chirwa et al. 2015; Kuyah et al. 2023). Of note, trees, such as *Parkia biglobosa* (Jacq.) G. Don (locust beans) and *Vitellaria paradoxa* C.F.Gaertn. (shea-butter trees) are recognized for their capacity to store substantial amounts of carbon in their biomass and canopies (Chap. 8). One key strategy involves promoting ecosystem-based approaches to adaptation, such as restoring degraded landscapes, conserving biodiversity hotspots, and enhancing natural carbon sequestration processes. Additionally, adopting sustainable land management practices, such as agroforestry, soil conservation, and water harvesting, can improve ecosystem health and enhance resilience to climate-related stresses. Strengthening institutional capacity and governance structures, investing in early warning systems and climate-smart infrastructure, and fostering community-based adaptation initiatives are also essential components of resilience-building strategies. Furthermore, promoting knowledge-sharing and collaboration among stakeholders, integrating traditional ecological knowledge with scientific expertise, and fostering adaptive learning processes can enhance the effectiveness and inclusivity of resilience-building efforts. By conserving and sustainably managing these NTFP-rich resources, the carbon storage potential is enhanced thus contributing to global efforts to combat climate change, while at the same time improving food security.

14.2.2.3 Rehabilitation of Degraded Landscapes

The SSA landscapes are a rich mosaic of different ecological diversity and integrity, with potential to support several restoration and rehabilitation activities. However, they are being degraded at a fast rate and their conservation values seem only more apparent in protected areas (IPBES 2018). As a result, the need to conserve and restore these landscapes is crucial if they have to continue supporting rural livelihood systems and the provisioning of other ecological services. Rehabilitation is currently based on the general understanding that the main tree species in the SSA are clonal, i.e., they use root suckers to spread (Chap. 12). This makes the forests and woodlands in the SSA landscapes resilient to aboveground disturbances like fire, wood harvesting and shifting cultivation (Syampungani et al. 2016, 2020). Nonetheless, rehabilitation models and techniques in this regard are relatively few, with the most common ones being natural regeneration, either assisted or enrichment planting, fire management and agroforestry systems as detailed in Chap. 12. Other major techniques utilized for landscape rehabilitation in SSA involve the role of trees in improving soil fertility. Evidence-based knowledge is given notably for the role of fertilizer trees in soil nitrogen enrichment through biological fixation, as well as the increase in soil carbon stock (Kuyah et al. 2020). This rationale is increasingly being proposed in the prospect of rethinking soil health management in SSA, notably in the light of declining crop yield responses to inorganic fertilizers in the region (Burke et al. 2016). In fact, Chap. 5 elaborates on the role of trees in enhancing soil physical properties as well as providing micro-climates conducive to improved growth and productivity of neighbouring plant species notably in the Tree-crop farming system and the Highland perennial farming system practised in different areas across SSA. Some evidence in the region also highlights the role of trees in reducing soil erosion, underscoring the important role of root systems of trees in soil particle retention, thereby preserving the quality of both the terrestrial and aquatic ecosystems (Chap. 12). Surprisingly however, despite being a growing environmental concern in Africa, to date, environmental pollution seems overlooked in trees-based technologies for improved environmental quality of African landscapes. Yet, this provides opportunities for reclamation and sustainable use of abandoned land, such as the salinized lands from large-scale irrigated farming systems (See Chap. 5), or metal-contaminated agricultural lands.

14.3 How Do We Manage a Multifunctional Landscape?

The management of multifunctional landscapes in Africa presents a complex interplay of ecological, social, and economic factors, posing numerous challenges alongside opportunities for sustainable development. Addressing these challenges necessitates precision control strategies and effective management approaches. In parallel, the socio-ecological dynamics of African landscapes highlight the need to balance conservation objectives with the diverse needs and aspirations of local communities. This underscores the importance of integrating scientific knowledge, indigenous knowledge, community engagement, and adaptive governance into management approaches. Collaborative initiatives such as ecological restoration and biodiversity conservation integration emphasize the critical role of international collaboration and landscape-scale interventions in achieving sustainable outcomes. In addition, the increasing alteration of ecosystems, mostly led by urbanization and agricultural land expansion to provide for a growing population, have substantially affected the capacity of landscapes to provide ecosystem services and sustain human well-being globally (IPBES 2018). Africa has seen a significant shift in its land use and land cover changes within the last few decades, which affects most ecological processes (Chap. 7). In SSA where the population growth rate is estimated to be high (2.5% per annum), the demand on the landscape is high. Therefore, the need for effective landscape management in SSA cannot be overemphasized.

14.3.1 Trees Management in Agricultural Landscapes

As a result of the important roles of trees on agricultural production and their overall contribution to community livelihoods, the presence of trees on agricultural landscapes is increasingly being encouraged. Across the SSA region, Farmer Managed Natural Regeneration (FMNR) has been identified as one of the most popular strategies to encourage the growth of native trees and promote their conservation on agricultural lands (Chap. 3). Chapter 12 demonstrates that tree species targeted in this endeavour are selected mostly for (i) their ability to access water resources efficiently, tolerate dry spells, and adapt to changing precipitation patterns (i.e. Senegalia, Vachellia, Faidherbia, Adansonia species), and (ii) their ability to resprout and recover after fires (i.e. thick bark and some eucalyptus species). These functional traits are thought to contribute significantly to the resilience of the African Savanna system (Syampungani et al. 2020), which provides a basis for designing sustainable multifunctional landscape approaches. While several authors support the view that FMNR provides multiple ecosystem services (Kuyah et al. 2023; Sileshi et al. 2023c), others feel that quantitative evidence on FMNR benefits is still sparse in relation to both its ecosystem function and its contribution to rural livelihoods (Chomba et al. 2020).

Another crucial aspect of trees management on agricultural lands, which has been explored from time immemorial, consists of traditional agroforestry practices guaranteeing a favourable interaction of trees with annual crops and other components of the agricultural landscape (Chap. 5). In this regard, research outcomes and experiences gained over decades in the SSA region have generated an appreciable body of knowledge to create settings that optimize ecosystem benefits of agroforestry practices across the landscape. The establishment of cocoa under the shade of larger native forest trees or under plantain and banana as practised in West Africa is an excellent example of a successful trees crop interaction (Chaps. 3 and 5). Other tree management practices tested with successful results across the SSA region include multi-storey systems, multi-purpose trees on farmlands, improved fallows, intercropping, parkland, rotational woodlots, taungya systems, and fodder banks, where detailed description is provided in Chaps. 5 and 12.

Other contemporary approaches of tree management discussed in this book, include tree domestication programs that have helped in reforestation projects across the continent by ensuring the planting of tree varieties adapted to local agro-ecological conditions (Chap. 10). In addition, more sophisticated approaches such as precision agriculture technologies are also increasingly considered in agroforestry systems in Africa to optimize resource use in the tree-crop interactions (Chaps. 5 and 12).

There is a growing consensus that both the roles and uses of trees in multifunctional landscapes must be appropriately contextualized to edaphic and climatic settings, in addition to cultural and socioeconomic considerations, which are variable across the SSA region. In Chap. 5, it is argued that the well-established farming systems in SSA (*sensu* Dixon et al. 2020) provide an opportunity in this regard. It is also demonstrated that northern SSA is dominated by agropastoral, Southern Africa by Maize mixed farming system and Perennial mixed farming system while West Africa is dominated by Cereal root crop farming system and Tree crop farming system, and Central Africa by Root and tuber crop farming system as well as Cereal-root crop farming system in which trees occupy important niches (Chap. 5).

14.3.2 Utilization and Nature Based Solution for Managing Resources in the Wild

While agricultural landscapes provide enormous benefits to society, other types of landscapes i.e., forests, woodlands, rocky outcrops also remain key to the well-being of communities in SSA as they provide both food and non-food ecosystem services. However, as the natural habitat is increasingly threatened by various natural and anthropogenic disturbances (Chap. 7), a decline in the provision of these services is being observed (Wessels et al. 2021), hence the need to extend sustainable landscape management practices beyond agricultural lands.

In many areas in the SSA, overharvesting of wild resources particularly for Non-Timber Forest Products (NTFPs) including edible insects, edible mushroom, Indigenous Fruits Trees (IFTs), and honey hunting has also been identified as a major threat to ecosystem integrity and therefore ecosystem service provisioning (Sileshi et al. 2023c). For this reason, regulating the collection of these resources from the wild and promoting their domestication, have been identified as an ecologically sound approach for their sustainable management. Across SSA, indigenous knowledge has contributed to sustainable harvesting of forest resources, such as *sopoodu* practice (partial pruning) in west African region (Gaoue and Ticktin 2009). Nonetheless, due to variation in the nature of NTFPs, developing sustainable harvesting systems based on scientific approaches is required. Local communities would be trained for the implementation of such systems (Fromentin et al. 2023) namely; regulating the rate of harvesting based on the functional attributes of a system whose components are being harvested (van Itterbeeck and van Huis 2012). Successful experiences with tree species include the fire-aided coppice regeneration in Mozambique and the stimulated abundance of *Ischyrolepis eleocharis* by harvesting in South Africa (Shackleton et al. 2007, Chap. 8), while host tree manipulation, wetland restoration, creation of insect-friendly corridor (Chap. 12) hold potential for improved management of edible insects in African landscapes (Dzerefos 2018). Additionally, the adoption of agroecological practices such as intercropping, and crop rotation in agricultural landscape (Chap. 5) are seen as potential opportunities for contributing towards creating environmental conditions to increase the abundance of edible insects.

14.3.3 Managing Germplasm Systems of Multipurpose Indigenous Trees in Africa

Multi-purpose indigenous trees play pivotal roles in SSA as evidenced in several chapters (Chaps. 3, 4, 5, 6, 8, and 9). To effectively harness the benefits of trees in the multifunctional landscapes of Africa, availability of germplasm, in the right volume, quality and timing, is important. Domestication of desired tree species has emerged as an important approach to providing alternative sources of NTFP for communities, especially for commodities with prospective increased demand on the international market, which are documented mainly for West, South and Eastern Africa, and the Sahel (Chaps. 8, 9 and 10). This approach is a contemporary strategy, where efforts are made to cultivate and breed tree species that are well-adapted to specific environments. Although still in its infancy, the domestication of some indigenous fruit trees documented across the Sub- Saharan Africa highlights its close association with commercialization perspective that is seen as a crucial incentive. This has brought about the concept of priority species (Leakey et al. 2022), which, in the context of SSA, are best represented by five fruit tree species namely: *Adansonia digitata* L., *Allanblackia spp, Sclerocarya birrea* (A.Rich.) Hochst., *Vitellaria paradoxa and Ziphiphus mauritiana* Lam., whose products have reached the international market and are already contributing to the community and national incomes (Chap. 9). Like for indigenous fruit trees, the cultivation of host trees for edible insects presents potential for providing alternative sources for wild resources thus reducing pressure on resources in the wild (Chap. 8).

Many African countries are committed to implementing large-scale landscape restoration initiatives to address the challenges of climate change, reduce land degradation, halt desertification and loss of biodiversity (Akinnifesi 2018; Giacomini et al. 2023; Sileshi et al. 2023a). Agroforestry has been identified as a key part of most developing country's Nationally Determined Contribution (NDC) targets, hence a potentially important contributor to global climate objectives. In a study by

Duguma et al. (2023), out of the 22 non-Annex 1 countries used for the analysis of NDC commitments, 15 were African countries that prioritized agroforestry as their strategies for meeting the unconditional targets of their NDC commitments. From Chap. 10, it is apparent that the success of these large-scale initiatives heavily hinges on steady access to good tree germplasm supply from both formal and informal sources. This warrants the development of a robust tree germplasm system geared towards improving timely access to good quality germplasm. Access to quality germplasm has continued to be a major limitation to most large-scale programmes. Currently, this is a daunting task in many African countries despite the presence of several actors including government institutions. As clearly presented in Chap. 10, ways to overcome these challenges have been proposed in this book, i.e. (i) increased research on long-term germplasm trials; (ii) developing capacity of germplasm actors along the supply chain, including institutional capacity strengthening of national tree seed centres (NTSCs), decentralized formal and informal seed sectors; (iii) scaling up community-led germplasm approaches, such as the Community Agroforestry Tree Seeds Banks model (CATS Banks) and Rural Resource Centers (RRCs) (these may include or be reinforced with the promotion of contract tree seed producers and seed out-grower schemes, with a caveat that, the seed quality control would be critical); and (iv) putting in place improved tree germplasm quality controls, certification, regulation and enforcement.

It is well established that effective policies on germplasm can promote tree germplasm value chain by ensuring good quality supply and promoting tree planting among communities; and much has been written about the importance of enabling policy on tree germplasm supply and delivery in Africa. To ensure sustainable delivery of quality tree germplasm, it is emphasized from Chap. 10, that there is need to improve current policies, regulations, governance, and the implementation of tree germplasm systems, putting appropriate measures in place to strengthen the capacity of decentralized germplasm supply system. This includes recognizing the role of communities, seed agri-businesses, smallholder producers and nursery operators, and capacitate them. Along with the germplasm policy, the legal framework is also a vital instrument for an effective germplasm sector. Such policies, legal and institutional frameworks can stimulate the efficiency of the delivery of quality tree germplasm and address current gaps in law enforcement in many African countries.

14.3.4 Management of Biological Invasion in African Multifunctional Landscape

Beside the negative impacts of climate changes and land deterioration, one of the prominent challenges faced by African multifunctional landscapes arises from the threats posed by invasive plant and insect species, which jeopardize biodiversity, food security, and local economies. For example, their negative impacts in Africa are so profound that economic losses averaging 1.2 billion US dollars are estimated per annum (Pratt et al. 2017). Alien invasive plant and insect species have been introduced into Africa through different ways and have established throughout the continent, providing a new host for native and non-native insect pests and pathogens in natural and plantation forests. Their transportation and dispersion are facilitated through international trade, which makes its management a daunting task especially in the contemporary time, characterized by globalization and increasing movement of people and goods. Hence, several management strategies are required in a holistic manner. This involves a series of resilience approaches categorized as (i) pre-border pathway management (i.e. transportation and introduction pathway), targeting the arrival stage and involving implementing measures to improve biosecurity surveillance at the ports of entry, and (ii) the post-border pathway management, which consists of containing or eradicating an invading population, through creation of habitat conditions unsuitable for the establishment and reproduction of the invasive species (Robertson et al. 2020). In the context of Africa, available management strategies include sentinel planting, mixed species forest, silviculture, breeding for resistance, biological control using biopesticides (i.e parasitoids, entomopathogens) and semiochemical lures/traps notably (Chap. 11).

Due to complexity inherent to its association to international trade, the management of invasive species and pests requires international and national phytosanitary policies and legislations. While the African Union-Inter-African Phytosanitary Council (AU-IAPSC) is mandated to play a key role in this regard, biosecurity policies and capacity to enforce them have been minimal in Africa, virtually lacking in almost 90% African countries (Sileshi et al. 2019). There is therefore an obvious need to enforce regional phytosanitary policies that will bolster efforts to control these biological invasions in African multifunctional landscapes. Besides, the knowledge on potential impacts of invasive species and pests on biodiversity conservation, agroecosystem viability and rural livelihoods is still limited in SSA. Such lack of assessments can undermine management options (Sileshi et al. 2019; Sileshi and Gebeyehu 2021; Vimercati et al. 2022). Hence, the integration of public awareness must be considered for effective management of biological invasions, for which indigenous knowledge within communities will be valuable.

14.3.5 Integrated Management Approaches for Sustainable Development in African Landscapes

In African landscapes, the importance of socioeconomic and ecological considerations are obvious, highlighting the need to balance conservation objectives with the diverse needs and aspirations of local communities. This underscores the importance of integrating scientific knowledge, indigenous knowledge, community engagement, and adaptive governance into management approaches.

Changes in land use patterns and ecosystem functions resulting from population growth and economic activities have implications for local communities dependent on natural resources for their well-being as highlighted in Chap. 7. For example, Chap. 13 discusses how displacement of indigenous populations, loss of access to forest resources, changes in agricultural productivity and socio-economic factors such as land tenure systems, market incentives, and policy frameworks can shape the composition, spatial arrangement, and ecological functions of trees within these landscapes, thereby affecting livelihood strategies and exacerbate poverty and vulnerability. Therefore, understanding and addressing the influence of socioeconomic drivers on landscapes are crucial for promoting sustainable land management practices and ensuring the well-being of both human communities and ecosystems.

Both Chaps. 12 and 13 show that management of multifunctional landscapes in Africa requires integrated approaches that embrace the diverse views, expertise, and interests among stakeholders. The two chapters delineate that cultural factors play a significant role in shaping socio-ecological management practices within landscapes. Local communities possess valuable traditional knowledge and practices that have sustained ecosystems for generations. Integrating these indigenous practices into management strategies not only fosters cultural resilience and enhances the relevance and effectiveness of conservation efforts, but also promotes cultural diversity, social cohesion and harmony between human communities and ecosystems. Hence, meaningful engagement of local communities is essential for the success of management initiatives in multifunctional landscapes. By involving communities in decision-making processes, project planning, and implementation, management efforts become more inclusive, transparent, and accountable.

Above all, the book underlines the paramount importance of collaboration and partnerships between relevant stakeholders, which are essential for addressing multifaceted challenges in landscape management (see Chaps. 5, 6, 7, 8, 9, 10, 11, 12 and 13). Recognizing the central role of local communities, it becomes imperative to involve them in decision-making processes, project planning, and implementation. By forging alliances between government agencies, NGOs, academia, local communities, and private sector entities, management initiatives can leverage complementary expertise, resources, and networks. The power of this endeavour lies in the fact that local communities, local government, and central authorities share rights and obligations under community-based natural resource management (CBNRM), which entails some degree of co-management of resources, ensuring management strategies align with local needs and aspirations. Through shared goals and mutual trust, partnerships amplify the impact of conservation efforts and facilitate collective action towards common objectives.

In the realm of policy, the book highlights that there are several strategic actions that policymakers should prioritize to facilitate the sustainable management of multifunctional landscapes in Africa. Policymakers need to develop and implement supportive frameworks that incentivize and reward sustainable land management practices. This may involve providing financial incentives, subsidies, or tax breaks for practices that promote soil conservation, water management, agroforestry, and biodiversity protection. By creating an enabling environment for sustainable land management, policymakers can encourage widespread adoption of practices that enhance ecosystem resilience and productivity.

14.4 Opportunities and Challenges for Provisioning of Ecosystem Services in the African Multifunctional Landscapes

The various chapters in this book have documented the potential of trees in African multifunctional landscapes. They have also established that trees hold promise to significantly enhance the resilience of African multifunctional landscapes and contribute to the well-being of local communities. Trees have historically been an integral part of the farming systems in SSA and indeed some (e.g., cocoa, coffee, commercial fruit trees, rubber) form the back-bone of the national economies of many countries (Chap. 5). The high value of products from trees is largely recognized across the region, as trees are identified as the main source of commodities in most high value farming systems such as the Highland perennial farming system in Eastern Africa and Madagascar, the Perennial mixed farming system in South Africa and North Africa and the Tree crop farming system in West Africa (Chap. 5). Particularly, African indigenous fruit trees (IFTs) can be singled out for their rich species diversity, which represents a great asset to contribute towards dietary diversification, and their adaptability for cultivation in a wider range of agroecological conditions across the SSA (Chaps. 5 and 9). Most importantly, the progressive changes in the global food market leaning towards preference for organically produced products, entails increased value for IFTs products, hence a flourishing market is foreseen for the future.

Like IFTs, edible insects and mushrooms, two major classes of tree-associated products, have been part of diets in the African rural communities for years, suggesting they are well imbedded in the culture of local populations (Chap. 8); therefore, they are easy to promote. Good prospects are foreseen for African mushrooms. The recognized nutritional and medicinal value, the wide diversity of identified edible species and the low cost and easy implementation of its farming, are expected to significantly boost the industry of African mushroom, which is currently cultivated only on small-scale basis in different areas across the SSA. Farming (with trading) of edible insects is also a rapidly growing business, notably in Eastern Africa (Chap. 8), holding promises not only to improve the provisioning services of African multifunctional landscape, but also to drive interest of the internal market as it provides access to the so-called 'climate-smart' proteins (Tanga and Kababu 2023).

Business potential is also evident for African rocky outcrop plant species, which are driving increasing interest in recent years (Chap. 4). Their superior ability for adaptation to extremely unfriendly environmental conditions make them a potential source of valuable genetic resources to contribute towards building a more resilient agri-food system in a region faced by challenges of emerging environmental issues such as climate changes and land degradation. Additionally, several epiphytic orchids, strongly related to tree species and used in the African food system are receiving more interest from food industries.

Beyond the enormous benefits provided by trees in African multifunctional landscapes, their management presents, however, a complex set of challenges that undermines exploitation of their potential to provide services. The future direction for research, management, and policy in multifunctional landscapes in Africa necessitates holistic and collaborative approaches that recognize the interconnectedness of ecological, social, and economic factors of landscape management. By adopting principles that prioritize a balance between human well-being and environmental management, stakeholders can ensure the resilience and vitality of multifunctional landscapes for future generations. Through these concerted efforts, Africa can realize the full potential of these landscapes while safeguarding the well-being of its ecosystems and communities, thus contributing to the resilience, sustainability and vitality of landscapes.

14.4.1 Research and Development

Future research endeavours should be geared towards advancing our understanding of socio-ecological interactions, fostering innovation in management strategies, and evaluating the efficacy of policy interventions. In order to drive advancements in this field, future research and development should prioritize several key areas, as follows:

(i) Improvement in our understanding of factors and mechanisms underlying the provision of ecosystem services by trees and related resources, which is currently limited;
(ii) Raising the awareness among policy-makers and the general public of the wide array of benefits that can be derived from trees and associated natural resources, particularly the medicinal potential of mushrooms and dietary attributes of epiphytic orchids that remains largely untapped in Africa;
(iii) Improvement of the fundamental knowledge pertaining to innovative practices for sustainable management of trees such as sustainable harvesting in the wild, domestication and agroecological practices to create conditions conducive to increased availability of trees-associated resources;
(iv) Development of market opportunities notably for indigenous fruit trees and non-timber forest products, which is currently limited due to the scarcity of information on their uses, a poor development of products, and a lack of adequate knowledge on postharvest handling of fruits;
(v) Formulation and enforcement of regulations and policies that integrate non-timber forest products in land management strategies;
(vi) Conduct more studies to build better insights about factors underlying the tolerance of rocky outcrop plant species which, together with the limited knowledge of their successful cultivation, currently hinders the prospect to explore their potential in enhancing the resilience of agricultural landscapes and other ecosystems;

(vii) Develop in-depth understanding of the complex interactions between socio-ecological systems within multifunctional landscapes, and elucidate the intricate relationships between anthropogenic activities, ecosystem dynamics, and landscape functionality, which can help researchers in developing more nuanced management approaches that balance ecological conservation with socio-economic development;
(viii) Integration of indigenous approaches to landscape restoration, identifying novel approaches for land use planning, resource allocation, and ecosystem restoration, for which the potential of emerging technologies such as remote sensing, GIS, and machine learning can be harnessed.
(ix) Develop biocontrol methods and technologies as environmentally-friendly alternatives to chemical pesticides for managing invasive species; and
(x) Foster climate-smart agriculture practices, such as conservation agriculture, precision farming, and water-saving irrigation techniques, to enhance resilience to climate variability and extreme weather events.

14.4.2 Policy Matters

Policymakers play a crucial role in shaping the management of multifunctional landscapes in Africa, as the effectiveness of these efforts is contingent upon supportive policy frameworks that incentivize sustainable land management. Despite strides made, existing policies in Africa often fall short, undermining efforts to harness the potential of regenerative agriculture, agroforestry and multifunctional landscapes. Thus, there is an identifiable gap and the need to consolidate management of African multifunctional landscapes by, among others:

(i) Formulating and implementing policies that integrate biodiversity conservation objectives into land use planning and decision-making processes, which is required to safeguard the ecological integrity and resilience of multifunctional landscapes;
(ii) Establishing incentives and support mechanisms for landowners and communities to actively participate in biodiversity conservation efforts within multifunctional landscapes, such as through payments for ecosystem services or conservation easements;
(iii) Investing in the capacity building of stakeholders involved in landscape management through education and training programs, as many of them lack the necessary skills, knowledge, and resources to address complex conservation challenges effectively;
(iv) Facilitating fair trade practices, reducing illegal harvesting and trade, and ensuring equitable distribution of benefits among local stakeholders.
(v) Reforming existing subsidies for chemical fertilizers and increase the competitiveness of organic fertilizers, including fertilizer trees, which are seen as crucial assets in rethinking soil health in the scenario of the African continent;

(vi) Raising public awareness and enforcement of strong regulations and controls of invasive species' movement at ports of entry, which currently jeopardizes biosecurity pertaining to invasive species across the continent;
(vii) Enhancing regional cooperation and information sharing mechanisms to facilitate early detection and rapid response to invasive species outbreaks, leveraging existing frameworks such as the African Union's African Plant Protection Convention;
(viii) Addressing the weak quality control and supply systems for tree germplasm that characterizes African countries, by fostering the development of technologies promoting production and delivery of germplasm to the users i.e. strengthening the capacity of decentralized germplasm supply system, and enforcing standards and certification of seeds, including phytosanitary standards; and
(ix) Evaluating the effectiveness of policy interventions through rigorous assessments of existing policies and initiatives aimed at managing multifunctional landscapes, which is paramount for informing evidence-based decision-making and guiding future policy development.

References

Akinnifesi FK (2018) Africa's Forest landscape restoration gathers momentum. Nature Faune 32(1):3–10

Akinnifesi FK, Kwesiga F, Mhango J, Chilanga T, Mkonda A, Kadu CAC, Kadzere I, Mithöfer D, Saka JDK, Sileshi G, Ramadhani T, Dhliwayo P (2006) Towards the development of miombo fruit trees as commercial tree crops in southern Africa. For Trees Livelihoods 16:103–121

Allen K (2020) Conservation values and the value of conservation. In: Goldstein MI, DellaSala DA (eds) Encyclopedia of the world's biomes. Elsevier, pp 303–310

Arnold JEM, Dewees PA (1995) Tree Management in Farmer Strategies: responses to agricultural intensification. Oxford University Press, Oxford

Baah SKT, Lakner C (2023) Fragility and poverty in sub-Saharan Africa: two sides of the same coin. World Bank Data Blog. https://blogs.worldbank.org/en/opendata/fragility-and-poverty-sub-saharan-africa-two-sides-same-coin#. Accessed on 8th May 2023

Bastos C, Liberal Â, Moldão M, Catarino L, Barros L (2023) Ethnomycological prospect of wild edible and medicinal mushrooms from central and southern Africa – a review. Food Front 4:549–575. https://doi.org/10.1002/fft2.215

Buechley ER, Şekercioğlu ÇH, Atickem A, Gebremichael G, Ndungu JK, Mahamued BA et al (2015) Importance of Ethiopian shade coffee farms for forest bird conservation. Biol Conserv 188:50–60. https://doi.org/10.1016/j.biocon.2015.01.011

Burke W, Frossard E, Kabwe S, Jayne T (2016) Understanding fertilizer effectiveness and adoption in Zambia. In: MSU international development working paper, vol No. 147. Michigan State University, East Lansing

Chirwa PW, Larwanou M, Syampungani S, Babalola FD (2015) Management and restoration practices in degraded landscapes of eastern Africa and requirements for up-scaling. Int For Rev 17(3):20–30

Chomba S, Sinclair F, Savadogo P, Bourne M, Lohbeck M (2020) Opportunities and constraints for using farmer managed natural regeneration for land restoration in sub-Saharan Africa. Front For Glob Change 3:571679. https://doi.org/10.3389/ffgc.2020.571679

Clough Y, Faust F, Tscharntke T (2009) Cacao boom and bust: sustainability of agroforestsand opportunities for biodiversity conservation. Conserv Lett 2:197–205
Duguma LA, Minang PA, Watson C, Nath AJ, Muthee KW, van Noordiwijk M, Mutune JM, Sileshi GW (2023) Agroforestry as a key intervention to achieve nationally determined contribution (NDC) targets. In: Dagar JC, Gupta SR, Sileshi GW (eds) Agroforestry for sustainable intensification of agriculture in Asia and Africa. Sustainability sciences in Asia and Africa. Springer, Singapore. https://doi.org/10.1007/978-981-19-4602-8_19
Dzerefos CM (2018) Conservation of edible insects in sub-Saharan Africa. In: Halloran A, Flore R, Vantomme P, Roos N (eds) Edible insects in sustainable food systems. Springer, Cham. https://doi.org/10.1007/978-3-319-74011-9_12
Fromentin JM, Emery MR, Donaldson J, Balachander G, Barron ES, Chaudhary RP et al (2023) Status, challenges and pathways to the sustainable use of wild species. Glob Environ Chang 81:102692
Gaoue OG, Ticktin T (2009) Fulani knowledge of the ecological impacts of *Khaya senegalensis* (Meliaceae) foliage harvest in Benin and its implications for sustainable harvest. Econ Bot 63:256–270. https://doi.org/10.1007/s12231-009-9091-6
Giacomini FL, Prempeh JA, Jalonen R, Vinceti B, Ekue M, Bosshard E, Burslem DFRP, Kettle CJ (2023) Seeding African Forest and landscape restoration: evaluating native tree seed systems in four African countries. Diversity 15(9):981. https://doi.org/10.3390/d15090981
Imarhiagbe O, Onyeukwu I, Egboduku WO, Mukah FE, Ogwu MC (2022) Forest conservation strategies in Africa: historical perspective, status and sustainable avenues for Progress. In: Chibueze IS (ed) Biodiversity in Africa: potentials, threats and conservation. Sustainable development and biodiversity, vol 29. Springer, Singapore. https://doi.org/10.1007/978-981-19-3326-4_21
IPBES (2018) Summary for policymakers of the assessment report on land degradation and restoration of the intergovernmental science policy platform on biodiversity and ecosystem services. In: Scholes R, Montanarella L, Brainich A, Barger N, ten Brink B, Cantele M et al (eds) Secretariat of the intergovernmental science-policy platform on biodiversity and ecosystem services, Bonn. 44p
Kraybill D (2013) Rural development in sub-Saharan Africa. In: Green GP (ed) Handbook of rural development. Edward Elgar Publisher, pp 270–293
Kuyah S, Sileshi G, Luedeling E, Akinnifesi F et al (2020) Potential of agroforestry to enhance livelihood security in Africa. In: Dagar JC, Gupta SR, Teketay D (eds) Agroforestry for degraded landscapes. Springer, Singapore. https://doi.org/10.1007/978-981-15-4136-0_4
Kuyah S, Buleti S, Dimobe K et al (2023) Farmer-managed natural regeneration in Africa: evidence for climate change mitigation and adaptation in drylands. In: Dagar JC, Gupta SR, Sileshi GW (eds) Agroforestry for sustainable intensification of agriculture in Asia and Africa. Sustainability sciences in Asia and Africa. Springer, Singapore. https://doi.org/10.1007/978-981-19-4602-8_3
Leakey RR, Tientcheu-Avana ML, Awazi NP, Assogbadjo AE, Mabhaudhi T, Hendre PS et al (2022) The future of food: domestication and commercialization of indigenous food crops in Africa over the third decade (2012–2021). Sustain For 14(4):2355
Loison SA (2015) Rural livelihood diversification in sub-Saharan Africa: a literature review. J Dev Stud 51(9):1125–1138. https://doi.org/10.1080/00220388.2015.1046445
Lovell ST, Johnston DM (2009) Designing landscapes for performance based on emerging principles in landscape ecology. Ecol Soc 14(1):44. [online]. http://www.ecologyandsociety.org/vol14/iss1/art44/
Maisharou A, Chirwa PW, Larwanou M, Babalola FD, Ofoegbu C (2015) Sustainable land management practices in the Sahel: review of practises, techniques and technologies for land restoration and strategy for up-scaling. Int For Rev 17(S3):1–19
Malkamäki A, D'Amato D, Hogarth NJ, Kanninen M et al (2018) A systematic review of the socio-economic impacts of large-scale tree plantations, worldwide. Glob Environ Chang 53:90–103
Maney C, Sassen M, Hill SLL (2022) Modelling biodiversity responses to land use in areas of cocoa cultivation. Agric Ecosyst Environ 324:107712

Martens JEJ, Emsens WJ, Jocque M, Geeraert L, de Beenhouwer M (2020) From natural forest to coffee agroforest: implications for communities of large mammals in the Ethiopian highlands. Oryx 54:715–722. https://doi.org/10.1017/S0030605318000844

Mensah ER, Shinde N, Kakpo NT, Djenontin NS (2024) The human Well-being outcomes of tree plantations in sub-Saharan Africa: a reassessment of evidence using longitudinal subnational-year data. Forest Policy Econ 160:103141

Neyret M, Peter S, Le Provost G et al (2023) Landscape management strategies for multifunctionality and social equity. Nat Sustain 6:391–403. https://doi.org/10.1038/s41893-022-01045-w

Pratt CF, Constantine KL, Murphy ST (2017) Economic impacts of invasive alien species on African smallholder livelihoods. Glob Food Secur 14:31–37. https://doi.org/10.1016/j.gfs.2017.01.011

Robertson PA, Mill A, Novoa A, Jeschke JM, Essl F, Gallardo B, Geist J, Jarić I, Lambin X, Musseau C, Pergl J (2020) A proposed unified framework to describe the management of biological invasions. Biol Invasions 22:2633–2645. https://doi.org/10.1007/s10530-020-02298-2

Scherr SJ, McNeely JA (2008) Multifunctional landscapes: interconnecting sustainable agriculture, rural development, and ecosystem services. Island Press. ISBN: 978-1-59726-131-2

SETAC (2019) Nature-based solutions are gaining momentum from international initiatives promoting environmental, social and economic benefits. Integr Environ Assess Manag 15(6):830–831

Shackleton CM, McGarry D, Fourie S, Gambiza J, Shackleton SE, Fabricius C (2007) Assessing the effects of invasive alien species on rural livelihoods: case examples and a framework from South Africa. Hum Ecol 35:113–127

Sileshi GB, Gebeyehu S (2021) Emerging infectious diseases threatening food security and economies in Africa. Glob Food Secur 28:100479. https://doi.org/10.1016/j.gfs.2020.100479

Sileshi GW, Gebeyehu S, Mafongoya PL (2019) The threat of alien invasive insect and mite species to food security in Africa and the need for a continent-wide response. Food Secur 11:763–775. https://doi.org/10.1007/s12571-019-00930-1

Sileshi GW, Dagar JC, Akinnifesi FK, Mng'omba SA (2023a) Potentials of indigenous fruit trees in enhancing nutrition, income and biodiversity conservation in African agroforestry, agroforestry for sustainable intensification of agriculture in Asia and Africa, sustainable agriculture and food security. Springer Nature Singapore Pte Ltd. pp. 321362. https://doi.org/10.1007/978-981-19-4602-8_11

Sileshi GW, Tibuhwa DD, Mlambo A (2023b) Underutilized wild edible fungi and their undervalued ecosystem services in Africa. CABI Agric Biosci 4:3. https://doi.org/10.1186/s43170-023-00145-7

Sileshi GW, Dagar JC, Kuyah K, Datta A (2023c) The great Green Wall initiatives and opportunities for integration of dryland agroforestry to mitigate desertification. In: Dagar JC, Gupta SR, Sileshi GW (eds) Agroforestry for sustainable intensification of agriculture in Asia and Africa, sustainable agriculture and food security. Springer Nature Singapore Pte Ltd., pp 175–206. https://doi.org/10.1007/978-981-19-4602-86

Suich H, Howe C, Mac G (2015) Ecosystem services and poverty alleviation: a review of the empirical links. Ecosyst Serv 12:37–147

Syampungani S, Geldenhuys CJ, Chirwa PWC (2016) Regeneration dynamics of miombo woodland in response to different anthropogenic disturbances: Forest characterisation for sustainable management. Agrofor Syst 90(4):563–576

Syampungani S, Sikanwe AN, Chirwa PW (2020) Ecology of natural regeneration of tropical dry forests of Africa and its implications for their sustainable man. In: Bhadouria R, Tripathi S, Srivastava P, Singh P (eds) Handbook of research on the conservation and restoration of tropical dry forests, IGI Global, pp 346–358

Tanga CM, Ekesi S (2024) Dietary and therapeutic benefits of edible insects: a global perspective. Annu Rev Entomol 69:303–331. https://doi.org/10.1146/annurev-ento-020123-013621

Tanga CM, Kababu M (2023) New insights into the emerging edible insect industry in Africa. Anim Front 13(4):26–40. https://doi.org/10.1093/af/vfad039

Tscharntke T, Tylianakis JM, Rand TA, Didham RK, Fahrig L, Batáry P, Bengtsson J, Clough Y, Crist TO et al (2012) Landscape moderation of biodiversity patterns and processes – eight hypotheses. Biol Rev Camb Philos Soc 87(3):661–685. https://doi.org/10.1111/j.1469-185X.2011.00216.x

Van Itterbeeck J, van Huis A (2012) Environmental manipulation for edible insect procurement: a historical perspective. J Ethnobiol Ethnomed 8:1–7

Vimercati G, Probert AF, Volery L, Bernardo-Madrid R, Bertolino S, Céspedes V, Essl F, Evans TT, Gallardo B, Gallien L, González-Moreno P (2022) The EICAT+ framework enables classification of positive impacts of alien taxa on native biodiversity. PLoS Biol 20(8):e3001729. https://doi.org/10.1371/journal.pbio.3001729

Wessels C, Merow C, Trisos CH (2021) Climate change risk to southern African wild food plants. Reg Environ Chang 21:29. https://doi.org/10.1007/s10113-021-01755-5

Yengoh GT (2020) Impact of land degradation on rural livelihoods in sub Saharan Africa. In: Leal Filho W et al (eds) Life on land, encyclopedia of the UN sustainable development goals. Springer Nature Switzerland AG, Cham

MIX
Papier aus verantwortungsvollen Quellen
Paper from responsible sources
FSC® C105338

If you have any concerns about our products,
you can contact us on
ProductSafety@springernature.com

In case Publisher is established outside the EU,
the EU authorized representative is:
Springer Nature Customer Service Center GmbH
Europaplatz 3, 69115 Heidelberg, Germany

Printed by Libri Plureos GmbH
in Hamburg, Germany